T0344348

C. Oliver Kappe, Doris Dallinger,
and S. Shaun Murphree
Practical Microwave Synthesis for Organic Chemists

Related Titles

Wyatt, P. / Warren, S.

Organic Synthesis

Strategy and Control

2007
ISBN: 987-0-471-92963-8

Loupy, A. (ed.)

Microwaves in Organic Synthesis

2006
ISBN: 978-3-527-31452-2

Kappe, C. O. / Stadler, A.

Microwaves in Organic and Medicinal Chemistry

2006
ISBN: 978-3-527-606831

Tierney, Jason / Lidström, Pelle (Eds.)

Microwave Assisted Organic Synthesis

2005
ISBN: 978-1-4051-1560-5

Wuts, P. G. M., Greene, T. W.

Greene's Protective Groups in Organic Synthesis

2006
ISBN: 978-0-471-69754-1

C. Oliver Kappe, Doris Dallinger, and S. Shaun Murphree

Practical Microwave Synthesis for Organic Chemists

Strategies, Instruments, and Protocols

WILEY-VCH
VCH

WILEY-VCH Verlag GmbH & Co. KGaA

The Authors

Prof. C. Oliver Kappe
Karl-Franzens-University Graz
Christian Doppler Laboratory for Microwave
Chemistry and Institute of Chemistry
Heinrichstrasse 28
8010 Graz
Austria

Dr. Doris Dallinger
Karl-Franzens-University Graz
Christian Doppler Laboratory for Microwave
Chemistry and Institute of Chemistry
Heinrichstrasse 28
8010 Graz
Austria

Prof. S. Shaun Murphree
Allegheny College
Department of Chemistry
520 N. Main Street
Meadville, PA 16335
USA

Library of Congress Card No.:
applied for

British Library Cataloguing-in-Publication Data
A catalogue record for this book is available from the
British Library.

**Bibliographic information published by
the Deutsche Nationalbibliothek**
The Deutsche Nationalbibliothek lists this
publication in the Deutsche Nationalbibliografie;
detailed bibliographic data are available on the
Internet at http://dnb.d-nb.de.

© 2009 WILEY-VCH Verlag GmbH & Co. KGaA,
Weinheim

Typesetting Thomson Digital, Noida, India
Printing betz-druck GmbH, Darmstadt
Binding Litges & Dopf GmbH, Heppenheim
Cover Design Grafik-Design Schulz, Fußgönheim

Printed on acid-free paper

ISBN: 978-3-527-32097-4

Contents

Practical Microwave Synthesis for Organic Chemists: Strategies, Instruments, and Protocols
C. Oliver Kappe, Doris Dallinger, and S. Shaun Murphree
Copyright © 2009 WILEY-VCH Verlag GmbH & Co. KGaA, Weinheim
ISBN: 978-3-527-32097-4

Preface

Today, a large body of work on microwave-assisted organic synthesis is available in the published and patent literature. Close to one hundred review articles, several books, and online databases already provide extensive coverage of the subject for the specialist reader. The goal of the present book is to provide an introductory treatise for beginners, a sort of "How To Get Started" guide. Apart from a few articles in the *Journal of Chemical Education*, very little introductory and practical hands-on information on controlled microwave chemistry has been presented in a textbook style format.

This fact has prompted the publication of the present work "*Practical Microwave Synthesis for Organic Chemists – Strategies, Instruments, and Protocols*" which serves both the beginner and the more experienced microwave user. A major motivation for writing this treatise has been the continuous and enthusiastic feedback obtained from scientists during conferences and short courses on microwave synthesis organized by the authors. In particular, the very popular MAOS conference series organized since 2003, in combination with practical hands-on or classroom-style training courses, has led to a collection of questions and comments from the attendees and has stimulated the design and concept for this book. It has been written mainly with the microwave novice in mind. Several chapters are specifically designed for beginners, such as undergraduate or graduate students in academia, or industrial scientists getting started in microwave-assisted organic synthesis.

Following a brief introduction (Chapter 1), Chapter 2 details the basic concepts of microwave dielectric heating theory and provides insight into the current understanding of microwave effects. In Chapter 3 a comprehensive review of most of the commercially available single-mode and multimode microwave reactors for organic synthesis is presented. Chapter 4 provides an extensive overview of the different microwave processing techniques that are available today, while Chapter 5 contains many useful tips for the microwave novice, including a "Frequently Asked Questions" Section. The last chapter of the book (Chapter 6) contains a collection of carefully selected and documented microwave experiments that may also be used by scientists in academia to design a course on microwave-assisted organic synthesis. All examples have been tested by advanced undergraduate students in the course of a

Practical Microwave Synthesis for Organic Chemists: Strategies, Instruments, and Protocols
C. Oliver Kappe, Doris Dallinger, and S. Shaun Murphree
Copyright © 2009 WILEY-VCH Verlag GmbH & Co. KGaA, Weinheim
ISBN: 978-3-527-32097-4

special "Microwave Chemistry Lab Course" during the spring of 2007 at the University of Graz.

Writing this book would have been impossible without considerable assistance from all the members of the Christian Doppler Laboratory for Microwave Chemistry (CDLMC) in Graz. All past and present members are acknowledged for their invaluable contributions over the years. Special thanks go to Jennifer M. Kremsner, Toma N. Glasnov, Hana Prokopcová, Bernadett Bacsa, Jamshed Hashim, Tahseen Razzaq and Florian Reder for their help in assembling the practical examples presented in Chapter 6. The students taking part in the "Microwave Chemistry Lab Course" are particularly acknowledged for their engagement and valuable input. In this context the authors wish to thank the Fulbright Commission and the Austrian–American Educational Commission for supporting the international collaboration leading to the development of this practical course. Thanks are also due to the microwave equipment vendors for providing detailed information and high resolution graphics of their instruments. In this context, the support of Alexander Stadler (Anton Paar GmbH), Martin Keil (Biotage AB), Axel Schöner (CEM Corp.) and Mauro Ianelli (Milestone s.r.l.) is acknowledged.

Graz   *C. Oliver Kappe and Doris Dallinger,*
Meadville   *S. Shaun Murphree*
May 2008

1
Microwave Synthesis – An Introduction

While fire is now rarely used in synthetic chemistry, it was not until Robert Bunsen invented the burner in 1855 that the energy from this heat source could be applied to a reaction vessel in a focused manner. The Bunsen burner was later superseded by the isomantle, oil bath or hot plate as a source of applying heat to a chemical reaction. In the past few years, heating chemical reactions by microwave energy has been an increasingly popular theme in the scientific community. Since the first published reports on the use of microwave irradiation to carry out organic chemical transformations by the groups of Gedye and Giguere/Majetich in 1986 [1], more than 3500 articles have been published in this fast moving and exciting field, today generally referred to as microwave-assisted organic synthesis (MAOS) [2, 3]. In many of the published examples, microwave heating has been shown to dramatically reduce reaction times, increase product yields and enhance product purities by reducing unwanted side reactions compared to conventional heating methods. The advantages of this enabling technology have, more recently, also been exploited in the context of multistep total synthesis [4] and medicinal chemistry/drug discovery [5], and have additionally penetrated related fields such as polymer synthesis [6], material sciences [7], nanotechnology [8] and biochemical processes [9]. The use of microwave irradiation in chemistry has thus become such a popular technique in the scientific community that it might be assumed that, in a few years, most chemists will probably use microwave energy to heat chemical reactions on a laboratory scale. The statement that, in principle, any chemical reaction that requires heat can be performed under microwave conditions has today been generally accepted as a fact by the scientific community.

The short reaction times provided by microwave synthesis make it ideal for rapid reaction scouting and optimization of reaction conditions, allowing very rapid progress through the "hypotheses–experiment–results" iterations, resulting in more decision points per unit time. In order to fully benefit from microwave synthesis one has to be prepared to fail in order to succeed. While failure could cost a few minutes, success would gain many hours or even days. The speed at which multiple variations of reaction conditions can be performed allows a morning discussion of "What should we try?" to become an after lunch discussion of "What were the results?" Not

Practical Microwave Synthesis for Organic Chemists: Strategies, Instruments, and Protocols
C. Oliver Kappe, Doris Dallinger, and S. Shaun Murphree
Copyright © 2009 WILEY-VCH Verlag GmbH & Co. KGaA, Weinheim
ISBN: 978-3-527-32097-4

surprisingly, therefore, many scientists, both in academia and in industry, have turned to microwave synthesis as a frontline methodology for their projects.

Arguably, the breakthrough in the field of MAOS on its way from laboratory curiosity to standard practice started in the pharmaceutical industry around the year 2000. Medicinal chemists were among the first to fully realize the true power of this enabling technology. Microwave synthesis has since been shown to be an invaluable tool for medicinal chemistry and drug discovery applications since it often dramatically reduces reaction times, typically from days or hours to minutes or even seconds [5]. Many reaction parameters can therefore be evaluated in a few hours to optimize the desired chemistry. Compound libraries can then be rapidly synthesized in either a parallel or (automated) sequential format using microwave technology [5]. In addition, microwave synthesis often allows the discovery of novel reaction pathways, which serve to expand "chemical space" in general, and "biologically-relevant, medicinal chemistry space", in particular.

In the early days of microwave synthesis, experiments were typically carried out in sealed Teflon or glass vessels in a domestic household microwave oven without any temperature or pressure measurements [1]. Kitchen microwave ovens are not designed for the rigors of laboratory usage: acids and solvents corrode the interiors quickly and there are no safety controls. The results were often violent explosions due to the rapid uncontrolled heating of organic solvents under closed vessel conditions. In the 1990s several groups started to experiment with solvent-free microwave chemistry (so-called dry-media reactions), which eliminated the danger of explosions [10]. Here, the reagents were pre-adsorbed onto either a more or less microwave transparent inorganic support (i.e., silica, alumina or clay) or a strongly absorbing one (i.e., graphite), that additionally may have been doped with a catalyst or reagent. Particularly in the beginning of MAOS, the solvent-free approach was very popular since it allowed the safe use of domestic microwave ovens and standard open vessel technology. While a large number of interesting transformations using dry-media reactions have been published in the literature [10], technical difficulties relating to non-uniform heating, mixing, and the precise determination of the reaction temperature remained unsolved, in particular when scale-up issues needed to be addressed.

Alternatively, microwave-assisted synthesis was, in the past, often carried out using standard organic solvents under open vessel conditions. If solvents are heated by microwave irradiation at atmospheric pressure in an open vessel, the boiling point of the solvent typically limits the reaction temperature that can be achieved. In order to nonetheless achieve high reaction rates, high-boiling microwave absorbing solvents were frequently used in open-vessel microwave synthesis [11]. However, the use of these solvents presented serious challenges during product isolation and recycling of solvent. In addition, the risks associated with the flammability of organic solvents in a microwave field and the lack of available dedicated microwave reactors allowing adequate temperature and pressure control were major concerns. The initial slow uptake of microwave technology in the late 1980s and 1990s has often been attributed to its lack of controllability and reproducibility, coupled with a general lack of understanding of the basics of microwave dielectric heating.

In particular, the use of kitchen microwave ovens in combination with non-reliable temperature monitoring devices led to considerable confusion in the microwave chemistry community in the late 1990s and has given MAOS a bad reputation and the stigma of a "black box" science. The majority of organic chemists at that time were not taking microwave chemistry seriously and the discussion and irritation around the topic of "microwave effects" has probably contributed to this situation [12]. Historically, since the early days of microwave synthesis, the observed rate-accelerations and sometimes altered product distributions compared to oil-bath experiments led to speculation on the existence of so-called "specific" or "non-thermal" microwave effects [13]. Such effects were claimed when the outcome of a synthesis performed under microwave conditions was different from the conventionally heated counterpart at the same apparent temperature. Reviewing the present literature it appears that today most scientists agree that, in the majority of cases, the reason for the observed rate enhancements is a purely thermal/kinetic effect, that is, a consequence of the high reaction temperatures that can rapidly be attained when irradiating polar materials in a microwave field, although clearly effects that are caused by the uniqueness of the microwave dielectric heating mechanism ("specific microwave effects") must also be considered. While for the chemist in industry this discussion may seem futile, the debate on "microwave effects" is undoubtedly going to continue for many years in the academic world. Because of the recent availability of modern dedicated microwave reactors with on-line accurate monitoring of both temperature and pressure, some of the initial confusion on microwave effects has subsided. This can also be attributed, to some extent, to the fact that microwave synthesis today is mostly carried out in solution phase using organic solvents, where the temperature of the reaction mixture can generally be adequately monitored.

Controlled MAOS in sealed vessels using standard solvents – a technique pioneered by Strauss in the mid 1990s [14] – has thus celebrated a steady comeback since the year 2000 and today clearly is the method of choice for performing microwave-assisted reactions. This is evident from surveying the recently published literature in the area of microwave-assisted organic synthesis (Figure 1.1). In addition to the primary and patent literature, many review articles [3–19], several books [2], special issues of journals [20], feature articles [21], online databases [22], information on the world-wide-web [23], and educational publications [24, 25] provide extensive coverage of the subject. Among the about 850 original publications that appeared in 2007 describing microwave-assisted reactions under controlled conditions, a careful analysis demonstrates that in about 90% of all cases sealed vessel processing (autoclave technology) in dedicated single-mode microwave instruments has been employed. A recent survey has, however, found that as many as 30% of all published MAOS papers still employ kitchen microwave ovens [26], a practice banned by most of the respected scientific journals today. For example, the American Chemical Society (ACS) organic chemistry journals will typically not consider manuscripts describing the use of kitchen microwave ovens or the absence of a reaction temperature, as specified in the relevant publication guidelines [27].

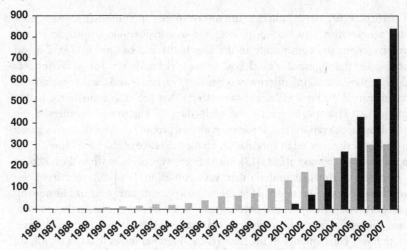

Figure 1.1 Publications on microwave-assisted organic synthesis (1986–2007). Gray bars: Number of articles involving MAOS for seven selected synthetic organic chemistry journals (*J. Org. Chem., Org. Lett., Tetrahedron, Tetrahedron Lett., Synth. Commun., Synthesis, Synlett*. SciFinder Scholar keyword search on "microwave"). The black bars represent the number of publications (2001–2007) reporting MAOS experiments in dedicated reactors with adequate process control (about 50 journals, full text search: microwave). Only those articles dealing with synthetic organic chemistry were selected.

Recent innovations in microwave reactor technology now allow controlled parallel and automated sequential processing under sealed vessel conditions, and the use of continuous or stop-flow reactors for scale-up purposes. In addition, dedicated vessels for solid-phase synthesis, for performing transformations using pre-pressurized conditions and for a variety of other special applications, have been developed. Today there are four major instrument vendors that produce microwave instrumentation dedicated to organic synthesis. All these instruments offer temperature and pressure sensors, built-in magnetic stirring, power control, software operation and sophisticated safety controls. The number of users of dedicated microwave reactors is therefore growing at a rapid rate and it appears only to be a question of time until most laboratories will be equipped with suitable microwave instrumentation.

In the past, microwave chemistry was often used only when all other options to perform a particular reaction had failed, or when exceedingly long reaction times or high temperatures were required to complete a reaction. This practice is now slowly changing and, due to the growing availability of microwave reactors in many laboratories, routine synthetic transformations are now also being carried out by microwave heating. One of the major drawbacks of this relatively new technology remains the equipment cost. While prices for dedicated microwave reactors for organic synthesis have come down considerably since their first introduction in the late 1990s, the current price range for microwave reactors is still many times higher than that of conventional heating equipment. As with any new technology, the current situation is bound to change over the next several years and less expensive equipment should become available. By then, microwave reactors will have truly become the "Bunsen burners of the twenty first century" and will be standard equipment in every chemical laboratory.

References

1 Gedye, R., Smith, F., Westaway, K., Ali, H., Baldisera, L., Laberge, L. and Rousell, J. (1986) *Tetrahedron Letters*, **27**, 279–282; Giguere, R.J., Bray, T.L., Duncan, S.M. and Majetich, G. (1986) *Tetrahedron Letters*, **27**, 4945–4948.

2 Loupy, A.(ed.) (2002) *Microwaves in Organic Synthesis*, Wiley-VCH, Weinheim; Hayes, B.L. (2002) *Microwave Synthesis: Chemistry at the Speed of Light*, CEM Publishing, Matthews, NC; Lidström, P. and Tierney, J.P.(eds) (2005) *Microwave-Assisted Organic Synthesis*, Blackwell Publishing, Oxford; Kappe, C.O. and Stadler, A. (2005) *Microwaves in Organic and Medicinal Chemistry*, Wiley-VCH, Weinheim; Loupy, A.(ed.) (2006) *Microwaves in Organic Synthesis*, 2nd edn, Wiley-VCH, Weinheim; Larhed, M. and Olofsson, K.(eds) (2006) *Microwave Methods in Organic Synthesis*, Springer, Berlin; Van der Eycken, E. and Kappe, C.O.(eds) (2006) *Microwave-Assisted Synthesis of Heterocycles*, Springer, Berlin.

3 Abramovitch, R.A. (1991) *Organic Preparations and Procedures International*, **23**, 685–711; Caddick, S. (1995) *Tetrahedron*, **51**, 10403–10432; Lidström, P., Tierney, J., Wathey, B. and Westman, J. (2001) *Tetrahedron*, **57**, 9225–9283; Kappe, C.O. (2004) *Angewandte Chemie-International Edition*, **43**, 6250–6284; Hayes, B.L. (2004) *Aldrichimica Acta*, **37**, 66–77; Ondruschka, B. and Bonrath, W. (2006) *Chimia*, **60**, 326–329; Kappe, C.O. (2006) *Chimia*, **60**, 308–312; Kappe, C.O. (2008) *Chemical Society Reviews*, **37**, 1127–1139.

4 Baxendale, I.R., Ley, S.V., Nessi, M. and Piutti, C. (2002) *Tetrahedron*, **58**, 6285–6304; Artman, D.D., Grubbs, A.W. and Williams, R.M. (2007) *Journal of the American Chemical Society*, **129**, 6336–6342; Appukkuttan, P. and Van der Eycken, E. (2006) *Topics in Current Chemistry*, **266**, 1–47.

5 Krstenansky, J.L. and Cotterill, I. (2000) *Current Opinion in Drug Discovery & Development*, **4**, 454–461; Larhed, M. and Hallberg, A. (2001) *Drug Discovery Today*, **6**, 406–416; Wathey, B., Tierney, J., Lidström, P. and Westman, J. (2002) *Drug Discovery Today*, **7**, 373–380; Wilson, N.S. and Roth, G.P. (2002) *Current Opinion in Drug Discovery & Development*, **5**, 620–629; Dzierba, C.D. and Combs, A.P. (2002) in *Annual Reports in Medicinal Chemistry* (ed. A.M. Doherty), Academic Press, New York, Vol. 37, pp. 247–256; Al-Obeidi, F., Austin, R.E., Okonya, J.F. and Bond, D.R.S. (2003) *Mini-Reviews in Medicinal Chemistry*, **3**, 449–460; Ersmark, K., Larhed, M. and Wannberg, J. (2004) *Current Opinion in Drug Discovery & Development*, **7**, 417–427; Shipe, W.D., Wolkenberg, S.E. and Lindsley, C.W. (2005) *Drug Discovery Today: Technologies*, **2**, 155–161; Kappe, C.O. and Dallinger, D. (2006) *Nature Reviews Drug Discovery*, **5**, 51–64; Mavandadi, F. and Pilotti, A. (2006) *Drug Discovery Today*, **11**, 165–174; Wannberg, J., Ersmark, K. and Larhed, M. (2006) *Topics in Current Chemistry*, **266**, 167–198; Chighine, A., Sechi, G. and Bradley, M. (2007) *Drug Discovery Today*, **12**, 459–464; Larhed, M., Wannberg, J. and Hallberg, A. (2007) *The QSAR & Combinatorial Science*,

26, 51–68; Alcázar, J., Diels, G. and Schoentjes, B. (2007) *Mini-Reviews in Medicinal Chemistry*, **7**, 345–369; Shipe, W.D., Yang, F., Zhao, Z., Wolkenberg, S.E., Nolt, M.B. and Lindsley, C.W. (2006) *Heterocycles*, **70**, 655–689.

6 Bogdal, D., Penczek, P., Pielichowski, J. and Prociak, A. (2003) *Advances in Polymer Science*, **163**, 193–263; Wiesbrock, F., Hoogenboom, R. and Schubert, U.S. (2004) *Macromolecular Rapid Communications*, **25**, 1739–1764; Hoogenboom, R. and Schubert, U.S. (2007) *Macromolecular Rapid Communications*, **28**, 368–386; Bogdal, D. and Prociak, A. (2007) *Microwave-Enhanced Polymer Chemistry and Technology*, Blackwell Publishing, Oxford; Zhang, C., Liao, L. and Gong, S. (2007) *Green Chemistry*, **9**, 303–314; Sinwell, S. and Ritter, H. (2007) *Australian Journal of Chemistry*, **60**, 729–743.

7 Barlow, S. and Marder, S.R. (2003) *Advanced Functional Materials*, **13**, 517–518; Zhu, Y.-J., Wang, W.W., Qi, R.-J. and Hu, X.-L. (2004) *Angewandte Chemie-International Edition*, **43**, 1410–1414; Perelaer, J., de Gans, B.-J. and Schubert, U.S. (2006) *Advanced Materials*, **18**, 2101–2104; Jhung, S.H., Jin, T., Hwang, Y.K. and Chang, J.-S. (2007) *Chemistry – A European Journal*, **13**, 4410–4417; Millos, C.J., Whittaker, A.G. and Brechin, E.K. (2007) *Polyhedron*, **26**, 1927–1933.

8 Tsuji, M., Hashimoto, M., Nishizawa, Y., Kubokawa, M. and Tsuji, T. (2005) *Chemistry – A European Journal*, **11**, 440–452; Tompsett, G.A., Conner, W.C. and Yngvesson, K.S. (2006) *ChemPhysChem*, **7**, 296–319; Langa, F. and de la Cruz, P. (2007) *Combinatorial Chemistry & High Throughput Screening*, **10**, 766–782.

9 Collins, J.M. and Leadbeater, N.E. (2007) *Organic and Biomolecular Chemistry*, **5**, 1141–1150; Lill, J.R., Ingle, E.S., Liu, P.S., Pham, V. and Sandoval, W.N. (2007) *Mass Spectrometry Reviews*, **26**, 657–671; Sandoval, W.N., Pham, V., Ingle, E.S., Liu, P.S. and Lill, J.R. (2007) *Combinatorial Chemistry & High Throughput Screening*, **10**, 751–765; Rejasse, B., Lamare, S., Legoy, M.-D. and Besson, T. (2007) *Journal of Enzyme Inhibition and Medicinal Chemistry*, **22**, 518–526.

10 Loupy, A., Petit, A., Hamelin, J., Texier-Boullet, F., Jacquault, P. and Mathé, D. (1998) *Synthesis*, 1213–1234; Varma, R.S. (1999) *Green Chemistry*, **1**, 43–55; Kidawi, M. (2001) *Pure and Applied Chemistry*, **73**, 147–151; Varma, R.S. (2001) *Pure and Applied Chemistry*, **73**, 193–198; Varma, R.S. (2002) *Tetrahedron*, **58**, 1235–1255; Varma, R.S. (2002) *Advances in Green Chemistry: Chemical Syntheses Using Microwave Irradiation*, Kavitha Printers, Bangalore; Varma, R.S. and Ju, Y. (2006) in *Microwaves in Organic Synthesis*, 2nd edn (ed. A. Loupy), Wiley-VCH, Weinheim, pp. 362–415 (Chapter 8); Besson, T., Thiéry, V. and Dubac, J. (2006) in *Microwaves in Organic Synthesis*, 2nd edn (ed. A. Loupy), Wiley-VCH, Weinheim, pp. 416–455 (Chapter 9).

11 Bose, A.K., Banik, B.K., Lavlinskaia, N., Jayaraman, M. and Manhas, M.S. (1997) *Chemtech*, **27**, 18–24; Bose, A.K., Manhas, M.S., Ganguly, S.N., Sharma, A.H. and Banik, B.K. (2002) *Synthesis*, 1578–1591.

12 Kuhnert, N. (2002) *Angewandte Chemie-International Edition*, **41**, 1863–1866; Strauss, C.R. (2002) *Angewandte Chemie-International Edition*, **41**, 3589–3591.

13 Perreux, L. and Loupy, A. (2001) *Tetrahedron*, **57**, 9199–9223; Perreux, L. and Loupy, A. (2002) in *Microwaves in Organic Synthesis* (ed. A. Loupy), Wiley-VCH, Weinheim, pp. 61–114 (Chapter 3); de La Hoz, A., Diaz-Ortiz, A. and Moreno, A. (2004) *Current Organic Chemistry*, **8**, 903–918; Perreux, L. and Loupy, A. (2006) in *Microwaves in Organic Synthesis*, 2nd edn (ed. A. Loupy), Wiley-VCH, Weinheim, pp. 134–218 (Chapter 4); Loupy, A. and Varma, R.S. (2006) *Chimica Oggi-Chemistry Today*, **24** (3), 36–39; de La Hoz, A., Díaz-Ortiz, A. and Moreno, A. (2005) *Chemical Society Reviews*, **34**,

164–178; de La Hoz, A., Díaz-Ortiz, A. and Moreno, A. (2005) *Advances in Organic Synthesis*, **1**, 119–171; de La Hoz, A., Díaz-Ortiz, A. and Moreno, A. (2006) in *Microwaves in Organic Synthesis*, 2nd edn (ed. A. Loupy), Wiley-VCH, Weinheim, pp. 219–277 (Chapter 5).

14 Strauss, C.R. and Trainor, R.W. (1995) *Australian Journal of Chemistry*, **48**, 1665–1692; Strauss, C.R. (1999) *Australian Journal of Chemistry*, **52**, 83–96; Strauss, C.R. (2002) in *Microwaves in Organic Synthesis* (ed. A. Loupy), Wiley-VCH, Weinheim, pp. 35–60 (Chapter 2); Roberts, B.A. and Strauss, C.R. (2005) *Accounts of Chemical Research*, **38**, 653–661.

15 For more technical reviews, see: Nüchter, M., Müller, U., Ondruschka, B., Tied, A. and Lautenschläger, W. (2003) *Chemical Engineering & Technology*, **26**, 1207–1216; Nüchter, M., Ondruschka, B., Bonrath, W. and Gum, A. (2004) *Green Chemistry*, **6**, 128–141; Nüchter, M., Ondruschka, B., Weiß, D., Bonrath, W. and Gum, A. (2005) *Chemical Engineering & Technology*, **28**, 871–881; Kremsner, J.M., Stadler, A. and Kappe, C.O. (2006) *Topics in Current Chemistry*, **266**, 233–278; Kirschning, A., Solodenko, W. and Mennecke, K. (2006) *Chemistry – A European Journal*, **12**, 5972–5990; Stankiewicz, A. (2006) *Chemical Engineering Research & Design*, **84**, 511–521; Schwalbe, T. and Simons, K. (2006) *Chimica Oggi-Chemistry Today*, **24** (3), 56–61; Baxendale, I.R. and Pitts, M.R. (2006) *Chimica Oggi-Chemistry Today*, **24** (3), 41–45; Leveque, J.-M. and Cravotto, G. (2006) *Chimia*, **60**, 313–320; Glasnov, T.N. and Kappe, C.O. (2007) *Macromolecular Rapid Communications*, **28**, 395–410; Cravotto, G. and Cintas, P. (2007) *Chemistry – A European Journal*, **13**, 1902–1909; Petricci, E. and Taddei, M. (2007) *Chimica Oggi-Chemistry Today*, **25**, 40–45; Matloobi, M. and Kappe, C.O. (2007) *Chimica Oggi-Chemistry Today*, **25**, 26–31; Kappe, C.O. and Matloobi, M. (2007) *Combinatorial Chemistry & High Throughput Screening*, **10**, 735–750;

Baxendale, I.R., Hayward, J.J. and Ley, S.L. (2007) *Combinatorial Chemistry & High Throughput Screening*, **10**, 802–836.

16 Organic synthesis: de la Hoz, A., Díaz-Ortiz, A., Moreno, A. and Langa, F. (2000) *European Journal of Organic Chemistry*, 3659–3673; Larhed, M., Moberg, C. and Hallberg, A. (2002) *Accounts of Chemical Research*, **35**, 717–727; Corsaro, A., Chiacchio, U., Pistara, V. and Romeo, G. (2004) *Current Organic Chemistry*, **8**, 511–538; Xu, Y. and Guo, Q.-X. (2004) *Heterocycles*, **63**, 903–974; Romanova, N.N., Gravis, A.G. and Zyk, N.V. (2005) *Russian Chemical Reviews*, **74**, 969–1013; de la Hoz, A., Díaz-Ortiz, A. and Moreno, A. (2005) *Advances in Organic Synthesis*, **1**, 119–171; Rakhmankulov, D.L., Shavshukova, S.Yu. and Latypova, F.N. (2005) *Chemistry of Heterocyclic Compounds*, **41**, 951–961; Kuznetsov, D.V., Raev, V.A., Kuranov, G.L., Arapov, O.V. and Kostikov, R.R. (2005) *Russian Journal of Organic Chemistry*, **41**, 1719–1749; Molteni, V. and Ellis, D.A. (2005) *Current Organic Synthesis*, **2**, 333–375; Leadbeater, N.E. (2005) *Chemical Communications*, **23**, 2881–2902;Bougrin, K., Loupy, A. and Soufiaoui, M. (2005) *Journal of Photochemistry and Photobiology C: Photochemistry Reviews*, **6**, 139–167; Strauss, C.R. and Varma, R.S. (2006) *Topics in Current Chemistry*, **266**, 199–231; Nilsson, P., Olofsson, K. and Larhed, M. (2006) *Topics in Current Chemistry*, **266**, 103–144; Suna, E. and Mutule, I. (2006) *Topics in Current Chemistry*, **266**, 49–101; Zhang, W. (2006) *Topics in Current Chemistry*, **266**, 145–166; Dallinger, D. and Kappe, C.O. (2007) *Chemical Reviews*, **107**, 2563–2591; Appukkuttan, P. and Van der Eycken, E. (2008) *European Journal of Organic Chemistry*, 1133–1155; Coquerel, Y. and Rodriguez, J. (2008) *European Journal of Organic Chemistry*, 1125–1132; Polshettiwar, V. and Varma, R.S. (2008) *Accounts of Chemical Research*, **41**, 629–639.

17 Radiochemistry: Elander, N., Jones, J.R., Lu, S.-Y. and Stone-Elander, S. (2000)

Chemical Society Reviews, **29**, 239–250; Stone-Elander, S. and Elander, N. (2002) *Journal of Labelled Compounds & Radiopharmaceu-ticals*, **45**, 715–746; Stone-Elander, S., Elander, N., Thorell, J.-O. and Fredriksson, A. (2007) *Ernst Schering Research Foundation Workshop*, **62**, pp. 243–269.

18 **Combinatorial chemistry**: Kappe, C.O. (2002) *Current Opinion in Chemical Biology*, **6**, 314–320; Lew, A., Krutznik, P.O., Hart, M.E. and Chamberlin, A.R. (2002) *Journal of Combinatorial Chemistry*, **4**, 95–105; Lidström, P., Westman, J. and Lewis, A. (2002) *Combinatorial Chemistry & High Throughput Screening*, **5**, 441–458; Blackwell, H.E. (2003) *Organic and Biomolecular Chemistry*, **1**, 1251–1255; Swamy, K.M.K., Yeh, W.-B., Lin, M.-J. and Sun, C.-M. (2003) *Current Medicinal Chemistry*, **10**, 2403–2423; Desai, B. and Kappe, C.O. (2004) *Topics in Current Chemistry*, **242**, 177–208; Santagada, V., Frecentese, F., Perissutti, E., Favretto, L. and Caliendo, G. (2004) *The QSAR & Combinatorial Science*, **23**, 919–944; Bhattacharyya, S. (2005) *Molecular Diversity*, **9**, 253–257; Lange, T. and Lindell, S. (2005) *Combinatorial Chemistry & High Throughput Screening*, **8**, 595–606; Martinez-Palou, R. (2006) *Molecular Diversity*, **10**, 435–462.

19 **Dielectric heating**: Gabriel, C., Gabriel, S., Grant, E.H., Halstead, B.S. and Mingos, D.M.P. (1998) *Chemical Society Reviews*, **27**, 213–224; Mingos, D.M.P. and Baghurst, D.R. (1991) *Chemical Society Reviews*, **20**, 1–47; Leadbeater, N.E., Torenius, H.M. and Tye, H. (2004) *Combinatorial Chemistry & High Throughput Screening*, **7**, 511–528; Habermann, J., Ponzi, S. and Ley, S.V. (2005) *Mini-Reviews in Organic Chemistry*, **2**, 125–137.

20 Kappe, C.O.(ed.) (2003) Microwaves in Combinatorial and High-Throughput Synthesis (a special issue). *Molecular Diversity*, **7**, 95–307; Van der Eycken, E. and Van der Eycken, J.(eds) (2004) Microwaves in Combinatorial and High-Throughput

Synthesis, (a special issue). *The QSAR & Combinatorial Science*, **23** (10), 823–986; Leadbeater, N.E.(ed.) (2006) Microwave-Assisted Synthesis (a special issue). *Tetrahedron*, **62** (19), 4623–4732; de la Hoz, A. and Díaz-Ortiz, A.(eds) (2007) The Use of Microwaves in High Throughput Synthesis (a special issue). *Combinatorial Chemistry and High Throughput Screening*, **10** (9/10), 773–934.

21 Stinson, S. (1996) *Chemical & Engineering News*, **74**, 45–46; Dagani, R. (1997) *Chemical & Engineering News*, **75**, 26–33; Whittaker, G. (1998) *New Scientist*, **2123**, 34; Cresswell, S.L. and Haswell, S.J. (1999) *Chemistry & Industry*, 621–624; Edwards, P. (2001) *Drug Discovery Today*, **6**, 614; Cresswell, S.L. and Haswell, S.J. (2001) *Journal of Chemical Education*, **78**, 900–904; Dutton, G. (2002) *Genetic Engineering News*, **22**, 13–17; Watkins, K.J. (2002) *Chemical & Engineering News*, **80** (6), 17–18; Adam, D. (2003) *Nature*, **421**, 571–572; Leadbeater, N.E. (2004) *Chemistry World*, **1**, 38–41; Minkel, J.R. (2004) *Drug Discovery & Development*, **7** (3), 47–52; Marx, V. (2004) *Chemical & Engineering News*, **82** (50), 14–19; Yarnell, A. (2007) *Chemical & Engineering News*, **85** (21), 32–33.

22 Biotage Pathfinder Database (http://www.biotagepathfinder.com); Microwave-Assisted Organic Synthesis Database (http://www.mwchemdb.com); Microwave Chemistry Literature Highlights (http://www.organic-chemistry.org/Highlights/microwave.shtm).

23 Websites on microwave synthesis: http://www.maos.net; http://microwavesynthesis.net.

24 Katritzky, A.R., Cai, C., Collins, M.D., Scriven, E.F.V., Singh, S.K. and Barnhardt, E.K. (2006) *Journal of Chemical Education*, **83**, 634–636; Murphree, S.S. and Kappe, C.O. (2008) *Journal of Chemical Education*, **85**, in press.

25 McGowan, C. and Leadbeater, N.E. (2006) *Clean, Fast Organic Chemistry: Microwave-*

Assisted Laboratory Experiments, CEM Publishing, Matthews, NC; Bogdal, D. (2005) *Microwave-assisted Organic Synthesis One Hundred Reaction Procedures*, Elsevier, Oxford.

26 Moseley, J.D., Lenden, P., Thomson, A.D. and Gilday, J.P. (2007) *Tetrahedron Letters*, **48**, 6084–6087 (Ref. 13).

27 *Journal of Organic Chemistry*, 73, 2008 Issue 1.

2
Microwave Theory

The physical principles behind and the factors determining the successful application of microwaves in organic synthesis are not widely familiar to chemists. Nevertheless, it is essential for the synthetic chemist involved in microwave-assisted organic synthesis to have at least a basic knowledge of the underlying principles of microwave–matter interactions and of the nature of microwave effects. The basic understanding of macroscopic microwave interactions with matter was formulated by von Hippel in the mid-1950s [1]. In this chapter a brief summary of the current understanding of microwaves and their interactions with matter is given. For more in-depth discussion on this quite complex field, the reader is referred to recent review articles [2–5].

2.1
Microwave Radiation

Microwave irradiation is electromagnetic irradiation in the frequency range 0.3 to 300 GHz, corresponding to wavelengths of 1 mm to 1 m. The microwave region of the electromagnetic spectrum (Figure 2.1) therefore lies between infrared and radio frequencies. The major use of microwaves is either for transmission of information (telecommunication) or for transmission of energy. Wavelengths between 1 and 25 cm are extensively used for RADAR transmissions and the remaining wavelength range is used for telecommunications. All domestic "kitchen" microwave ovens and all dedicated microwave reactors for chemical synthesis that are commercially available today operate at a frequency of 2.45 GHz (corresponding to a wavelength of 12.25 cm) in order to avoid interference with telecommunication, wireless networks and cellular phone frequencies. There are other frequency allocations for microwave heating applications (ISM (industrial, scientific and medical) frequencies, see Table 2.1) [6], but these are not generally employed in dedicated reactors for synthetic chemistry. Indeed, published examples of organic synthesis carried out with microwave heating at frequencies other than 2.45 GHz are extremely rare [7].

From comparison of the data presented in Table 2.2 [8], it is obvious that the energy of the microwave photon at a frequency of 2.45 GHz (0.0016 eV) is too low to cleave

Practical Microwave Synthesis for Organic Chemists: Strategies, Instruments, and Protocols
C. Oliver Kappe, Doris Dallinger, and S. Shaun Murphree
Copyright © 2009 WILEY-VCH Verlag GmbH & Co. KGaA, Weinheim
ISBN: 978-3-527-32097-4

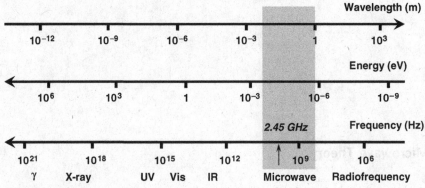

Figure 2.1 The electromagnetic spectrum.

Table 2.1 ISM microwave frequencies (data from Ref. [6]).

Frequency (MHz)	Wavelength (cm)
$433.92 \pm 0.2\%$	69.14
915 ± 13	32.75
2450 ± 50	12.24
5800 ± 75	5.17
$24\,125 \pm 125$	1.36

Table 2.2 Comparison of radiation types and bond energies (data from Refs. [6, 8]).

Radiation type	Frequency (MHz)	Quantum energy (eV)	Bond type	Bond energy (eV)
Gamma rays	3.0×10^{14}	1.24×10^{6}	C–C	3.61
X-rays	3.0×10^{13}	1.24×10^{5}	C=C	6.35
Ultraviolet	1.0×10^{9}	4.1	C–O	3.74
Visible light	6.0×10^{8}	2.5	C=O	7.71
Infrared light	3.0×10^{6}	0.012	C–H	4.28
Microwaves	2450	0.0016	O–H	4.80
Radiofrequencies	1	4.0×10^{-9}	hydrogen bond	0.04–0.44

molecular bonds and is also lower than Brownian motion. It is therefore clear that microwaves cannot "induce" chemical reactions by direct absorption of electromagnetic energy, as opposed to ultraviolet and visible radiation (photochemistry).

2.2
Microwave Dielectric Heating

Microwave chemistry is based on the efficient heating of materials by "microwave dielectric heating" effects [4, 5]. Microwave dielectric heating is dependent on the

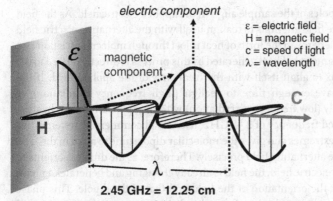

electric component

ε = electric field
H = magnetic field
c = speed of light
λ = wavelength

magnetic component

2.45 GHz = 12.25 cm

Figure 2.2 Electric and magnetic field components in microwaves.

ability of a specific material (e.g. a solvent or reagent) to absorb microwave energy and convert it into heat. Microwaves are electromagnetic waves which consist of an electric and a magnetic field component (Figure 2.2). For most practical purposes related to microwave synthesis it is the electric component of the electromagnetic field that is of importance for wave–material interactions, although in some instances magnetic field interactions (e.g. with transition metal oxides) can also be of relevance [9].

The electric component of an electromagnetic field causes heating by two main mechanisms: dipolar polarization and ionic conduction. The interaction of the electric field component with the matrix is called the dipolar polarization mechanism (Figure 2.3a) [4, 5]. For a substance to be able to generate heat when irradiated with microwaves it must possess a dipole moment. When exposed to microwave

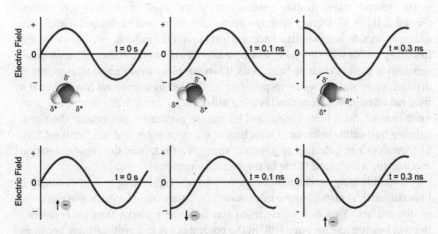

Figure 2.3 (a) Dipolar polarization mechanism. Dipolar molecules try to align with an oscillating electric field.
(b) Ionic conduction mechanism. Ions in solution will move in the electric field.

frequencies, the dipoles of the sample align in the applied electric field. As the field oscillates, the dipole field attempts to realign itself with the alternating electric field and, in the process, energy in the form of heat is lost through molecular friction and dielectric loss. The amount of heat generated by this process is directly related to the ability of the matrix to align itself with the frequency of the applied field. If the dipole does not have enough time to realign (high frequency irradiation) or reorients too quickly (low frequency irradiation) with the applied field, no heating occurs. The allocated frequency of 2.45 GHz, used in all commercial systems, lies between these two extremes and gives the molecular dipole time to align in the field but not to follow the alternating field precisely. Therefore, as the dipole reorients to align itself with the electric field, the field is already changing and generates a phase difference between the orientation of the field and that of the dipole. This phase difference causes energy to be lost from the dipole by molecular friction and collisions, giving rise to dielectric heating. In summary, field energy is transferred to the medium and electrical energy is converted into kinetic or thermal energy, and ultimately into heat. It should be emphasized that the interaction between microwave radiation and the polar solvent, which occurs when the frequency of the radiation approximately matches the frequency of the rotational relaxation process, is not a quantum mechanical resonance phenomenon. Transitions between quantized rotational bands are not involved and the energy transfer is not a property of a specific molecule but the result of a collective phenomenon involving the bulk [4, 5]. The heat is generated by frictional forces occurring between the polar molecules whose rotational velocity has been increased by the coupling with the microwave irradiation. It should also be noted that gases cannot be heated under microwave irradiation since the distance between the rotating molecules is too far. Similarly, ice is also (nearly) microwave transparent, since the water dipoles are constrained in a crystal lattice and cannot move as freely as in the liquid state.

The second major heating mechanism is the ionic conduction mechanism (Figure 2.3b) [4, 5]. During ionic conduction, as the dissolved charged particles in a sample (usually ions) oscillate back and forth under the influence of the microwave field, they collide with their neighboring molecules or atoms. These collisions cause agitation or motion, creating heat. Thus, if two samples containing equal amounts of distilled water and tap water, respectively, are heated by microwave irradiation at a fixed radiation power, more rapid heating will occur for the tap water sample due to its ionic content. Such ionic conduction effects are particularly important when considering the heating behavior of ionic liquids in a microwave field (see Section 4.5.2). The conductivity principle is a much stronger effect than the dipolar rotation mechanism with regard to the heat-generating capacity.

A related heating mechanism exists for strongly conducting or semiconducting materials such as metals, where microwave irradiation can induce a flow of electrons on the surface. This flow of electrons can heat the material through resistance (ohmic) heating mechanisms [10]. In the context of organic synthesis this becomes important for heating strongly microwave absorbing materials, such as thin metal films (Pd, Au) (see Section 4.8.4), graphite supports (see Section 4.1) or so-called passive heating elements made out of silicon carbide (see Section 4.6).

2.3
Dielectric Properties

The heating characteristics of a particular material (e.g. a solvent) under microwave irradiation conditions are dependent on the dielectric properties of the material. The ability of a specific substance to convert electromagnetic energy into heat at a given frequency and temperature is determined by the so-called loss tangent, $\tan \delta$. The loss factor is expressed as the quotient, $\tan \delta = \varepsilon''/\varepsilon'$, where ε'' is the dielectric loss, indicative of the efficiency with which electromagnetic radiation is converted into heat, and ε' is the dielectric constant describing the polarizability of molecules in the electric field. A reaction medium with a high $\tan \delta$ is required for efficient absorption and, consequently, for rapid heating. Materials with a high dielectric constant, such as water (ε' at $25\,^{\circ}C = 80.4$), may not necessarily also have a high $\tan \delta$ value. In fact, ethanol has a significantly lower dielectric constant (ε' at $25\,^{\circ}C = 24.3$) but heats much more rapidly than water in a microwave field due to its higher loss tangent ($\tan \delta$: ethanol $= 0.941$, water $= 0.123$). The loss tangents for some common organic solvents are summarized in Table 2.3 [11]. In general, solvents can be classified as high ($\tan \delta > 0.5$), medium ($\tan \delta$ 0.1–0.5), and low microwave absorbing ($\tan \delta < 0.1$). Other common solvents without a permanent dipole moment, such as carbon tetrachloride, benzene and dioxane, are more or less microwave transparent. It has to be emphasized that a low $\tan \delta$ value does not preclude a particular solvent from being used in a microwave-heated reaction. Since either the substrates or some of the reagents/catalysts are likely to be polar, the overall dielectric properties of the reaction medium will, in most cases, allow sufficient heating by microwaves. Furthermore, polar additives (such as alcohols or ionic liquids) or passive heating elements can be added to otherwise low-absorbing reaction mixtures in order to increase the absorbance level of the medium (see Sections 4.5.2 and 4.6).

The loss tangent values are both frequency and temperature dependent. Figure 2.4 shows the dielectric properties of distilled water as a function of frequency at

Table 2.3 Loss tangents ($\tan \delta$) of different solvents (2.45 GHz, 20 °C; data from Ref. [11]).

Solvent	tan δ	Solvent	tan δ
Ethylene glycol	1.350	N,N-dimethylformamide	0.161
Ethanol	0.941	1,2-dichloroethane	0.127
Dimethylsulfoxide	0.825	Water	0.123
2-propanol	0.799	Chlorobenzene	0.101
Formic acid	0.722	Chloroform	0.091
Methanol	0.659	Acetonitrile	0.062
Nitrobenzene	0.589	Ethyl acetate	0.059
1-butanol	0.571	Acetone	0.054
2-butanol	0.447	Tetrahydrofuran	0.047
1,2-dichlorobenzene	0.280	Dichloromethane	0.042
1-methyl-2-pyrrolidone	0.275	Toluene	0.040
Acetic acid	0.174	Hexane	0.020

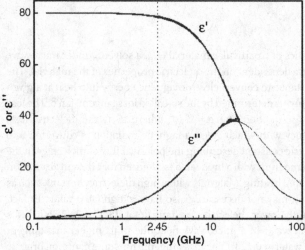

Figure 2.4 Dielectric properties of water as a function of frequency at 25 °C [12].

25 °C [1, 4, 5]. It is apparent that appreciable values of the dielectric loss ε'' exist over a wide frequency range. The dielectric loss ε'' goes through a maximum as the dielectric constant ε' falls. The heating, as measured by ε'', reaches its maximum around 18 GHz, while all domestic microwave ovens and dedicated reactors for chemical synthesis operate at a much lower frequency, 2.45 GHz. The practical reason for the lower frequency is the necessity to heat food efficiently throughout its interior. If the frequency is optimal for a maximum heating rate, the microwaves are absorbed in the outer regions of the food, and penetrate only a short distance ("skin effect") [4].

According to definition, the penetration depth is the point where 37% ($1/e$) of the initially irradiated microwave power is still present [6]. The penetration depth is inversely proportional to $\tan \delta$ and therefore depends critically on factors such as temperature and irradiation frequency. Materials with relatively high $\tan \delta$ values are thus characterized by low values of penetration depth and therefore microwave irradiation may be totally absorbed within the outer layers of these materials. For a solvent such as water ($\tan \delta = 0.123$ at 25 °C and 2.45 GHz), the penetration depth at room temperature is only of the order of a few centimeters (Table 2.4). Beyond this penetration depth, volumetric heating due to absorption of microwave energy becomes negligible. This means that, during microwave experiments on a larger scale, only the outer layers of the reaction mixture may be directly heated by microwave irradiation via dielectric heating mechanisms. The inner part of the reaction mixture will, to a large extent, be heated by conventional heat convection and/or conduction mechanisms. Issues relating to the penetration depth are therefore critically important when considering the scale-up of MAOS (see Section 4.8).

The dielectric loss and loss tangent of pure water and most other organic solvents decrease with increasing temperature (Figure 2.5). The absorption of microwave

Table 2.4 Penetration depth of some common materials (data from Ref. [10]).

Material	Temperature (°C)	Penetration depth (cm)
Water	25	1.4
Water	95	5.7
Ice	−12	1100
Polyvinylchloride	20	210
Glass	25	35
Teflon	25	9200
Quartz glass	25	16 000

radiation in water therefore decreases at higher temperatures. While it is relatively easy to heat water from room temperature to 100 °C by 2.45 GHz microwave irradiation, it is significantly more difficult to further heat water to 200 °C and beyond in a sealed vessel. In fact, supercritical water ($T > 374$ °C) is transparent to microwave irradiation (see Section 4.5.1).

Most organic materials and solvents behave similarly to water, in the sense that the dielectric loss ε'' will decrease with increasing temperature [2–5]. From the practical point of view this may be somewhat inconvenient, since microwave heating at higher temperatures may often be compromised. On the other hand, from the standpoint of safety, it should be stressed that the opposite situation may lead to a scenario where a material will become a stronger microwave absorber

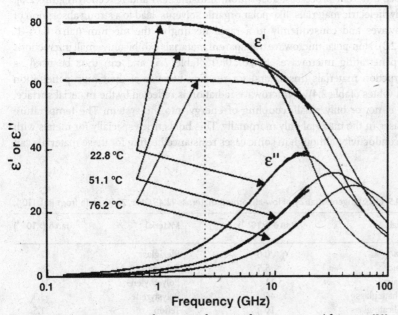

Figure 2.5 Dielectric properties of water as a function of temperature and frequency [12].

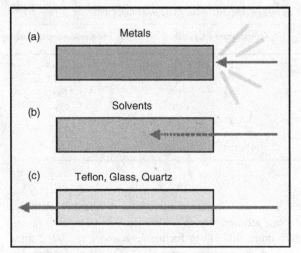

Figure 2.6 Interaction of microwaves with different materials.
(a) Electrical conductors, (b) absorbing materials (tan δ 0.05–1),
(c) insulators (tan δ < 0.01).

with increasing temperature. This is the case for some inorganic/polymeric materials [4], and will lead to the danger of a thermal runaway during microwave heating.

In summary, the interaction of microwave irradiation with matter is characterized by three different processes: absorption, transmission and reflection (Figure 2.6). Highly dielectric materials, like polar organic solvents, lead to a strong absorption of microwaves and consequently to a rapid heating of the medium (tan δ 0.05–1, Table 2.3). Non-polar microwave transparent materials exhibit only small interactions with penetrating microwaves (tan δ < 0.01, Table 2.5) and can thus be used as construction materials (insulators) for reactors because of their high penetration depth values (Table 2.4). If microwave radiation is reflected by the material surface, there is no, or only small, coupling of energy into the system. The temperature increases in the material only marginally. This holds true especially for metals with high conductivity, although in some cases resistance heating for these materials can occur.

Table 2.5 Loss tangents (tan δ) of low-absorbing materials (2.45 GHz, 25 °C; data from Ref. [10]).

Material	tan δ (×10⁻⁴)	Material	tan δ (×10⁻⁴)
Quartz	0.6	Plexiglass	57
Ceramic	5.5	Polyester	28
Porcelain	11	Polyethylene	31
Phosphate glass	46	Polystyrene	3.3
Borosilicate glass	10	Teflon	1.5

2.4
Microwave versus Conventional Thermal Heating

Traditionally, organic synthesis is carried out by conductive heating with an external heat source (e.g. an oil-bath or heating mantle). This is a comparatively slow and inefficient method for transferring energy into the system since it depends on convection currents and on the thermal conductivity of the various materials that must be penetrated, and generally results in the temperature of the reaction vessel being higher than that of the reaction mixture (Figure 2.7). This is particularly true if reactions are performed under reflux conditions, whereby the temperature of the bath fluid is typically kept at 10–30 °C above the boiling point of the reaction mixture in order to ensure an efficient reflux. In addition, a temperature gradient can develop within the sample and local overheating can lead to product, substrate or reagent decomposition.

In contrast, microwave irradiation produces efficient internal heating (in core volumetric heating) by direct coupling of microwave energy with the molecules (solvents, reagents, catalysts) that are present in the reaction mixture. Microwave irradiation, therefore, raises the temperature of the whole volume simultaneously (bulk heating) whereas in the conventionally heated vessel, the reaction mixture in contact with the vessel wall is heated first (Figure 2.7a). Since the reaction vessels employed in modern microwave reactors are typically made out of (nearly) microwave transparent materials such as borosilicate glass, quartz or Teflon (Table 2.5), the radiation passes through the walls of the vessel and an inverted temperature gradient as compared to conventional thermal heating results. If the microwave cavity is well designed, the temperature increase will be uniform throughout the sample (see Section 2.5.1). The very efficient internal heat transfer results in minimized wall effects (no hot vessel surface) which may lead to the observation of so-called specific microwave effects (see Section 2.5.3), for example in the context of diminished catalyst deactivation. It should be emphasized that microwave dielectric heating and

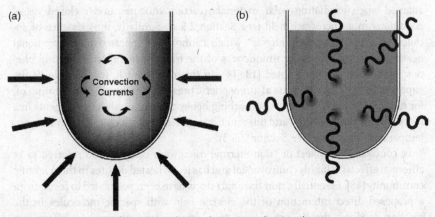

Figure 2.7 Comparison of conventional (a) and microwave heating (b).

thermal heating by convection are totally different processes, and that any comparison between the two is inherently difficult (Box 2.1).

Box 2.1
Summary: Characteristics of microwaves and microwave–matter interaction.

- Electromagnetic waves
- Low energy photon (does not break chemical bonds)
- Causes movement of molecules (dipole rotation)
- Causes movement of ions (ionic conduction)
- Will be reflected, transmitted or absorbed
- Volumetric heating throughout an absorbing material.

2.5
Microwave Effects

Despite the relatively large body of published work on microwave-assisted chemistry (Figure 1.1), and the basic understanding of high-frequency electromagnetic irradiation and microwave–matter interactions, the exact reasons why and how microwaves enhance chemical processes are still not fully understood. Since the early days of microwave synthesis, the observed rate-accelerations and sometimes altered product distributions compared to conventionally heated experiments have led to speculations on the existence of so-called "specific" or "non-thermal" microwave effects [13, 14]. Such effects have been claimed when the outcome of a synthesis performed under microwave conditions was different from the conventionally heated counterpart at the same measured reaction temperature. Today it is generally agreed that, in most standard cases, the observed enhancements in microwave-heated reactions are in fact the result of purely thermal/kinetic effects, in other words, are a consequence of the high reaction temperatures that can rapidly be attained when irradiating polar materials/reaction mixtures under closed vessel conditions in a microwave field (see Section 2.5.2). Similarly, the existence of so-called "specific microwave effects" which cannot be duplicated by conventional heating and result from the uniqueness of the microwave dielectric heating phenomenon is largely undisputed [13, 14]. In this category fall, for example (i) the superheating effect of solvents at atmospheric pressure, (ii) the selective heating of, for example, strongly microwave absorbing heterogeneous catalysts or reagents in a less polar reaction medium, and (iii) the elimination of wall effects caused by inverted temperature gradients (see Section 2.5.3).

In contrast, the subject of "non-thermal microwave effects" (also referred to as athermal effects) is highly controversial and has led to heated debates in the scientific community [15]. Essentially, non-thermal effects have been postulated to result from a proposed direct interaction of the electric field with specific molecules in the reaction medium that is not related to a macroscopic temperature effect (see

Section 2.5.4) [13, 14]. It has been argued, for example, that the presence of an electric field leads to orientation effects of dipolar molecules or intermediates and hence changes the pre-exponential factor A or the activation energy (entropy term) in the Arrhenius equation for certain types of reactions. Furthermore, a similar effect has been proposed for polar reaction mechanisms, where the polarity is increased on going from the ground state to the transition state, resulting in an enhancement of reactivity by lowering of the activation energy. Significant non-thermal microwave effects have been suggested for a wide variety of synthetic transformations [13, 14].

It should be obvious from a scientific standpoint that the question of non-thermal microwave effects needs to be addressed in a serious manner, given the rapid increase in the use of microwave technology in the chemical sciences, in particular in organic synthesis. There is an urgent need to provide a scientific rationalization for the observed effects and to investigate the general influence of the electric field (and therefore of the microwave power) on chemical transformations. This is even more important if one considers engineering and safety aspects once this technology moves from small scale laboratory work to pilot or production scale instrumentation. Although the detailed discussion of microwave effects lies outside the scope of this introductory book, the present chapter provides a summary of the basic concepts of relevance to the microwave chemistry practitioner.

Historically, microwave effects were claimed when the outcome of a synthesis performed under microwave conditions was different from the conventionally heated counterpart at the same apparent temperature. An extreme example is highlighted in Scheme 2.1. Here, Soufiaoui and coworkers [16] have synthesized a series of 1,5-aryldiazepin-2-ones in high yield in only 10 min by the condensation of *ortho*-aryldiamines with β-ketoesters in xylene under microwave irradiation in an open vessel at reflux temperature, utilizing a conventional domestic microwave oven. Surprisingly, they observed that no reaction occurred when the same reactions were heated conventionally for 10 min at the same temperature. In their publication, the authors specifically point to the involvement of "specific effects (which are not necessarily thermal)" in rationalizing the observed product yields. These results could be taken as clear evidence for a specific microwave effect. Interestingly, Gedye and Wei have later reinvestigated the exact same reaction under thermal and microwave conditions and found that there is virtually no difference in the rate of the microwave and the conventionally heated reactions, leading to similar product

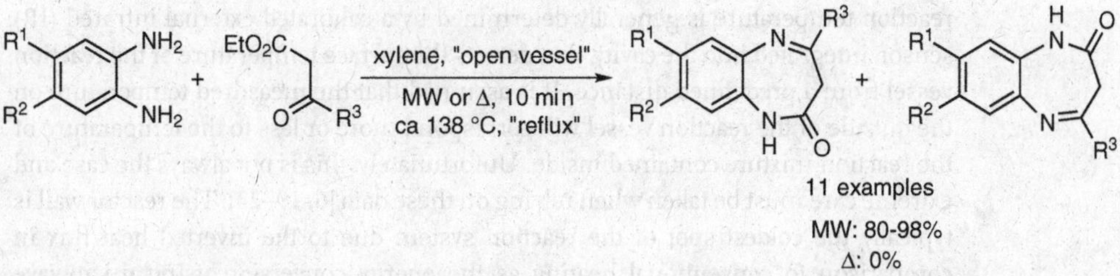

11 examples

MW: 80–98%

Δ: 0%

Scheme 2.1 Molecular magic with microwaves?

yields [7, 17]. The literature is full of examples like the one highlighted above, with conflicting reports on the involvement or non-involvement of "specific" or "non-thermal" microwave effects for a wide variety of different types of chemical reactions [13–15]. Microwave effects are the subject of considerable current debate and controversy and it is evident that extensive research efforts are necessary in order to truly understand these and related phenomena.

Essentially, one can envision three different possibilities for rationalizing rate-enhancements observed in a microwave-assisted chemical reaction [18]:

- Thermal effects (kinetics)
- Specific microwave effects
- Non-thermal (athermal) microwave effects.

Clearly, a combination of two or all three contributions may be responsible for the observed phenomena, which makes the investigation of microwave effects an extremely complex subject. Before discussing the above-mentioned effects in detail, it is important to have an understanding of how the reaction temperature in a microwave-heated reaction can be adequately determined. In order to obtain reproducible and reliable results from a microwave-assisted reaction, it is absolutely essential to have an accurate way of directly measuring the temperature of the reaction mixture online during the irradiation process. This is even more important if a comparison with conventionally heated experiments is performed.

2.5.1
Temperature Monitoring in Microwave Chemistry

Dedicated microwave reactors for organic synthesis are in most cases operated in "temperature control" mode, which means that the desired reaction temperature is selected by the user (see Chapter 3). By coupling the feedback from a suitable temperature probe to the modulation of magnetron output power the reaction mixture is heated and kept at the pre-selected value (see e.g. Figures 2.12 and 2.13). This process requires a reliable way of rapidly monitoring the reaction temperature online during the microwave irradiation process. The correct temperature measurement in microwave-assisted reactions, however, often presents a problem since classical temperature sensors such as thermometers or metal-based thermocouples will fail as they will couple with the electromagnetic field [6]. In the most popular single-mode microwave reactors (Biotage Initiator, CEM Discover, see Chapter 3) the reaction temperature is generally determined by a calibrated external infrared (IR) sensor, integrated into the cavity, that detects the surface temperature of the reaction vessel from a predefined distance. It is assumed that the measured temperature on the outside of the reaction vessel will correspond more or less to the temperature of the reaction mixture contained inside. Unfortunately, this is not always the case and extreme care must be taken when relying on these data [6, 19–23]. The reactor wall is typically the coldest spot of the reaction system due to the inverted heat flux in comparison to conventional heating as the energy conversion using microwave irradiation takes place directly in the reaction mixture (Figure 2.7) [6].

A more accurate way is to determine the temperature of the reaction mixture directly by an internal probe such as a fiber-optic sensor or a gas-balloon thermometer [19–23]. Fiber optic probes are more accurate than IR sensors but are far more expensive. Another disadvantage, compared to other temperature measurement systems, is the generally more narrow operating range of 0 to 300 °C. In addition, permanent aging phenomena can already be observed above 250 °C after a few hours [6]. These probes are also very sensitive toward mechanical stress and one reason for the lower temperature resistance is the unavoidable use of polymers during their fabrication, for example for gluing the sensor crystal to the optical fiber. The routine use of fiber-optic probes in microwave-assisted synthesis is therefore often not practical. Fiber-optic probes are available to monitor internal reaction temperatures in the CEM Discover system and are also used in the multimode reactors from Milestone and CEM (see Chapter 3). A gas balloon thermometer is employed for internal temperature measurement with the Synthos 3000 multimode reactor from Anton Paar. In most multimode microwave instruments, the surface temperatures of the reaction vessels are additionally controlled by an external IR sensor. The lead sensor that controls the power input in this case is either a fiber-optic probe (Milestone, CEM) or a gas balloon thermometer (Anton Paar). In certain instances, it can also be of interest to investigate the temperature of a microwave-heated reaction mixture or vessel surface with the aid of a thermovision camera [23–25].

All the measurement systems described above have certain characteristics and specific advantages and disadvantages, as shown in Box 2.2. For routine synthetic

---

Box 2.2
Temperature sensors for microwave chemistry.

- Infrared sensors (external).
  temperature measured at the outside of reactor,
  typically registers a lower temperature,
  low cost,
  temperature range −40 °C to 400 °C.

- Fiber-optic probes (internal).
  accurate but costly and fragile,
  temperature range 0–330 °C,
  aging phenomena >250 °C.

- Shielded thermocouples (internal).
  inexpensive,
  act as antenna, self-absorbing.

- Gas balloon thermometer (internal).
  temperature range 0–300 °C,
  accurate but fragile (accuracy ±1 °C).

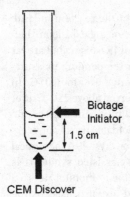

Figure 2.8 Position of infrared temperature sensors in single-mode microwave cavities from Biotage and CEM (10 mL reaction vessel).

applications in single-mode microwave reactors (Biotage Initiator, CEM Discover), the use of standard IR probes is acceptable, mainly because of convenience, the robust nature and low cost of these types of probes. However, the user should be aware of the limitations of these devices and should recognize situations where the use of these external probes is not appropriate. In general, external IR sensors will only represent the internal reaction temperature properly if efficient agitation of the homogeneous reaction mixture is ensured. Inefficient agitation can lead to temperature gradients within the reaction mixture due to field inhomogeneities in the high-density single-mode microwave cavities [23, 25, 26]. Extreme care must therefore be taken with heterogeneous reactions, such as solvent free, dry-media or highly viscous systems (see Section 4.1).

Additionally, it has to be emphasized that in the two most popular single-mode microwave reactors the temperature is measured at different positions of the otherwise more or less identical microwave vessels (Figure 2.8). Taking into account inherent field inhomogeneities that exist in both cavities [23], this fact in itself can lead to discrepancies when comparing the results obtained from running the exact same chemical reaction in the two systems [26]. It has to be noted that in the Biotage microwave systems a certain minimum filling volume must be used in order to ensure a proper temperature reading (see Section 5.3.3). These differences are aggravated for biphasic mixtures where one of the phases is strongly microwave absorbing and the other is only weakly absorbing. A case in point are, for example, unstirred biphasic mixtures of ionic liquids and nonpolar organic solvents where a strong differential heating (see Section 2.5.3) of the ionic liquid phase will occur [21]. Depending on the microwave system used, either the temperature of the very hot ionic liquid phase (IR from the bottom) or the temperature of the cooler organic layer (IR from the side) will be recorded (see also Figure 4.6).

Importantly, external IR sensors should never be used in conjunction with simultaneous external cooling of the reaction vessel (see Section 2.5.4). It has been demonstrated by several research groups that by using this technique the internal reaction temperatures will be significantly higher than recorded by the IR sensor on the outside [6, 19, 20, 22, 23]. When using simultaneous external cooling, an internal fiber-optic probe device must therefore be employed. Even without using external cooling, one should be aware of the fact that the IR sensor will need some time until it reflects the actual internal reaction temperature. This is because it will take a certain time for the reaction vessel, made of glass, to be warmed "from the inside" by microwave dielectric heating of its polar contents. Although this delay is typically only of the order of a few seconds, it may already suffice to lead to an undetected small overshooting of the internal reaction temperature, in particular in the case of strongly microwave-absorbing reaction mixtures that are rapidly heated by microwave irradiation.

In the case of low-absorbing or nearly microwave transparent reaction mixtures the opposite phenomenon may occur. Since the glass used for making the comparatively low-cost microwave process vials used in single-mode reactors is not completely microwave transparent (for loss tangents of different types of glasses, see Table 2.5), significant heating of the reaction vessel, rather than of the reaction mixtures, will occur under these circumstances (Figure 2.9). In contrast, no detectable heating of the microwave transparent reaction mixture is seen when a custom-made reaction vessel made of high-purity quartz is employed (Figure 2.9). Heating of the microwave transparent solvent when using the standard glass vessel is the result of indirect heating by conduction and convection phenomena via the hot surface of the self-absorbing glass. Since an IR sensor directly monitors the surface temperature of the glass (rather than of its contents), the observed effects are more pronounced using this type of monitoring method [21]. It is important to note, however, that in the case of medium or strongly microwave-absorbing reaction mixtures, the heating of the glass reaction vessel can be considered negligible and is therefore of little practical concern in microwave synthesis [21].

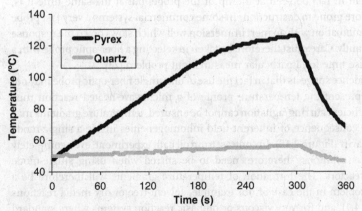

Figure 2.9 Heating profiles for microwave-transparent CCl_4 in Pyrex and quartz reaction vessels at constant 150 W magnetron output power (CEM Discover, IR sensor). Reproduced with permission from Ref. [21].

From a practical point of view, it should be highlighted that IR sensors need to be re-calibrated from time to time against internal probes, and that the path between the actual sensor and the reaction vessel must be unobstructed in order to ensure a proper temperature measurement. This is particularly important when the IR sensor is housed at the bottom of the microwave cavity where debris can more easily accumulate (Figure 2.8).

Based on the information provided above, it is evident that 'more accurate temperature measurements in conjunction with microwave-assisted reactions can be obtained using internal fiber-optic probes. In contrast to thermocouples, fiber-optic sensors are immune to electromagnetic interference and high voltage, do not require shielding and do not spark or transmit current. Although different types of sensing technologies exist, most microwave reactor manufacturers that provide fiber-optic temperature sensor's (CEM, Milestone) rely on probes that use semiconductor band-gap technology. This method uses the band-gap shift principle to deduce the temperature from a semiconductor crystal (normally gallium arsenide) attached to the end of the fiber-optic probe. In essence, a light source sends light to the semiconductor crystal sensor. Depending on the temperature, light with different wavelengths is absorbed by the crystal. This wavelength shift occurs due to a corresponding shift in the sensors energy band gap with temperature. The probes typically have an accuracy of $\pm 1.5\,°C$.

Although internal fiber-optic temperature probes are more accurate than external IR sensors their use is also not without complications. This is, in part, because the mechanically sensitive sensor crystal needs to be protected, requiring the use of appropriate protective immersion wells for the fiber-optic probes. In some fiber-optic probes the actual sensor crystal (GaAs) is additionally protected by a polymer coating. This increases the lifetime of the probe but slows down the response time. Delay times of up to 13 s have been measured for some commercially available fiber-optic probes/immersion wells [23]. In other, "faster", probes the GaAs crystal is unprotected and can in fact be seen at the tip of the probe, but at the same time it is, therefore, more prone to destruction. In some commercial systems a very fast probe is used in combination with an inert immersion well which slows down the response time significantly. Care must therefore be taken in selecting a fiber-optic probe with a short response time for a particular measurement problem [23].

Recent evidence suggests that in fact the use of one single fiber-optic probe may not suffice to represent the temperature profile of a microwave-heated reaction mixture [23]. If efficient stirring/agitation cannot be ensured, temperature gradients may develop as a consequence of inherent field inhomogeneities inside a single-mode microwave cavity (Figure 2.10). In contrast to an oil bath experiment, even completely homogeneous solutions, therefore, need to be stirred when using single-mode microwave reactors. The formation of temperature gradients will therefore be a particular problem in the case of, for example, solvent-free or dry-media reactions (see Section 4.1), and for very viscous or biphasic reaction systems where standard magnetic stirring is not effective.

The temperature monitoring studies shown in Figure 2.10 demonstrate that microwave heating in high field density single-mode cavities is in fact not as

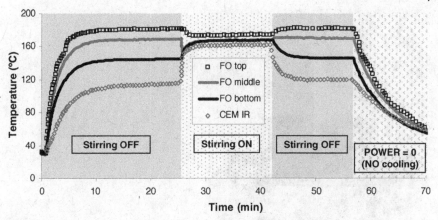

Figure 2.10 Temperature profiles for a sample of 5 mL of NMP contained in a 10 mL quartz vessel equipped with three internal fiber-optic sensors positioned at different heights. The sample was irradiated with constant 50 W magnetron output power (CEM Discover). Shown are the profiles for the three internal fiber-optic probes and the external IR sensor. Magnetic stirring reduces the temperature differences between the individual fiber-optic probes from a maximum of 36 °C to less than 6 °C. The temperature of the IR sensor deviates by 12 °C from the top fiber-optic probe with stirring. Adapted from Ref. [23].

homogeneous as often portrayed, and that extreme care must be taken in determining the proper reaction temperature in these experiments, especially in those cases where adequate mixing cannot be assured.

It should be stressed that, when studying differences between microwave heating and conventional heating (microwave effects), it is particularly important to use highly accurate and fast responding temperature monitoring devices. In order to accurately compare the results obtained by direct microwave heating with the outcome of a conventionally heated reaction, a reactor system should be used that allows one to perform both types of transformations *in the identical reaction vessel* and to monitor the internal reaction temperature in both experiments directly with the same fiber-optic probe device. Such a reactor set-up, originally introduced by Maes and coworkers for the CEM Discover reactor (Figure 2.11) [27], can be either immersed into the cavity of the microwave reactor or into a preheated and temperature equilibrated oil or metal bath placed on a magnetic stirrer/hotplate. In both cases, the software of the microwave instrument is recording the internal temperature. Such a system has the advantage that the same reaction vessel and the same method of temperature measurement are used. In this way all parameters apart from the mode of heating are identical and, therefore, a fair comparison between microwave heating and thermal heating can generally be made [22, 23, 27, 28].

Unfortunately, at the time when most of the early work on microwave effects was published, many of the complications and subtleties of accurate online temperature measurement under microwave irradiation conditions were not known [13, 14]. The conclusions of these studies should therefore in some cases be treated with

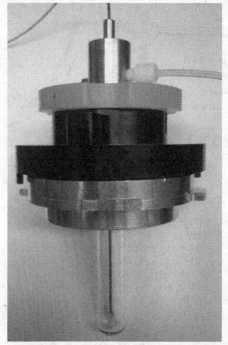

Figure 2.11 Set-up for monitoring internal reaction temperatures with fiber-optic probes in microwave and oil-bath experiments.

skepticism. The following sections provide an overview of the currently existing hypotheses on different types of microwave effects.

2.5.2
Thermal Effects (Kinetics)

Reviewing the present literature, it appears that today many scientists would agree that, in the majority of cases, the reason for the observed rate enhancements seen in microwave chemistry is a purely thermal/kinetic effect, that is, a consequence of the high reaction temperatures that can rapidly be attained when irradiating polar materials in a microwave field. As shown in Figure 2.12, even a moderately strong microwave absorbing solvent such as 1-methyl-2-pyrrolidone (NMP, bp 202–204 °C, $\tan \delta = 0.275$, see Table 2.3) can be heated very rapidly ("microwave flash heating") in a microwave cavity. As indicated in Figure 2.12, a sample of NMP can be heated to 200 °C within about 40 s, depending on the maximum output power of the magnetron [29].

Today, most of the published microwave-assisted reactions are performed under sealed vessel conditions in relatively small, so-called single mode microwave reactors with high power density (see Chapter 3). Under these autoclave-type conditions, microwave absorbing solvents with a comparatively low boiling point such as methanol (bp 65 °C, $\tan \delta = 0.659$), can be rapidly superheated to temperatures of more

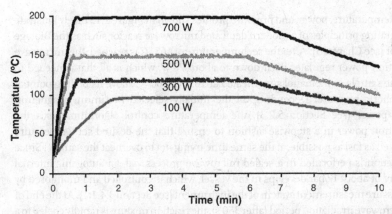

Figure 2.12 Temperature profiles for a 30 mL sample of 1-methyl-2-pyrrolidone heated under open-vessel microwave irradiation conditions [29]. Multimode microwave heating (MicroSYNTH, Milestone) at different maximum power levels for 6 min with temperature control using the feedback from a fiber-optic probe. After the set temperatures of 200 °C (700 W), 150 °C (500 W), 120 °C (300 W), and 100 °C (100 W) are reached, the power regulates itself down to an appropriate level (not shown). Reproduced with permission from Ref. [29].

than 100 °C in excess of their boiling points when irradiated under microwave conditions (Figure 2.13). The rapid increase in temperature can be even more pronounced for media with extreme loss tangents such as, for example, ionic liquids (see Section 4.5.2), where temperature jumps of 200 °C within a few seconds are not uncommon. Naturally, such temperature profiles are very difficult if not impossible to reproduce by standard thermal heating. Therefore, comparisons with conventionally heated processes are inherently troublesome.

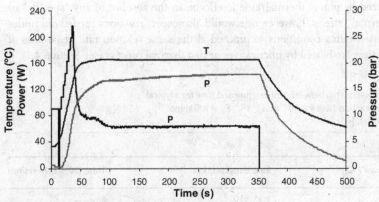

Figure 2.13 Temperature (*T*), pressure (*p*), and power (*P*) profile for a 3 mL sample of methanol heated under sealed-vessel microwave irradiation conditions. Biotage Initiator 2.0 (400 W maximum power, absorbance level: high). Reproduced with permission from Ref. [30].

The temperature, power, and pressure profiles shown in Figure 2.13 nicely illustrate the operating principles of a modern dedicated microwave reactor such as the Biotage Initiator (see Chapter 3). After the set temperature of 165 °C is reached, the microwave magnetron power regulates itself down to about 70 W, which is all that is needed to keep the sample of methanol superheated at 165 °C, 100 °C above its boiling point at atmospheric pressure. Depending on the initially selected maximum magnetron output power (see Section 5.2.3), the temperature control algorithm raises the magnetron power in a stepwise fashion to ensure that the desired set temperature is reached as fast as possible, at the same time trying not to overheat the sample. Since the reaction is performed in a sealed microwave process vial, an autogenic internal pressure of about 17 bar develops in the vessel, which is controlled and monitored by the pressure measurement system of the instrument (see Section 3.4.1.1). At the end of the microwave irradiation period (after 350 s), the reaction mixture is rapidly cooled to a temperature of typically 40–50 °C by a stream of compressed air (active gas jet cooling). This allows the user to remove the processed microwave reaction vial from the cavity in a reasonably short time period, which is particularly important when processing several reactions in sequence using robotic vial handling (see Section 4.7.1).

It appears obvious, that a specific reaction performed under the conditions depicted in Figure 2.13 utilizing superheated methanol as solvent at 165 °C, will occur at a much faster rate than when carried out in refluxing methanol at 65 °C. Dramatic rate-enhancements when comparing reactions that are performed under standard oil-bath conditions (heating under reflux) with high-temperature microwave-heated processes are therefore not uncommon. As Mingos and Baghurst [4] have pointed out, based on simply applying the Arrhenius law [$k = A \exp(-E_a/RT)$], a transformation that requires 68 days to reach 90% conversion at 27 °C, will show the same degree of conversion within 1.61 s (!) when performed at 227 °C (Table 2.6). Due to the very rapid heating and extreme temperatures observable in microwave chemistry, it appears obvious that many of the reported rate-enhancements can be rationalized by purely thermal/kinetic effects. In the absence of any "specific" or "non-thermal" effects, however, one would also expect reactions carried out under open vessel, reflux conditions to proceed at the same reaction rate, regardless of whether they are heated by microwaves or in a thermal process (see Section 4.3). It

Table 2.6 Relationship between temperature and time for a typical first order reaction ($A = 4 \times 10^{10}$ mol^{-1} s^{-1}, $E_a = 100$ kJ mol^{-1}; data from Ref. [4]).

$k = Ae^{-E_a/RT}$		
Temperature (°C)	Rate constant, k (s^{-1})	Time (90% conversion)
27	1.55×10^{-7}	68 d
77	4.76×10^{-5}	13.4 h
127	3.49×10^{-3}	11.4 min
177	9.86×10^{-2}	23.4 s
227	1.43	1.61 s

should be emphasized that, for these strictly thermal effects, the pre-exponential factor A and the energy term (activation energy E_a) in the Arrhenius equation are not affected, only the temperature term changes (see Section 2.5.4).

It should also be noted that the rapid heating and cooling typical of small scale microwave-assisted transformations (see Figure 2.13) may lead to altered product distributions compared to a conventional oil bath reflux experiment, where heating (and cooling) typically is not as fast and the reaction temperature is generally lower. It has been argued that the very different heating profiles experienced in microwave and conventional heating can actually lead to different reaction products if the product distribution is controlled by complex temperature-dependent kinetic profiles [3]. This may be the reason why, in many cases, microwave-assisted reactions have been found to be cleaner, leading to less by-products compared to the conventionally heated processes.

At the same time, it is obvious that microwave heating will not always favor the desired reaction pathway, and that there may be cases where, because of the higher reaction temperatures, unwanted reaction products (e.g. isomers), not seen during a conventionally heated experiment performed at a lower temperature, will be formed.

2.5.3
Specific Microwave Effects

In addition to the above-mentioned thermal/kinetic effects, microwave effects that are caused by the uniqueness of the microwave dielectric heating mechanisms (see Section 2.2) must also be considered. These effects should be termed "specific microwave effects" and shall be defined as *accelerations of chemical transformations in a microwave field that can not be achieved or duplicated by conventional heating, but essentially are still thermal effects*. In this category falls, for example, the superheating effect of solvents at atmospheric pressure [31–33]. The question of the boiling point of a liquid undergoing microwave irradiation is one of the basic problems facing microwave heating. Several groups have established that the enthalpy of vaporization is the same under both microwave and conventional heating [34]. These studies have also shown that the rate of evaporation, as well as the temperature of both vapor and liquid at the interface, depend strongly on the experimental conditions. Initial studies of boiling phenomena related to microwave chemistry appeared in 1992 by Baghurst and Mingos [31], and later by Saillard and coworkers [32]. It was established that microwave-heated liquids boil at temperatures above the equilibrium boiling point at atmospheric pressure. For several solvents, the superheating temperature can be up to 40 °C above the classical boiling point [33]. Therefore, in a microwave-heated reactor, the average temperature of the solvent can be significantly higher than the atmospheric boiling point. This is because the microwave power is dissipated over the whole volume of the solvent. The most significant way to lose excess thermal energy is by boiling. However, this will only occur at the existing liquid–gas interfaces, in contrast to a thermally heated solvent where boiling typically occurs at nucleation points (cavities, pits and scratches) on the glass reactor surface [31]. The

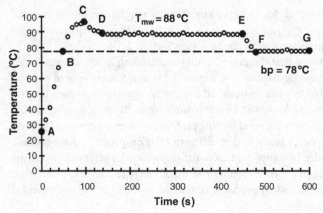

Figure 2.14 Microwave heating (single mode reactor) of ethanol under open-vessel conditions. Initially the temperature rises during the heating phase (AB), above the normal boiling point of ethanol (78 °C) to a point C where the solvent bumps and starts to boil at the vapor/liquid interface. At this point the temperature drops to a plateau region (D) and can be maintained at this temperature for many hours within ±1 °C. Addition of a boiling chip (E) brings the temperature to the normal boiling point of ethanol (FG). Reproduced with permission from Ref. [33].

bulk temperature of a microwave irradiated solvent under boiling depends on many factors, such as the physical properties of the solvent, reactor geometry, mass flow, heat flow, and electric field distribution. It should be emphasized that practically all super-heating can be removed by adding boiling chips or stirring (Figure 2.14) [33]. Importantly, however, the kinetics of homogeneous organic reactions shows an extension of Arrhenius behavior into the superheated temperature region [33]. Therefore, 10–100-fold reaction rate enhancements can be achieved, which is normally only possible under pressure. Since all dedicated microwave reactors offer a stirring option (see Chapter 3), and most of the current microwave chemistry is performed under sealed vessel conditions (for exceptions, see Section 4.3), the microwave superheating effect under atmospheric pressure conditions is of little practical relevance and concern.

Closely related to the superheating effect under atmospheric pressure are wall effects, more specifically the elimination of wall effects caused by inverted temperature gradients (Figure 2.7). With microwave heating, the surface of the wall is generally not heated since the energy is dissipated inside the bulk liquid. Therefore, the temperature at the inner surface of the reactor wall is lower than that of the bulk liquid. It can be assumed that in a conventional oil-bath experiment (hot vessel surface, Figure 2.7a) temperature-sensitive species, for example, catalysts, may decompose at the hot reactor surface (wall effects). The elimination of such a hot surface will increase the lifetime of the catalyst and will, therefore, lead to better conversions in a microwave-heated as compared to a conventionally heated process. However, no dedicated study of the reduction of wall effects in microwave chemistry has been published to date.

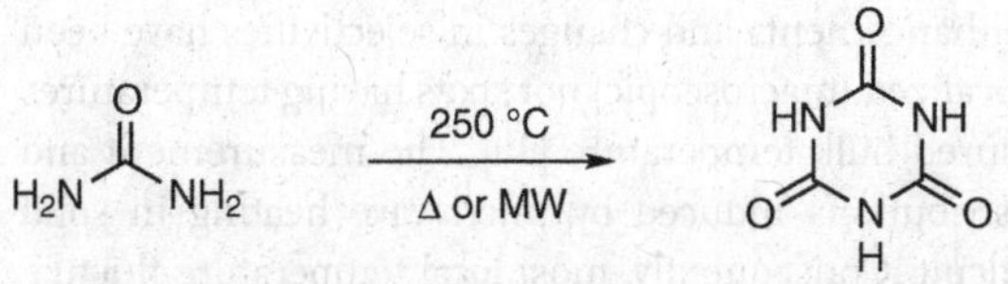

Scheme 2.2 Decomposition of urea to cyanuric acid.

Another phenomenon characteristic of microwave dielectric heating is mass heating, that is, the rapid and even heating of the whole reaction mixture by microwaves (volumetric heating). An example to illustrate this effect, involving the decomposition of urea to cyanuric acid (Scheme 2.2), was studied by Berlan [35]. Cyanuric acid is obtained by heating urea to temperatures around 250 °C. Under conventional heating the reaction is sluggish and chemical yields are low due to the formation of various side-products. The reason for this is that cyanuric acid, which is first formed as a solid at the walls of the reactor, is a poorly heat-conductive material (it decomposes without melting at 300 °C), and it forms an insulating crust which prevents heat transfer to the rest of the reaction mixture. Increasing the temperature of the wall (i.e. the oil-bath temperature) results in partial decomposition and does not improve the chemical yield of cyanuric acid significantly. In contrast, very good yields of cyanuric acid (83%) can be obtained under volumetric microwave heating on a 2 g scale for 2 min without any urea or biuret side product being detected at the end of the reaction. Based on the discussion on penetration depth issues in conjunction with microwave heating (Section 2.3) it should be stressed that these effects will only be seen on a comparatively small scale.

The same concept of volumetric *in situ* heating by microwaves was also exploited by Larhed and coworkers in the context of scaling-up a biochemical process such as the polymerase chain reaction (PCR) [36]. In PCR technology, strict control of temperature in the heating cycles is essential in order not to deactivate the enzymes involved. With classic heating of a milliliter scale sample, the time required for heat transfer through the wall of the reaction tube and to obtain an even temperature in the whole sample is still substantial. In practice, the slow distribution of heat (temperature gradients), together with the importance of short processing times and reproducibility, limits the volume for most PCR transformations in conventional thermocyclers to 0.2 mL. With microwave heating, the thermal gradients are eliminated since the full volume is heated simultaneously. Therefore, microwave heating under strict temperature control has been shown to be an extremely valuable tool for carrying out large scale PCR processing up to a 15 mL scale [36].

Probably one of the most important "specific microwave effects" results from the selective heating of strongly microwave-absorbing heterogeneous catalysts or reagents in a less polar reaction medium [37]. Selective heating generally means that, in a sample containing more than one component, only that component which couples with microwaves is selectively heated. The non-absorbing components are thus not heated directly, but only by heat transfer from the heated component. For heterogeneous mixtures, in particular for gas/solid systems involving heterogeneous gas-phase catalysis [37, 38], selective heating of the catalyst bed is of importance and here the sometimes observed rate enhancements and changes in selectivities have been attributed to the formation of localized (macroscopic) hot spots having temperatures of 100–150 °C above the measured bulk temperature [39]. The measurement and estimation of temperature distributions induced by microwave heating in solid materials is, however, very difficult. Consequently, most local temperature fluctuations are greater than those measured. Under stronger microwave irradiation it is, therefore, very easy to obtain local temperature gradients. Temperature measurements usually yield an average temperature, because temperature gradients induce convective motions. Despite these difficulties, some methods, for example IR thermography, can reveal surface temperature distribution without any contact with the sample under study.

Of greater importance for the organic chemist are microwave-assisted transformations in organic solvents catalyzed by a heterogeneous catalyst (liquid/solid systems) such as palladium-on-charcoal (Pd/C). Since here the catalyst is a very strong absorber of microwave energy (see Section 4.1), it can be assumed that the reaction temperature on the catalyst surface is significantly higher than the bulk temperature of the solvent, in particular when a solvent with a low tan δ value is chosen (Table 2.3). The selective heating/activation by microwave irradiation of a Pd/C catalyst was exploited by Vanier in the Pd-catalyzed hydrogenation of various carbon–carbon double bond systems [28]. In the example shown in Scheme 2.3 hydrogenation of the butadiene under single-mode microwave irradiation conditions at 80 °C gave complete conversion after 5 min of microwave irradiation. When the same process was performed in an oil-bath at the same measured bulk temperature under strictly comparable conditions (fiber-optic temperature measurement, see Figure 2.11), the conversion was only 55% [28]. Similar observations were also made by Holzgrabe and coworkers in related microwave-assisted hydrogenation reactions [40]. The selective absorption of microwave energy by a heterogeneous encapsulated palladium catalyst (PdEnCat) was also suggested to be responsible for very efficient Suzuki couplings of aryl bromides performed by the Ley group under both batch and flow microwave conditions, although no control experiments using conventional heating were presented [41].

The selective heating phenomenon cannot only be exploited for strongly absorbing catalysts, but also for reagents. In the example shown in Scheme 2.4 primary and secondary alcohols were oxidized with a chromium dioxide reagent under microwave conditions. Irradiation for 2 min of a neat sample of chromium dioxide led to surface temperatures of up to 360 °C, measured with an IR thermovision camera [24]. When the oxidant was suspended in the weakly microwave-absorbing toluene, the temperature of the chromium dioxide reached about 140 °C. Importantly, even though the temperature of the solid material was higher than the boiling point of toluene, no

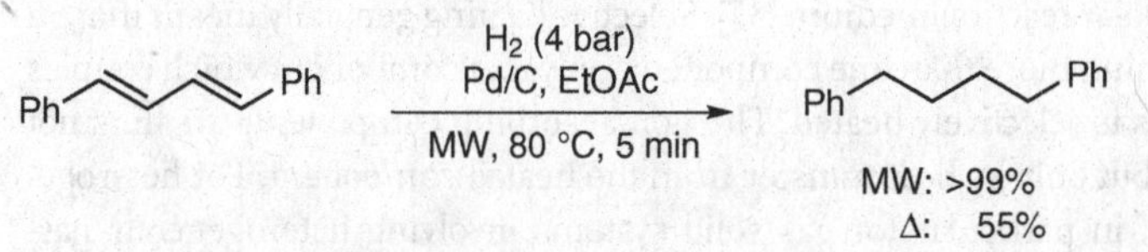

$$Ph\diagup\diagdown\diagup Ph \xrightarrow[\text{MW, 80 °C, 5 min}]{\substack{\text{H}_2 \text{ (4 bar)} \\ \text{Pd/C, EtOAc}}} Ph\diagup\diagdown\diagup Ph$$

MW: >99%
Δ: 55%

Scheme 2.3 Hydrogenation reactions using a heterogeneous palladium catalyst.

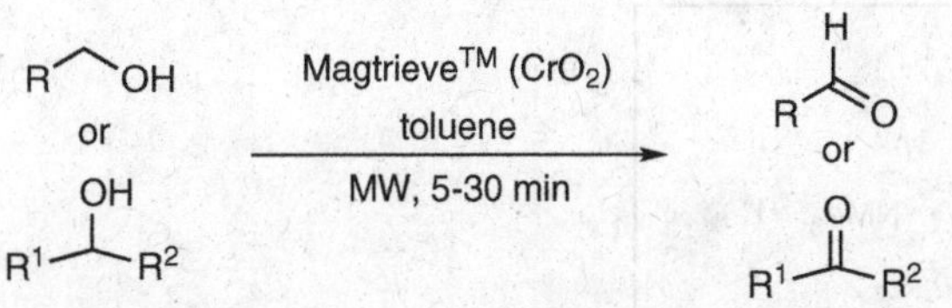

Scheme 2.4 Oxidation of primary and secondary alcohols with chromium dioxide.

boiling was observed in the reaction vessel. The direct interaction of the microwave field with magnesium metal turnings was recently reported for the formation of Grignard reagents under microwave conditions [42]. Irradiating a solution of an arylhalide in dry tetrahydrofuran together with magnesium turnings led to strong arcing of the magnesium turnings, as observed through the front glass door of the multimode microwave reactor. The formation of the Grignard reagent using microwave heating was significantly faster than by conventional heating at the same temperature of 65 °C (refluxing tetrahydrofuran) [42]. This could be due to a cleansing effect (electrostatic etching), that is removal of a layer of magnesium oxide from the magnesium, initiated by microwave irradiation. Recent evidence suggests that heterogeneity itself plays a major role in the enhancement of chemical processes by microwave irradiation. It has been demonstrated that microwave irradiation can change the energies and/or the "effective temperatures" of individual species at interfaces as the result of Maxwell–Wagner interfacial microwave polarization [43].

These studies provide clear evidence for the existence of selective heating effects in MAOS involving heterogeneous mixtures. It should be stressed that the standard methods for determining the temperature in microwave-heated reactions, namely with an IR pyrometer from the outside of the reaction vessel, or with a fiber-optic probe on the inside, would here only allow measurement of the average bulk temperature of the solvent, not the "true" reaction temperature on the surface of the solid reagent, or on an interface.

For homogeneous mixtures, for example polar reagents in a microwave transparent solvent (liquid/liquid systems), in principle the same arguments about selective heating can be made. However, the existence of such "molecular radiators" [44] is experimentally difficult to prove and it would have to be assumed that the energy of these "hot" molecules would be instantaneously dissipated to the surrounding "cooler" solvent molecules [2–5]. It should also be stressed that it is not possible to selectively "activate" polar functional groups (so-called antenna groups [45]) within a larger molecule by microwave irradiation. It is tempting for a chemist to give a chemical significance to the fact that localized rotations of such antenna groups are indeed possible [5], and to speculate that microwave dielectric heating of molecules containing these groups may result in an enhancement of reaction rates specifically at these groups. However, the dielectric heating process involves the rapid energy transfer from these groups to neighboring molecules and it is not possible to store the energy in a specific part of the molecule [5].

Another specific microwave effect not easily duplicated by conventional heating is the differential heating of bi- or multiphasic liquid/liquid systems. This type of selective heating was exploited by Strauss and coworkers in a Hofmann elimination

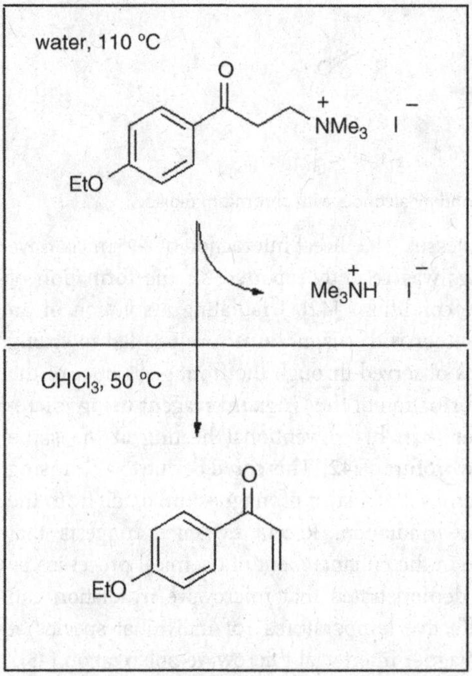

Figure 2.15 Selective dielectric heating of water/chloroform mixtures.

reaction using a two-phase water/chloroform system (Figure 2.15) [46]. The temperatures of the aqueous and organic phases under microwave irradiation were 110 and 50 °C, respectively, due to differences in the dielectric properties of the solvents (Table 2.3). This difference avoids decomposition of the final product which is soluble in the cooler organic chloroform phase. Comparable conditions would be difficult to obtain using traditional heating methods. A similar effect has been observed by Hallberg and coworkers in the preparation of β, β-diarylated aldehydes by hydrolysis of enol ethers in a two-phase toluene/aqueous hydrochloric acid system [47].

Differential heating phenomena will almost always be observed when heterogeneous liquid/liquid systems are irradiated by microwaves, since there is likely to be a difference in loss tangents between the two phases. The effects can be extreme, as in the case of an ionic liquid/hexane mixture (see Figure 4.6), or more moderate in nature. The user, however, should always be aware of the possibility of differential heating when dealing with non-homogeneous reaction mixtures under microwave irradiation conditions. Because of the potentially different temperatures in the phases, mass- and heat-transfer across the phase boundaries may be altered compared to conventional heating where both phases have the same temperature. Naturally, particular care must be given to the temperature measurement in these cases, as it will be critically important in which phase the temperature is measured. In any event, intensive stirring should always be applied when dealing with heterogeneous mixtures [23]. It is perhaps no coincidence that a recent study has found that microwave effects can often be eliminated when the heterogeneity of the reaction system is reduced by adding appropriate solvents [42].

In summary, all potential rate-enhancements discussed above falling under the category of "specific microwave effects", such as the superheating effect of solvents at atmospheric pressure, the selective heating of strongly microwave absorbing hetero-geneous catalysts or reagents in a less polar reaction medium, differential heating of liquid/liquid biphasic mixtures, and the elimination of wall effects caused by inverted temperature gradients, are essentially still a result of a *thermal* effect (i.e. a change in temperature compared to heating by standard convection methods), although it may be difficult to determine the exact reaction temperature experimentally.

2.5.4
Non-Thermal (Athermal) Microwave Effects

In contrast to the so-called "specific microwave effects" described in Section 2.5.3 some authors have suggested the possibility of "non-thermal microwave effects" (also referred to as athermal effects). These should be classified as *accelerations of chemical transformations in a microwave field that cannot be rationalized by either purely thermal/kinetic or specific microwave effects* [18]. Essentially, non-thermal effects result from a proposed direct interaction of the electric field with specific molecules in the reaction medium. It has been argued, for example, that the presence of an electric field leads to orientation effects of dipolar molecules and, hence, changes the pre-exponential factor A [48] or the activation energy (entropy term) [49] in the Arrhenius equation. Furthermore, it has been argued that a similar effect should be observed for polar reaction mechanisms, where the polarity is increased on going from the ground state to the transition state, resulting in an enhancement of reactivity by lowering of the activation energy (Figure 2.16).

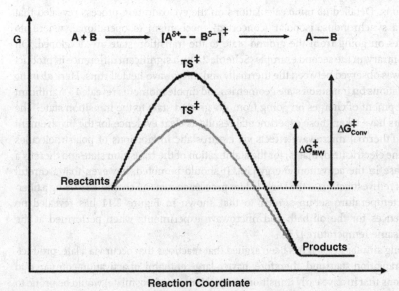

Figure 2.16 Proposed lowering of polar transition state energies by stabilization in the microwave field [13, 14].

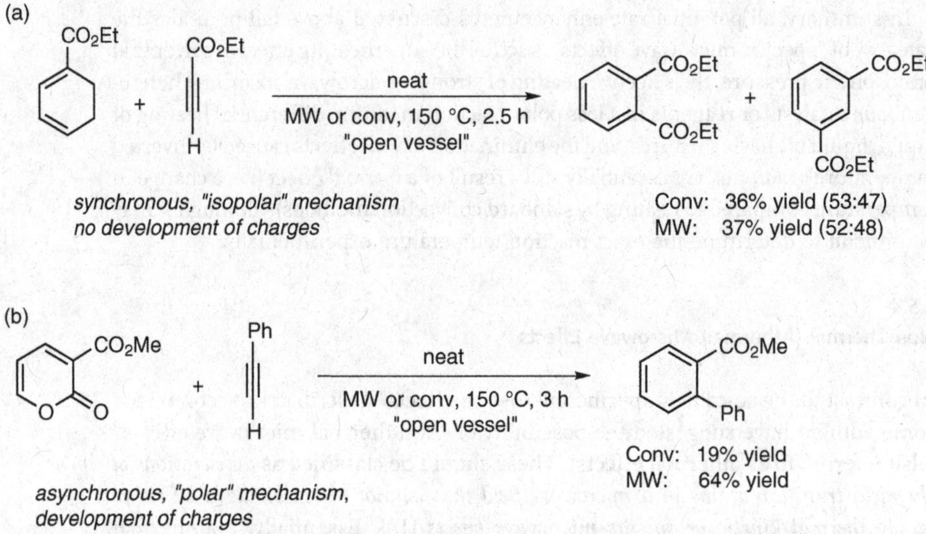

(a)

synchronous, "isopolar" mechanism
no development of charges

Conv: 36% yield (53:47)
MW: 37% yield (52:48)

(b)

Conv: 19% yield
MW: 64% yield

asynchronous, "polar" mechanism,
development of charges

Scheme 2.5 Proposed involvement of non-thermal microwave effects in Diels–Alder cycloaddition reactions.

Several publications in the literature use arguments like this to explain the outcome of a chemical reaction carried out under microwave irradiation conditions [13, 14]. A 2004 study by Loupy and coworkers provides a representative example [50]: as shown in Scheme 2.5, two irreversible Diels–Alder cycloaddition processes were compared. In the first example (Scheme 2.5a), no difference in either yield or selectivity was observed between the conventionally and microwave-heated reactions. Detailed *ab initio* calculations on the cycloaddition process revealed that here a synchronous, isopolar (concerted) mechanism is operational, where no charges on going from the ground state to the transition state are developed. On the contrary, in the second example (Scheme 2.5b), a significant difference in product yield was observed between the thermally and microwave-heated runs. Here *ab initio* calculations on transition state geometries and dipole moments revealed a significant development of charges on going from the ground state to the transition state. The authors have taken these experimental results as clear evidence for the involvement of non-thermal microwave effects via electrostatic interactions of polar molecules with the electric field, that is, for the stabilization of the transition state and thereby a decrease in the activation energy [50]. It should be noted, however, that a careful recent reinvestigation of the Diels–Alder process shown in Scheme 2.5b using a fiber-optic temperature set-up similar to that shown in Figure 2.11 has revealed no differences for the oil bath and microwave experiments when performed at the exact same temperature [23].

Along similar lines, it has been argued that reactions that occur via a late, product-like, transition state and, therefore, have a large enthalpy of activation compared to reactions that involve early transition states (Hammond postulate) would be prone to show large microwave effects (Figure 2.17) [13, 14]. Several authors have used

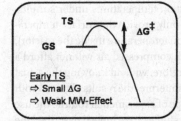 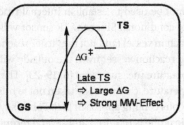

Figure 2.17 Proposed stabilization of transition states positioned late on the reaction coordinate by microwave irradiation [13, 14]. Adapted with permission from Ref. [13].

quantum mechanical methods to study the interaction between reaction molecules and the electromagnetic field [51, 52].

As already mentioned above, the issue of non-thermal microwave effects is highly controversial. Many scientists denounce the existence of dipolar orientation effects in electric fields on the grounds of overriding disorientation phenomena (thermal agitation) that should prevent any statistically significant orientation (alignment) of dipoles [2]. In this context it may be noted that it is probably no coincidence that non-thermal microwave effects have, in many cases, been claimed for processes involving solvent-free/dry-media reactions, and/or for transformations involving polar reaction intermediates or products (which will strongly absorb microwave energy) [13, 14]. Based on the difficulties of exact online temperature measurement under these experimental conditions (see Section 2.5.1) it can be argued that, in many of the published cases, the observed differences between microwave and conventional heating may be rationalized by inaccurate temperature measurements, often using external IR temperature probes, rather than being the consequence of a genuine non-thermal effect [6].

Related to the issue of non-thermal or specific microwave effects is the recent concept that simultaneous external cooling of the reaction mixture (or maintenance of sub-ambient reaction temperatures) while heating by microwaves leads to an enhancement of the overall process (PowerMax, "Enhanced Microwave Synthesis" [11, 53, 54]). Here, the reaction vessel is cooled from the outside by compressed air while being irradiated by microwaves. This allows a higher level of microwave power to be directly administered to the reaction mixture, but will prevent overheating by continuously removing latent heat. Some authors [11, 54] have made the argument that since microwave energy is transferred into the sample much faster than molecular kinetic relaxation occurs, non-equilibrium conditions will result in a microwave-heated reaction system. Therefore, the more power that is applied, the higher will be the "instantaneous" temperature relative to the measured bulk temperature [11, 54]. Published applications of the simultaneous air cooling technology are comparatively rare [20, 22, 23, 28, 41, 55, 56]. While it is tempting to speculate about the apparent benefits of this approach, one must be aware that the actual reaction temperature has not been determined in many cases [55]. As detailed in Section 2.5.1, the standard external IR pyrometer used in a single-mode microwave reactor such as the CEM

Discover cannot be used to establish internal reaction temperatures under simultaneous cooling conditions. Since the IR sensor will only provide the surface temperature of the reaction vessel (and not the "true" reaction temperature inside the reactor), cooling of the reaction vessel from the outside with compressed air will not afford a reliable temperature measurement [6, 19–23]. Therefore, without knowing the actual reaction temperature, care must be taken not to misinterpret the results obtained with the simultaneous cooling approach [22]. For this type of experimental routine the use of a fiber-optic probe is required and several recent studies have indeed described the application of the simultaneous cooling concept in conjunction with fiber-optic temperature measurements [20, 22, 23, 28, 41, 56, 57]. While, in some cases, the effects compared to conventional microwave heating were small [20, 22, 23], in other instances significant differences in conversion or purity between the "cooled" and "uncooled" experiments were seen at the same nominal bulk temperature (Scheme 2.6) [28, 41, 56, 57]. It should be noted that in all three cases presented in Scheme 2.6 a heterogeneous catalyst is involved which is probably superheated to temperatures significantly higher than the measured bulk temperature by the microwave irradiation (see Section 2.5.3).

A logical extension of the simultaneous air cooling idea is to execute microwave chemistry under sub-ambient temperature conditions. For this purpose, the CEM Discover Coolmate has been developed, featuring a jacketed low-temperature reaction vessel in combination with appropriate microwave transparent cooling media and chilling technology (see Section 3.4.2.1). Again, there exist conflicting reports about the usefulness of this instrument and the general concept [22, 57, 58]. In the

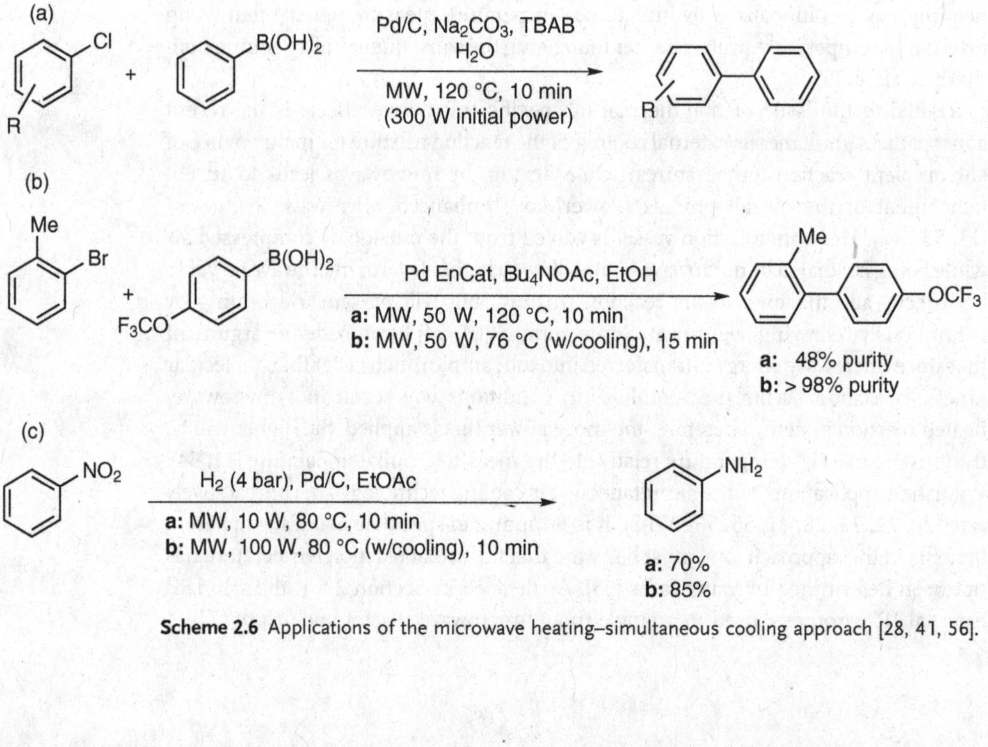

(a)

$$\text{R}\text{—}C_6H_3\text{—Cl} \; + \; C_6H_5\text{—B(OH)}_2 \xrightarrow[\text{MW, 120 °C, 10 min}]{\substack{\text{Pd/C, Na}_2\text{CO}_3\text{, TBAB} \\ \text{H}_2\text{O} \\ \text{(300 W initial power)}}} \text{R—biphenyl}$$

(b)

Pd EnCat, Bu$_4$NOAc, EtOH

a: MW, 50 W, 120 °C, 10 min
b: MW, 50 W, 76 °C (w/cooling), 15 min

a: 48% purity
b: > 98% purity

(c)

$$C_6H_5\text{—NO}_2 \xrightarrow[\substack{\text{a: MW, 10 W, 80 °C, 10 min} \\ \text{b: MW, 100 W, 80 °C (w/cooling), 10 min}}]{\text{H}_2 \text{ (4 bar), Pd/C, EtOAc}} C_6H_5\text{—NH}_2$$

a: 70%
b: 85%

Scheme 2.6 Applications of the microwave heating–simultaneous cooling approach [28, 41, 56].

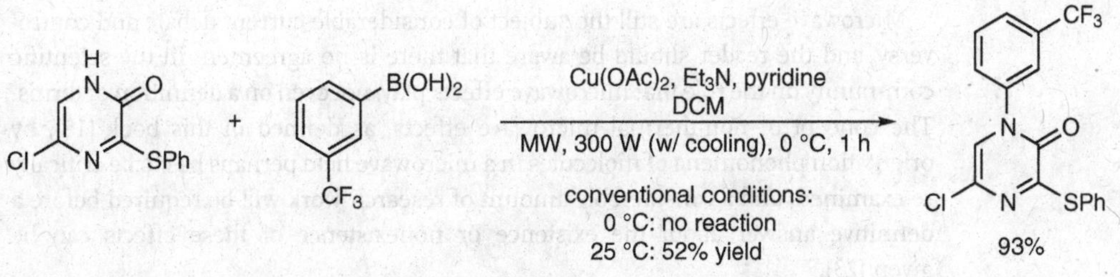

Scheme 2.7 Microwave chemistry at sub-ambient conditions.

example shown in Scheme 2.7, a dramatic improvement was seen in comparing the conversion of a transition metal-catalyzed reaction performed at 0 °C in an ice-bath with the same reaction carried out a 0 °C in the Coolmate using 300 W microwave irradiation power [57].

Recently, Hajek and coworkers have reported results on microwave-assisted chemistry performed by cooling of the reaction mixture to as low as −176 °C. Reaction rates were recorded under microwave and conventional conditions. The higher reaction rates under microwave heating at sub-ambient temperatures were attributed to a superheating of the heterogeneous K10 catalyst and, therefore, to a specific microwave effect [59]. Related to the simultaneous cooling approach is the concept of using pre-cooled reaction vessels [60].

Another unusual phenomenon in microwave synthesis was described by Ley and coworkers. In several examples [61, 62], the authors found that pulsed microwave irradiation gave higher conversions than irradiation of the reaction mixture continuously for the same amount of time. In the example shown in Scheme 2.8, the highest conversion in the Claisen rearrangement was obtained by exposing the reaction to three 15 min sequences of microwave heating to 220 °C (with intermittent cooling). Continuous microwave irradiation for 45 min at the same temperature provided a somewhat lower yield [61]. It is tempting to speculate that repeated overshooting of temperature of the strongly microwave-absorbing reaction mixture in the pulsed experiments may be responsible for the observed effects.

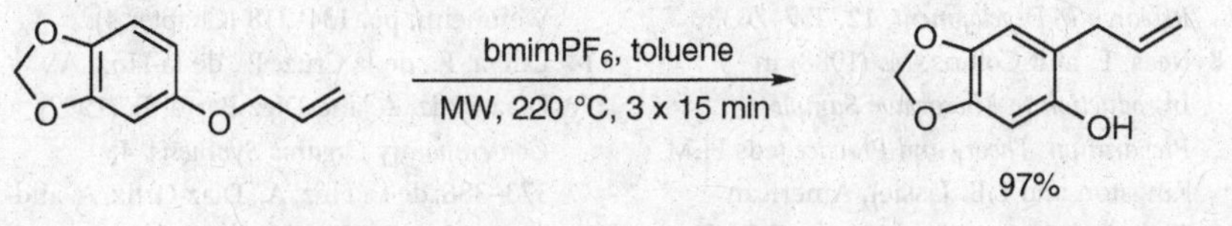

MW (220 °C)	Conversion (%)
3 x 15 min	97
30 min continuous	78
45 min continuous	86
2+2+2+1+1+15+15+15	85

Scheme 2.8 Pulsed versus continuous microwave irradiation (bmimPF$_6$ = 1-butyl-3-methylimidazolium hexafluorophosphate).

Microwave effects are still the subject of considerable current debate and controversy, and the reader should be aware that there is no agreement in the scientific community on the role that microwave effects play, not even on a definition of terms. The concept of non-thermal microwave effects, as defined in this book [18], by orientation phenomena of molecules in a microwave field perhaps has to be critically re-examined, and a considerable amount of research work will be required before a definitive answer about the existence or nonexistence of these effects can be given [23].

References

1 von Hippel, A.R. (1954) *Dielectric Materials and Applications*, MIT Press, Cambridge, MA, USA.

2 Stuerga, D. (2006) in *Microwaves in Organic Synthesis*, 2nd edn (ed. A. Loupy), Wiley-VCH, Weinheim, pp. 1–34 (Chapter 1).

3 Mingos, D.M.P. (2005) in *Microwave-Assisted Organic Synthesis* (eds P. Lidström and J.P. Tierney), Blackwell Publishing, Oxford, pp. 1–22 (Chapter 1).

4 Baghurst, D.R. and Mingos, D.M.P. (1991) *Chemical Society Reviews*, 20, 1–47.

5 Gabriel, C., Gabriel, S., Grant, E.H., Halstead, B.S. and Mingos, D.M.P. (1998) *Chemical Society Reviews*, 27, 213–223.

6 Nüchter, M., Ondruschka, B., Bonrath, W. and Gum, A. (2004) *Green Chemistry*, 6, 128–141.

7 Gedye, R.N. and Wei, J.B. (1998) *Canadian Journal of Chemistry*, 76, 525–532; Horikoshi, S., Iida, S., Kajitani, M., Sato, S. and Serpone, N. (2008) *Organic Process Research & Development*, 12, 257–263.

8 Neas, E. and Collins, M. (1988) in *Introduction to Microwave Sample Preparation: Theory and Practice* (eds H.M. Kingston and L.B. Jassie), American Chemical Society, Washington, DC.

9 Stass, D.V., Woodward, J.R., Timmel, C.R., Hore, P.J. and McLauchlan, K.A. (2000) *Chemical Physics Letters*, 329, 15–22; Timmel, C.R. and Hore, P.J. (1996) *Chemical Physics Letters*, 257, 401–408; Woodward, J.R., Jackson, R.J., Timmel, C.R., Hore, P.J. and McLauchlan, K.A.

(1997) *Chemical Physics Letters*, 272, 376–382.

10 Bogdal, D. and Prociak, A. (2007) *Microwave-Enhanced Polymer Chemistry and Technology*, Blackwell Publishing, Oxford.

11 Hayes, B.L. (2002) *Microwave Synthesis: Chemistry at the Speed of Light*, CEM Publishing, Matthews, NC.

12 Grice, D.D., Fletcher, P.D.I. and Haswell, S.J. Department of Chemistry, University of Hull, UK, unpublished data.

13 Bodgal, D. (2005) *Microwave-assisted Organic Synthesis. One Hundred Reaction Procedures*, Elsevier, Oxford; Perreux, L. and Loupy, A. (2001) *Tetrahedron*, 57, 9199–9223; Perreux, L. and Loupy, A. (2002) in *Microwaves in Organic Synthesis* (ed. A. Loupy), Wiley-VCH, Weinheim, pp. 61–114 (Chapter 3);Perreux, L. and Loupy, A. (2006) in *Microwaves in Organic Synthesis*, 2nd edn (ed. A. Loupy), Wiley-VCH, Weinheim, pp. 134–218 (Chapter 4).

14 Langa, F., de la Cruz, P., de la Hoz, A., Díaz-Ortiz, A. and Díez-Barra, E. (1997) *Contemporary Organic Synthesis*, 4, 373–386; de La Hoz, A., Díaz-Ortiz, A. and Moreno, A. (2004) *Current Organic Chemistry*, 8, 903–918; de La Hoz, A., Díaz-Ortiz, A. and Moreno, A. (2005) *Chemical Society Reviews*, 34, 164–178; de La Hoz, A., Díaz-Ortiz, A. and Moreno, A. (2006) in *Microwaves in Organic Synthesis*, 2nd edn (ed. A. Loupy), Wiley-VCH, Weinheim, pp. 219–277 (Chapter 5).

15 Kuhnert, N. (2002) *Angewandte Chemie-International Edition*, 41, 1863–1866; Strauss, C.R. (2002) *Angewandte Chemie-International Edition*, 41, 3589–3590; Panunzio, M., Campana, E., Martelli, G., Vicennati, P. and Tamanini, E. (2004) *Materials Research Innovations*, 8, 27–31.

16 Bougrin, K., Bannani, A.K., Tetouani, S.F. and Soufiaoui, M. (1994) *Tetrahedron Letters*, 35, 8373–8376.

17 Koizumi, H., Itoh, Y. and Ichikawa, T. (2006) *Chemistry Letters*, 35, 1350–1351.

18 Kappe, C.O. (2004) *Angewandte Chemie-International Edition*, 43, 6250–6284.

19 Nüchter, M., Ondruschka, B., Weiß, D., Bonrath, W. and Gum, A. (2005) *Chemical Engineering & Technology*, 28, 871–881.

20 Leadbeater, N.E., Pillsbury, S.J., Shanahan, E. and Williams, V.A. (2005) *Tetrahedron*, 61, 3565–3585.

21 Kremsner, J.M. and Kappe, C.O. (2006) *The Journal of Organic Chemistry*, 71, 4651–4658.

22 Hosseini, M., Stiasni, N., Bàrbieri, V. and Kappe, C.O. (2007) *The Journal of Organic Chemistry*, 72, 1417–1424.

23 Herrero, M.A., Kremsner, J.M. and Kappe, C.O. (2008) *The Journal of Organic Chemistry*, 73, 36–47.

24 Bogdal, D., Lukasiewicz, M., Pielichowski, J., Miciak, A. and Bednarz, Sz. (2003) *Tetrahedron*, 59, 649–653; Lukasiewicz, M., Bogdal, D. and Pielichowski, J. (2003) *Advanced Synthesis and Catalysis*, 345, 1269–1272.

25 Bogdal, D., Bednarz, S. and Lukasiewicz, M. (2006) *Tetrahedron*, 62, 9440–9445.

26 Moseley, J.D., Lenden, P., Thomson, A.D. and Gilday, J.P. (2007) *Tetrahedron Letters*, 48, 6084–6087.

27 Hostyn, S., Maes, B.U.W., Van Baelen, G., Gulevskaya, A., Meyers, C. and Smits, K. (2006) *Tetrahedron*, 62, 4676–4684.

28 Vanier, G.S. (2007) *Synlett*, 131–135.

29 Stadler, A. and Kappe, C.O. (2001) *European Journal of Organic Chemistry*, 919–925.

30 Kappe, C.O. (2008) *Chemical Society Reviews*, 37, 1127–1139.

31 Baghurst, D.R. and Mingos, D.M.P. (1992) *Journal of the Chemical Society. Chemical Communications*, 674–677.

32 Saillard, R., Poux, M., Berlan, J. and Audhuy-Peaudecerf, M. (1995) *Tetrahedron*, 51, 4033–4042.

33 Chemat, F. and Esveld, E. (2001) *Chemical Engineering & Technology*, 24, 735–744; Gilday, J.P., Lenden, P., Moseley, J.D. and Cox, B.G. (2008) *The Journal of Organic Chemistry*, 73, 3130–3134.

34 Abtal, E., Lallemant, M., Bertrand, G. and Watelle, G. (1985) *Journal of Chemical Physics*, 82, 381–399; Roussy, G., Thibaut, J.M. and Collin, P. (1986) *Thermochimica Acta*, 98, 57–62.

35 Berlan, J. (1995) *Radiation Physics and Chemistry*, 45, 581–589.

36 Orrling, K., Nilsson, P., Gullberg, M. and Larhed, M. (2004) *Chemical Communications*, 790–791.

37 Hajek, M. (2006) in *Microwaves in Organic Synthesis*, 2nd edn (ed. A. Loupy), Wiley-VCH, Weinheim, pp. 615–652 (Chapter 13).

38 Will, H., Scholz, P. and Ondruschka, B. (2002) *Chemie-Ingenieur-Technik*, 74, 1057–1067.

39 Zhang, X., Hayward, D.O. and Mingos, D.M.P. (2003) *Catalysis Letters*, 88, 33–38.

40 Heller, E., Lautenschläger, W. and Holzgrabe, U. (2005) *Tetrahedron Letters*, 46, 1247–1249.

41 Baxendale, I.R., Griffith-Jones, C.M., Ley, S.V. and Tranmer, G.K. (2006) *Chemistry – A European Journal*, 12, 4407–4416.

42 Dressen, M.H.C.L., van de Kruijs, B.H.P., Meuldijk, J., Vekemans, J.A.J.M. and Hulshof, L.A. (2007) *Organic Process Research & Development*, 11, 865–869.

43 Conner, W.C. and Tompsett, G.A. (2008) *The Journal of Physical Chemistry. B*, 112, 2110–2118.

44 Kaiser, N.-F.K., Bremberg, U., Larhed, M., Moberg, C. and Hallberg, A. (2000)

Angewandte Chemie-International Edition, **39**, 3596–3598.

45 Laurent, R., Laporterie, A., Dubac, J., Berlan, J., Lefeuvre, S. and Audhuy, M. (1992) *The Journal of Organic Chemistry*, **57**, 7099–7102.

46 Raner, K.D., Strauss, C.R., Trainor, R.W. and Thorn, J.S. (1995) *The Journal of Organic Chemistry*, **60**, 2456–2460.

47 Nilsson, P., Larhed, M. and Hallberg, A. (2001) *Journal of the American Chemical Society*, **123**, 8217–8225.

48 Jacob, J., Chia, L.H.L. and Boey, F.Y.C. (1995) *Journal of Materials Science*, **30**, 5321–5327; Binner, J.G.P., Hassine, N.A. and Cross, T.E. (1995) *Journal of Materials Science*, **30**, 5389–5393; Shibata, C., Kashima, T. and Ohuchi, K. (1996) *Japanese Journal of Applied Physics*, **35**, 316–319.

49 Berlan, J., Giboreau, P., Lefeuvre, S. and Marchand, C. (1991) *Tetrahedron Letters*, **32**, 2363–2366; Lewis, D.A., Summers, J.D., Ward, T.C. and McGrath, J.E. (1992) *Journal of Polymer Science Part A-Polymer Chemistry*, **30**, 1647–1653.

50 Loupy, A., Maurel, F. and Sabatié-Gogová, A. (2004) *Tetrahedon*, **60**, 1683–1691.

51 Kalhori, S., Minaev, B., Stone-Elander, S. and Elander, N. (2002) *The Journal of Physical Chemistry A*, **106**, 8516–8524.

52 Miklavc, A. (2001) *ChemPhysChem*, **2**, 552–555.

53 Hayes, B.L. and Collins, M.J. Jr. (2004) World Patent WO 04002617.

54 Hayes, B.L. (2004) *Aldrichimica Acta*, **37**, 66–77.

55 Chen, J.J. and Deshpande, S.V. (2003) *Tetrahedron Letters*, **44**, 8873–8876; Mathew, F., Jayaprakash, K.N., Fraser-Reid, B., Mathew, J. and Scicinski, J. (2003) *Tetrahedron Letters*, **44**, 9051–9054; Crawford, K., Bur, S.K., Straub, C.S. and Padwa, A. (2003) *Organic Letters*, **5**, 3337–3340; Humphrey, C.E., Easson,

M.A.M., Tierney, J.P. and Turner, N.J. (2003) *Organic Letters*, **5**, 849–852; Katritzky, A.R., Zhang, Y., Singh, S.K. and Steel, P.J. (2003) ARKIVOC, (xv) 47–65; Bennett, C.J., Caldwell, S.T., McPhail, D.B., Morrice, P.C., Duthie, G.G. and Hartley, R.C. (2004) *Bioorganic and Medicinal Chemistry*, **12**, 2079–2098; Villard, A.-L., Warrington, B. and Ladlow, M. (2004) *Journal of Combinatorial Chemistry*, **6**, 611–622; Bejugam, M. and Flitsch, S.L. (2004) *Organic Letters*, **6**, 4001–4004.

56 Arvela, R.K. and Leadbeater, N.E. (2005) *Organic Letters*, **7**, 2101–2104.

57 Appukkuttan, P., Husain, M., Gupta, R.K., Parmar, V.S. and Van der Eycken, E. (2006) *Synlett*, wdb 350Singh, B.K., Appukkuttan, P., Claerhout, S., Parmar, V.S. and Van der Eycken, E. (2006) *Organic Letters*, **8**, 1863–1866; Singh, B.K., Mehta, V.P., Parmar, V.S. and Van der Eycken, E. (2007) *Organic and Biomolecular Chemistry*, **5**, 2962–2965.

58 Leadbeater, N.E., Stencel, L.M. and Wood, E.C. (2007) *Organic and Biomolecular Chemistry*, **5**, 1052–1055.

59 Kurfürstová, J. and Hajek, M. (2004) *Research on Chemical Intermediates*, **30**, 673–681.

60 Bose, A.K., Ganguly, S.N., Manhas, M.S., He, W. and Speck, J. (2006) *Tetrahedron Letters*, **47**, 3213–3215; Bacsa, B., Desai, B., Dibo, G. and Kappe, C.O. (2006) *Journal of Peptide Science*, **12**, 633–638.

61 Baxendale, I.R., Lee, A.-I. and Ley, S.V. (2002) *Journal of the Chemical Society-Perkin Transactions*, 1850–1857.

62 Durand-Reville, T., Gobbi, L.B., Gray, B.L., Ley, S.V. and Scott, J.S. (2002) *Organic Letters*, **4**, 3847–3850; Baxendale, I.R., Ley, S.V., Nessi, M. and Piutti, C. (2002) *Tetrahedron*, **58**, 6285–6304; Yin, W., Ma, Y. and Zhao, Y.J. (2006) *The Journal of Organic Chemistry*, **71**, 4312–4315.

3
Equipment Review

3.1
Introduction

Although many of the early pioneering experiments in microwave-assisted organic synthesis were carried out in domestic microwave ovens, the current trend undoubtedly is to use dedicated instruments for chemical synthesis (Figure 1.1). In a domestic microwave oven the irradiation power is generally controlled by on–off cycles of the magnetron (pulsed irradiation), and it is typically not possible to monitor the reaction temperature reliably. Combined with the inhomogeneous field produced by the low-cost multimode designs and the lack of safety controls, the use of such equipment cannot be recommended for scientific purposes. In contrast, all of today's commercially available dedicated microwave reactors for synthesis feature built-in magnetic stirrers or alternative agitation devices, direct temperature control of the reaction mixture with the aid of IR sensors, fiber-optic probes or gas balloon thermometers, and software that enables on-line temperature/pressure control by regulation of microwave power output. Currently two different philosophies with respect to microwave reactor design are emerging: multimode and monomode (also referred to as single-mode). In the so-called multimode instruments (conceptually similar to a domestic microwave oven), the microwaves that enter the cavity are reflected by the walls and the load over the typically large cavity. In many multimode instruments a mode stirrer ensures that the field distribution is as homogeneous as possible. In the much smaller monomode cavities, only one mode is present and the electromagnetic irradiation is directed through a precision-designed rectangular or circular waveguide onto the reaction vessel mounted at a fixed distance from the radiation source, creating a standing wave. The key difference between the two types of reactor systems is that, whereas in multimode cavities several reaction vessels can be irradiated simultaneously in multi-vessel rotors (parallel synthesis), in monomode systems typically only one vessel can be irradiated at a time. In the latter case high throughput can be achieved by integrated robotics that moves individual reaction vessels in and out of the microwave cavity. Most instrument manufacturers offer a variety of diverse

Practical Microwave Synthesis for Organic Chemists: Strategies, Instruments, and Protocols
C. Oliver Kappe, Doris Dallinger, and S. Shaun Murphree
Copyright © 2009 WILEY-VCH Verlag GmbH & Co. KGaA, Weinheim
ISBN: 978-3-527-32097-4

reactor platforms with different degrees of sophistication with respect to automation, database capabilities, safety features, temperature and pressure monitoring, and vessel design. Importantly, single-mode reactors processing comparatively small volumes also have an efficient built-in cooling feature that allows rapid cooling of the reaction mixture by compressed air after completion of the irradiation period (see Figure 2.13). The dedicated single-mode instruments available today can process volumes ranging from about 60 μL to 50 mL under sealed-vessel conditions (300 °C, *circa* 20 bar), and somewhat higher volumes (125 mL) under open-vessel reflux conditions. In the much larger multimode instruments several liters can be processed under both open- and closed-vessel conditions. For both single-mode and multimode cavities continuous flow reactors are nowadays available that allow the preparation of kilograms of materials using microwave technology.

This chapter provides a detailed description of the various currently commercially available microwave reactors that are dedicated for microwave-assisted organic synthesis [1]. Publications describing the use of such equipment are discussed in Chapter 4. A comprehensive coverage of microwave reactor design, applicator theory, description of waveguides, magnetrons and microwave cavities can be found elsewhere [2–6]. An overview of experimental, non-commercial microwave reactors has recently been presented by Ondruschka *et al.* [5].

3.2
Domestic Microwave Ovens

At its beginning in the mid-1980s, microwave-assisted organic synthesis was carried out exclusively in conventional multimode domestic microwave ovens [7, 8]. The main drawback of these household appliances is the lack of control systems. In general, it is not possible to determine the reaction temperature accurately using domestic microwave ovens. The lack of pressure control and no possibility to stir the reaction mixture additionally makes performing chemical syntheses in domestic microwave ovens troublesome. Furthermore, the pulsed irradiation (on–off duty cycles) and the resulting inhomogeneity of the microwave field may lead to problems of reproducibility. Due to the non-homogenous distribution of the energy intensity throughout the cavity some areas receive higher amounts of energy (so-called hot spots) whereas others receive less energy (so-called cold spots). These hot and cold spots can be detected by "mapping experiments" of the cavity and visualized, for instance, with an IR camera that records images of a waterfilm on a glass plate after microwave irradiation (see Figures 3.1 and 3.2) [9]. If a load is located in the cavity the energy distribution is changed compared to the unloaded empty cavity (see Figure 3.1). To compensate for horizontal inhomogeneities that are also height-dependent (see Figure 3.2), the samples are generally rotated in the cavity to achieve a more even energy distribution.

Another concern is safety. Heating organic solvents in open vessels in a microwave oven can lead to violent explosions induced by electric arcs inside the cavity or sparking resulting from the switching of the magnetrons. On the other hand,

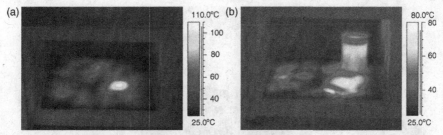

Figure 3.1 Inhomogeneities in domestic microwave ovens: formation of hot and cold spots in an empty cavity (a), and loading with a beaker filled with water (b). IR thermovision images of a waterfilm on a glass plate after irradiation at 800 W for 40 or 80 s, respectively. Reproduced with permission from Ref. [9].

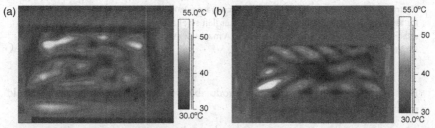

Figure 3.2 Inhomogeneities in domestic microwave ovens: formation of hot and cold spots in 3.5 cm (a), and 8 cm height (b). IR thermovision images of a waterfilm on a glass plate after 15 s irradiation at 800 W. Reproduced with permission from Ref. [9].

working with sealed vessels under pressure without real-time monitoring of pressure can lead to unexpected vessel failures in the case of a thermal runaway and to serious accidents. Early on, simple modifications of the available ovens with self-made accessories like mechanical stirrers or reflux condensers mounted through holes in the cavity were attempted in order to generate instrumentation useful for chemical synthesis (Figure 3.3). However, some safety risks remained as those instruments were not explosion-proof and leakage of microwaves harmful to the operator could occur. Clearly, the use of any microwave equipment not specifically designed for organic synthesis cannot be recommended and is generally banned from most industrial and academic laboratories today (see also Chapter 5).

3.3
Dedicated Microwave Reactors for Organic Synthesis

With growing interest in MAOS during the mid-1990s, the demand for more sophisticated microwave instrumentation, offering, for example, stirring of the reaction mixture, temperature measurement and power control features, increased.

Figure 3.3 Modified domestic household microwave oven. Inlets for temperature measurement by IR pyrometer (left side) and for attaching reflux condensers (top) are visible. A magnetic stirrer is situated below the instrument.

For scientifically valuable, safe and reproducible work, the microwave reactors should offer the following features:

- built-in magnetic or mechanical stirring
- accurate temperature measurement
- pressure control
- continuous power regulation
- efficient post-reaction cooling
- computer-aided method programming
- explosion proof cavities.

A particularly difficult problem in microwave processing is the correct measurement of the reaction temperature during the irradiation phase. Classical temperature sensors (thermometers, thermocouples) will fail since they will couple with the electromagnetic field. Temperature measurement can be achieved either by an immersed temperature probe (fiber-optic or gas balloon thermometer) or, on the outer surface of the reaction vessels, by a remote IR sensor. Due to the volumetric character of microwave heating, the surface temperature of the reaction vessel will not always reflect the actual temperature inside the reaction vessel (see Section 2.5.1 for details).

In general, a microwave reactor consists of a microwave power source (magnetron), a transmission line (waveguide) that delivers microwaves from the magnetron into an antenna or applicator, and a microwave applicator (cavity). Continuous microwaves are generated in the magnetron, which can be considered as a vacuum tube. A magnetron consists of a cylindrical cathode that is encircled by an anode block while a magnetic field is created parallel to the axis of the cathode by external magnets (Figure 3.4). The anode block additionally possesses small cavities. The

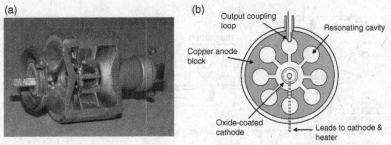

Figure 3.4 Magnetron shown without magnet (a) and cross-section of a magnetron (b).

electrons emitted from the cathode are deflected by the electric and magnetic field and rotate around the cathode before they can reach the anode. Electron bunches are generated due to the acceleration or deceleration of electrons in the cavity and thus transform their energy into microwave oscillation. Finally, microwave energy from one of the resonant cavities is coupled to the antenna that is connected to the waveguide. Many modes may be excited when the energy is coupled into the waveguide where the number is determined by the dimensions of the wave-guide cross-section. In the same way, an infinite number of modes can exist in microwave cavities. According to the geometry and dimensions of the cavities, multimode or single-mode reactors can be distinguished. A state-of-the-art micro-wave reactor design should ensure that all the incident power is absorbed by the load since the magnetron may be damaged when too much energy is reflected back. Several techniques to overcome this problem are incorporated in dedicated reactors (see Sections 3.4 and 3.5).

Since the early applications in microwave-assisted synthesis were based on the use of domestic multimode microwave ovens, the primary focus in the development of dedicated microwave instruments was on the improvement of multimode reactors. In multimode instruments, typically, one or two magnetrons create the microwave irradiation, which is, typically, directed into the cavity through a waveguide and distributed by a mode stirrer (Figure 3.5). The microwaves are reflected from the walls of the cavity, thus interacting with the sample in a chaotic manner. Multimode cavities therefore may show multiple energy pockets with different levels of energy intensity, thus resulting in hot and cold spots. To provide an equal energy distribu-tion, the samples are continuously rotated within the cavity. Therefore, multimode instruments offer convenient platforms for the increase of reaction throughput by utilizing multi-vessel rotors for parallel synthesis or scale-up. A general problem for multimode instruments is the weak performance for small-scale experiments (<3 mL). While the generated microwave power is high (1000–1600 W), the power density of the field is generally rather low, making heating of small individual samples rather difficult – a major drawback, especially for research and development purposes. Therefore, the general use of multimode instruments for small-scale synthetic organic chemistry is not so extensive in comparison with the much more popular single-mode cavities.

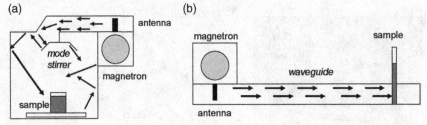

Figure 3.5 Multimode (a) versus single-mode cavities (b).

In contrast, single-mode (monomode) instruments generate a single, comparatively homogeneous energy field of high power intensity. Thus, these systems couple efficiently with small samples and the maximum output power is typically limited to 300 or 400 W. The microwave energy is created by a single magnetron and is typically directed through a rectangular waveguide to the sample, which is positioned at a maximized energy point (Figures 3.5b and 3.6b). To create optimum conditions for variations in microwave absorptivity, in some systems (Biotage) a tuning device provides proper adjustment of the microwave field. The homogenous field generally enables excellent reproducibility, although field inhomogeneities also do exist with these devices (see Section 2.5.1).

In addition to the rectangular waveguide applicator, instruments with a self-tuning circular waveguide are also available (Figure 3.6a). This single-mode cavity (CEM) features multiple entry points for the microwave energy to be delivered into the cavity (slots), compensating for variations in the coupling characteristics of the sample. Due to the cavity design, it is suitable for different vessel types and sizes, such as sealed 4–80 mL vials or up to 125 mL round-bottom flasks.

Recent advances and further improvements have led to a broad variety of applications for single-mode microwave instruments, offering flow-through systems

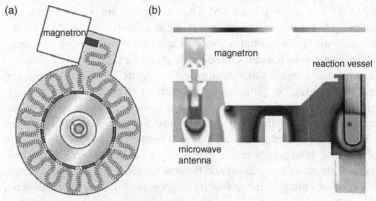

Figure 3.6 Circular waveguide (CEM Discover (a)) and rectangular waveguide (Biotage Initiator (b)).

as well as special features such as solid-phase peptide synthesis. The use of single-mode reactors has therefore increased tremendously since the year 2000 and these types of instruments have become very popular in many synthetic laboratories, both in industry and in academia. However, it should be pointed out that the type of instrumentation used is based on the desired application and scale rather than on the kind of chemistry to be performed. Both multimode and single-mode reactors are able to carry out chemical reactions efficiently and to improve classic heating protocols.

The following sections give a comprehensive description of all commercial multimode and single-mode microwave reactors available as of 2008, including various accessories and special application tools. Essentially, there are currently four major instrument manufacturers that produce microwave reactors for laboratory scale organic synthesis: Anton Paar GmbH (Graz, Austria) [10], Biotage AB (Uppsala, Sweden) [11], CEM Corporation (Matthews, NC, USA) [12], and Milestone s.r.l./MLS GmbH (Sorisole, Italy and Leutkirch, Germany) [13].

3.4
Single-Mode Instruments

The first microwave instrument company offering single-mode cavities ("focused microwaves") was the French company Prolabo [14]. In the early 1990s the Synthewave 402 was released, followed by the Synthewave 1000 [15]. The instruments were designed with a rectangular waveguide, providing focused microwaves with a maximum magnetron output power of 300 W. The cavity was designed for the use of cylindrical glass or quartz tubes with different diameters for reactions at atmospheric pressure only. Temperature measurement was performed at the bottom of the vessels by an IR sensor that required calibration by a fiber-optic probe. In 1999 all patents and microwave-based product lines were acquired by CEM Corporation. However, some instruments are still in use, mainly throughout the French scientific community, and several publications per year describing the use of this equipment still appear in the literature.

3.4.1
Biotage AB

3.4.1.1 Initiator Platform
The currently available instrumentation from Biotage is the Initiator reactor for small-scale reactions in a single-mode cavity (Figure 3.7). This instrument is closely related to the former Emrys Creator platform (2000–2004, see Section 3.4.1.2), but 45% smaller in footprint and now equipped with a touch-screen so that no external PC is needed. The Initiator is upgradeable from the single-sample manual format to the automated systems Initiator Eight or Initiator Sixty (Figure 3.7). Both are equipped with a vial rack and a robotic gripper, allowing one to perform up to 8 reactions sequentially for the Initiator Eight and up to 60 for the Initiator Sixty.

(a) (b) (c)

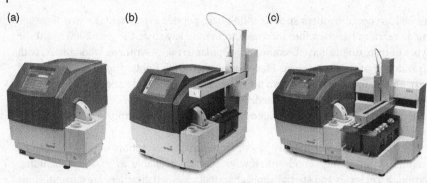

Figure 3.7 Biotage Initiator (a), Initiator Eight (b) and Initiator Sixty (c).

For the basic set-up two different vial types with reaction volumes from 0.5 to 5 mL can be used (Figure 3.8). Additional vessel sizes from very small scale (0.2–0.5 mL) to a low-level scale-up (20 mL vessel) are available with the EXP upgrade which offers a larger microwave cavity. In this way a direct scale-up from the mg to g range using different vessels from 0.2 to 20 mL operating volume (Figure 3.8) can be achieved without any system modifications or reoptimizations. The reaction vials, made of borosilicate glass, are permanently sealed by a cap and can withstand operating pressures up to 20 bar. For each vial type appropriate stir bars are provided. The vial cap consists of an aluminum crimp in which a Teflon septum is inserted. The advantage of this septum is the Reseal design which allows sample withdrawal for analysis or addition of reagents, albeit not while the reaction vessel is contained inside the microwave cavity. The sealing ring located in the cavity lid reseals the septum after penetration while the cavity is closed.

The Initiator microwave unit consists of a closed rectangular waveguide tube (see Figure 3.6b), combined with a deflector device that, via a power-sensor, physically maximizes the energy absorption by the reaction mixture (Dynamic Field Tuning). Dynamic Field Tuning allows the system to detect the absorbance characteristics of

Figure 3.8 Biotage Initiator vessel types (0.2–0.5 mL, 0.5–2.0 mL, 2.0–5.0 mL, 10.0–20.0 mL, left to right) and schematic description of maximum volume.

the reaction mixture and to optimize the coupling and quantity of microwave energy delivered. Thus, uniform and high density heating is ensured, resulting in fast and highly reproducible synthesis performances.

In 2006 the Initiator 2.0 was launched and all the following specifications refer to this model. The output power is maximized to continuously deliver 400 W, sufficient for rapid heating of most reaction mixtures. The Initiator series is equipped with built-in magnetic stirring with stirring speed variable from 300 to 900 rpm. Temperature measurement is achieved by an IR sensor perpendicular to the position of the vial in the waveguide, working in a measuring range of 40–250 °C. This arrangement requires a specific minimum filling height in each vessel type in order to obtain accurate temperature values. On the other hand, if the maximum filling volume of the vial (Figure 3.8) is exceeded then insufficient space for pressure build-up is left. The temperature is measured on the outer surface of the reaction vessels and no inside temperature measurement is currently available. The pressure limit for the Initiator instruments is 20 bar, imposed by the sealing mechanism which utilizes Teflon-coated silicon seals in aluminum crimp tops. Pressure control is achieved by means of a non-invasive sensor integrated into the closing lid of the cavity, which senses the deformation of the seal due to the pressure build-up. Efficient cooling is accomplished by means of a pressurized air supply with a rate of approximately 60 L min^{-1}, enabling cooling from 250 to 50 °C within approximately 1 min, depending on the heat capacity of the solvent used.

Reaction control is temperature-based, with the system trying to attain the adjusted maximum temperature as fast as possible by using the appropriate microwave magnetron output power up to 400 W. For polar reaction mixtures, for example ion-containing mixtures or good absorbing solvents like NMP, the absorption level can be set to "high" and the starting output power is limited to a maximum of *circa* 100 W to avoid an overshooting of the temperature. When concentrated ionic solutions or mixtures including, for example, ionic liquids are employed, a "very high" absorption level can be selected where the starting output power is limited to a maximum of *circa* 60 W to prevent overheating of the sample.

Regarding safety features, if the measured temperature exceeds 250 °C or the pressure exceeds 22 bar, respectively, the instrument switches off and the cooling mechanism is activated as the reaction is aborted. The same is true if the temperature increases by more than 30 °C s^{-1} or the pressure by more than 5 bar s^{-1}, respectively.

As no external PC is needed, the characteristics of the software package have changed compared to the Emrys series. Via the touch-screen the user is able to change the performed reaction protocols "on-the-fly" without aborting the experiment-temperature and time changes can be made immediately. Furthermore, reactions can be performed under "power control", that is, the power can be adjusted to a defined value over the process time. In addition, so-called "cooling-while-heating" can be applied, introducing more microwave energy into the reaction system as the cooling is activated during the irradiation process (see Section 2.5.4 for details).

For data management purposes, the Initiator 2.0 is equipped with a USB port for saving and transferring methods and results. Instruments that are connected to a

Biotage HUB can share data, methods and user profiles, furthermore, electronic lab books can be directly interfaced. For instruments that are connected to a network, online monitoring of the processing reaction via the office PC is possible. Additionally, the result files can be sent via e-mail.

An additional tool is the PathFinder database, representing a collection of detailed protocols for reactions performed with Biotage microwave instruments. As of 2008 more than 4500 entries are available in this web-based tool, where keyword and/or structure searches can be conducted.

3.4.1.2 Emrys Platform (2000–2004)

Up to 2004, Biotage offered the Emrys monomode reactor series, which is the predecessor to the Initiator instruments. Although no longer commercially available, many instruments are currently still in use. Therefore this line of products is also discussed in this chapter with particular attention focused on the Emrys Liberator instrument.

The Emrys Creator, similar to the Initiator, is the basic instrument platform for reactions on a 0.5–5.0 mL scale. The Emrys Optimizer, predecessor of the Initiator Sixty, allows automation with integrated robotics to run up to 60 reactions sequentially. The EXP upgrade offers a larger microwave cavity and additional vessel sizes from very small scale (0.2–0.5 mL) to a low-level scale-up (20 mL vessel).

The Emrys Liberator is a fully automated instrument with robotic sample handling and a liquid dispenser, especially designed for high-throughput library synthesis (Figure 3.9). Up to 120 reactions can be operated sequentially with this equipment

Figure 3.9 Emrys Liberator from Biotage.

using the small vial types which allow working volumes from 0.2 to 5.0 mL. With the liquid handler, stock solutions or liquid reagents can be dispensed into designated vials. The Liberator is equipped with a comprehensive software package for creation of protocols, including an ISIS draw surface for graphical description of the reactions. Furthermore, all the parameters, like weights/volumes of the reagents, needed for each experiment are calculated automatically.

Similar to the Initiator series, the operation limits for the Emrys system are 60–250 °C at a maximum pressure of 20 bar. Temperature control is achieved in the same way by an IR sensor rectangular to the sample position. In contrast to the Initiator single-mode reactors from Biotage all Emrys instruments work only with external computer control.

3.4.1.3 Chemspeed SWAVE

The SWAVE unattended microwave synthesizer station from Chemspeed (Figure 3.10) incorporates a Biotage Initiator with all the features which are necessary to enable a fully automated synthesis workflow platform [16]. Automation is given, starting from reaction preparation up to purification, via the robotic platform which consists of an XYZ robotic arm and 8 racks of up to 30 samples (maximum 240 samples). Addition of reagents under an inert atmosphere, solid weighing and direct dispensing, liquid handling (4 syringe pumps) is covered, as well as an automated single tool for capping, crimping and gripping of the microwave vials. For multi-step synthesis, reagents can be added directly through the septum while the vial is located in the microwave cavity. Solid phase extraction (SPE), automated filtration, liquid–liquid extraction and online chromatographic analysis are some of the available features for the work-up process. To increase the throughput, a second Initiator microwave reactor can be incorporated on the right-hand side of the platform (Figure 3.10).

Figure 3.10 Chemspeed SWAVE: fully automated microwave synthesizer station.

Figure 3.11 Available CEM Discover systems: (a) BenchMate, (b) LabMate and (c) S-Class.

3.4.2
CEM Corporation

3.4.2.1 Discover Platform

The CEM Discover system, introduced in 2001, is a single-mode instrument based on the self-tuning circular waveguide technique (see Figure 3.6a). This circular cavity automatically adjusts to ensure that the reaction receives the optimum amount of energy, regardless of the reaction volume. This concept provides modularity for automation, scale-up under closed-vessel, open-vessel and flow-through conditions, low-temperature chemistries and application in biosciences. All Discover instruments are equipped with a built-in keypad for programming the reaction procedures and allowing on-the-fly changes. The manual Discover reactor (Figure 3.11, Table 3.1) covers a variety of reaction conditions in open- (up to 125 mL) and closed-vessel systems (up to 50 mL filling volume).

The output power is maximized to continuously deliver 300 W (power can be set between 0 and 300 W), sufficient for rapid heating of most reaction mixtures. Routine temperature measurement within the Discover series is achieved by an IR sensor positioned at the bottom of the cavity, below the vessel. This allows accurate temperature control of the reaction while using minimum amounts of materials (60–70 µL for the 4 mL vials). The platform also accepts an optional fiber-optic temperature sensor system for internal temperature measurement that can be applied in cases where IR technology is not suitable, such as with sub-zero temperature reactions or with specialized reaction vessels (see Section 2.5.1 for details).

Table 3.1 Features of the manual Discover systems from CEM.

Features	BenchMate	LabMate	S-Class
Open vessel	✓	✓	✓
Closed vessel	✓	✓	✓
Pressure management and venting	✓	✓	✓
Pressure measurement	✓	✓	✓
Synergy included		✓	✓
Automated pressure device			✓
35 mL vessels			✓

Offering an economical choice in terms of footprint, the Discover BenchMate provides an entry-level system with basic reaction temperature and pressure management (Figure 3.11a). The pressure-management device does not measure pressure but, due to the "snap-on" cap design, automatic venting is feasible when the internal pressure exceeds 20 bar (IntelliVent technology). With the BenchMate system 4 and 10 mL reaction vials with a maximum filling volume of 2.5 or 7 mL, respectively, can be used.

The Discover LabMate (Figure 3.11b) is the next generation, possessing additional features like the pressure measurement device with the IntelliVent Pressure Control System. If the pressure in the vial exceeds 20 bar, the IntelliVent sensor allows a controlled venting of the pressure and subsequently automatically reseals the vial to maintain optimum safety. The same vial sizes as for the BenchMate (4, 10 mL) can be employed with this system.

In 2006, the newest Discover system, the Discover S-Class was introduced (Figure 3.11c). Supplementary to the features of the LabMate, the instrument incorporates a fully automated pressure control and a USB interface. Additionally to the 4 and 10 mL vial, a 35 mL reaction vial with working volumes from 2.5–25 mL is available, allowing low-level scale-up. An interesting and unique element is the optional integrated CCD camera for *in situ* reaction monitoring (see Section 5.3.8 for details).

All three Discover systems work with the Synergy software which is included in the base package for the LabMate and S-Class. The systems can be run and programmed via PC with the Synergy software which allows documentation, automated data handling and parameter control capabilities (e.g. variable stirring speed: off, low, medium, high). For achieving simultaneous cooling for "enhanced microwave synthesis" (EMS), (see Section 2.5.4 for details), the patented enhanced cooling system (PowerMAX) can be used during irradiation.

An extension to the common Discover reactors is the Discover CoolMate (Figure 3.12), a sub-zero cooling module that can be used on all Discover systems and which is designed to perform sub-ambient temperature chemistry. The reactor is equipped with a jacketed low-temperature vessel and the system's microwave

Figure 3.12 CEM Discover CoolMate: low-temperature cooling module.

transparent cooling media and chilling technology to keep the bulk temperature low (-80 to $+65\,°C$). Temperature measurement is performed by internal fiber optic temperature control. Thus, thermal degradation of materials in temperature-sensitive reactions may be prevented while microwave energy is introduced to the reaction mixture (see Section 2.5.4 for more details).

3.4.2.2 Explorer$_{PLS}$ Systems

The Explorer series provides an automation upgrade for the Discover platform by introducing sample racks with a robotic gripper on the top. The Discover BenchMate and LabMate can be converted to the Explorer-24, a 24-position system which can be used only with the 10 mL vials.

For the Discover S-Class three different autosampler rack sizes are optional. All three vial types (4, 10, 35 mL) can be employed for the Explorer-48, -72 and -96 instruments (Figure 3.13). Due to the rack design and software, the vial size and position is recognized automatically by the autosampler. The Explorer-48 holds 4 interchangeable racks and up to 48 positions (24 when 35 mL vials are used) that can be loaded and run in any combination. The same is true for the Explorer-72 and -96 which are useful for high-throughput syntheses and library production; the former holds 6 and the latter 8 vial racks.

3.4.2.3 Voyager Systems

The Voyager System converts the Discover BenchMate and LabMate reactors into an automated flow system (see Sections 4.8.2 and 4.8.3), designed to allow the scale-up of reactions from mg quantities to *circa* 1 kg while still maintaining the advantages of single-mode energy transfer. While the technology accommodates both continuous and stop-flow formats, the stop-flow technique better serves the majority of scale-up applications encountered in today's synthesis laboratory since it combines the advantages of a batch reactor with those of a continuous flow reactor. The Voyager$_{SF}$ system in stop-flow mode is operated with a special 80 mL vessel (see Figure 3.14), where the reaction mixture is pumped in and out by peristaltic pumps. Reaction limits are $250\,°C$ and 18 bar and the system is also applicable for heterogeneous

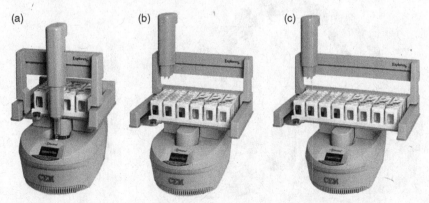

(a) (b) (c)

Figure 3.13 Available CEM Explorer systems: Explorer-48 (a), -72 (b), and -96 (c).

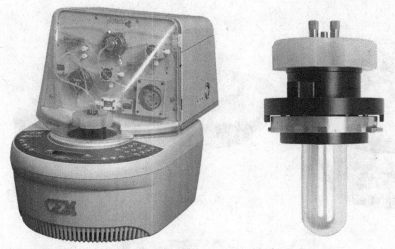

Figure 3.14 The CEM Voyager$_{SF}$ and its 80 mL reaction vessel.

mixtures, slurries and solid phase reactions. Uniform mixing is ensured by dynamic stirring. With this system, a direct scale-up of reactions that were performed on the Discover or Explorer units is possible.

Reaction scale-up in genuine continuous flow format is performed by the use of special flow-through cells in combination with two HPLC pumps in the Voyager$_{CF}$ reactor. Available are 10 mL and 80 mL flow cells (Figures 3.15b and 3.15c), which can be equipped with inert packing material like glass beads or catalysts and scavengers immobilized on a solid support. A fiber-optic temperature control module is standard for both the Voyager$_{CF}$ and Voyager$_{SF}$ for internal temperature measurement. In addition, coiled flow-through cells, made of Teflon or glass (Figures 3.15d and 3.15e), with a maximum flow rate of 20 mL min^{-1} can be obtained. Operating limits for working with these flow cells in the Voyager$_{CF}$ are 250 °C and 17 bar. The continuous flow format should only be used for homogeneous solution-phase chemistry as slurried mixtures may cause problems with the pumping system.

3.4.2.4 Peptide Synthesizers

For solid-phase peptide synthesis the Liberty, a fully automated microwave peptide synthesizer, is available (Figure 3.16). This instrument, based on the same Discover reactor core, enables the sequential synthesis of up to 12 peptides, unattended, using a fluidics module to allow the controlled addition of resins, amino acids, coupling, deprotection, and washing reagents, as well as cleavage cocktails. The system is equipped with up to 25 amino acid reservoirs with 125 or 250 mL capacity and 7 external bottle positions. Vessels with 30 and 125 mL volumes allow the synthesis of peptides on a 0.025 to 5 mmol scale applying either Fmoc or Boc strategies. A spraying system for top-down washing contained in the vessel ensures that the resin beads are properly washed from the vessel wall and covered with the reaction mixture. Appropriate mixing of the resin with reagents is ensured by low-pressure nitrogen bubbling. Typical cycle times with this system are 15 min, including all washings, for each residue addition. The Liberty also allows programmable cleavage from the resin,

(a)

(b)

(c)

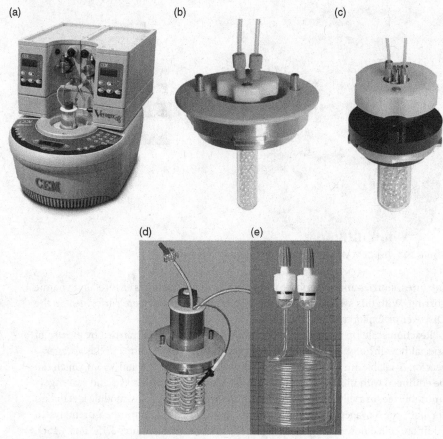

(d)

(e)

Figure 3.15 The CEM Voyager$_{CF}$ (a) and its flow cells: 10 mL (b) and 80 mL (c) flow cell, 5 mL Kevlar-enforced Teflon coil (d) and 10 mL glass coil (e).

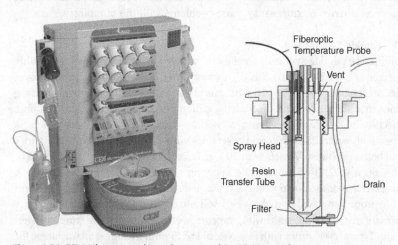

Fiberoptic
Temperature Probe

Vent

Spray Head

Resin
Transfer Tube

Drain

Filter

Figure 3.16 CEM Liberty peptide synthesizer and its 30 mL vessel.

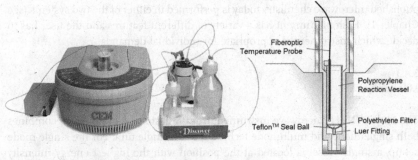

Figure 3.17 CEM Discover SPS and its 25 mL reaction vessel.

either immediately after synthesis or at a later programmed time. Potential issues with racemization have been addressed via internal temperature control by a fiber-optic probe (see Section 4.7.2.1).

In addition to the automated peptide synthesizer a manual microwave reactor, the Discover SPS system, is also available (Figure 3.17). Reactions are performed on a 0.025–1 mmol scale in a 25 mL polypropylene vessel that can be considered as a modified SPE cartridge with a filtration device at the bottom. Leakage is prevented by a special sealing technology: a Teflon seal ball at the bottom stops reagents from dripping out of the vessel. For washing and filtration steps, a vacuum manifold station, consisting of one waste container and one for product collection after cleavage, is used in combination with the Discover SPS. As for the Liberty system, agitation can be performed via inert gas bubbling and internal temperature measurement is via a fiber-optic sensor.

In Table 3.2 some of the most important features of the Biotage Initiator 2.0 in comparison to the CEM Discover systems are summarized as the majority of all

Table 3.2 Comparison of Biotage Initiator and CEM Discover instruments.

Features	Biotage Initiator 2.0	CEM Discover
Waveguide	rectangular	circular
Max. output power	400 W	300 W
Operation temperature	40–250 °C	rt–300 °C
Max. pressure	20 bar	20 bar
		15 bar (80 mL vessel)
Vessel sizes	0.2–20 mL	4–80 mL
		max. 125 mL round-bottom flask
Sealing mechanism	permanent with crimped caps	"Snap-on" IntelliVent caps
IR sensor	from the side at a defined height	from the bottom
Fiber optic	✕	✓
Simultaneous cooling	✓	✓
Closed vessel	✓	✓
Open vessel	✕	✓
Magnetic stirring	300–900 rpm	3 different speeds
Method programming	touch screen	touch pad or PC

published microwave chemistry today is performed in either of the two systems (see Chapter 1). Each instrument has a variety of different features and the user has to decide which instrument is appropriate for individual demands.

3.4.3
Milestone s.r.l.

The MultiSYNTH microwave instrument from Milestone is unique as it combines both single-mode and multimode techniques in a single unit. For the single-mode set-up, a single vessel is located at the position with the highest energy intensity allowing rapid heating of small reaction volumes (250 µL). Temperature reaction control is provided by both IR and fiber-optic sensors. Agitation is accomplished by either magnetic stirring or by shaking of the reaction vessel. For more detailed information on the MultiSYNTH system, see Section 3.5.4.1.

3.5
Multimode Instruments

The development of multimode reactors for organic synthesis occurred mainly from already available microwave acid digestion/solvent extraction systems. Instruments for this purpose were first designed in the 1980s and with the growing demand for synthesis systems, these reactors were subsequently adapted for organic synthesis applications. Today, a broad spectrum of instruments with a large range of different accessories for organic synthesis is available. In addition, the reactors are designed in such a way that a direct scale-up from small scales performed in single-mode platforms to larger scales in multimode instruments is possible without any change in already optimized reaction conditions.

3.5.1
Anton Paar GmbH

The Anton Paar Synthos 3000 (Figure 3.18, Table 3.3) is one of the most recent multimode instruments on the market. This microwave reactor dedicated for scale-up synthesis in quantities of up to 1 L reaction volume per run was initially designed for chemistry under high-pressure and high-temperature conditions. Very recently, new features that allow combinatorial and library synthesis under more moderate conditions and on a smaller scale have been introduced.

The use of two magnetrons (1400 W continuously delivered output power) allows mimicking of small-scale runs to produce large amounts of the desired compounds within a similar time frame. The homogeneous microwave field guarantees identical conditions at every position of the rotors, resulting in good reproducibility of experiments. Offering high operation limits (80 bar at 300 °C) the instrument facilitates the investigation of new reaction methods like near-critical water chemistry. The instrument can be operated with various rotor types ranging from an 8- to a

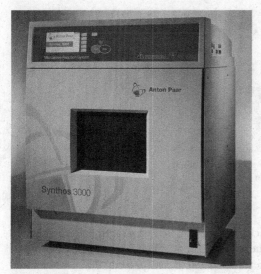

Figure 3.18 Anton Paar Synthos 3000.

96-position rotor, equipped with a choice of vessel types for different pressure and temperature conditions. For a combinatorial approach a rotor system that holds four 48-well microtiter plates made of silicon carbide is also available. Various accessories (Figure 3.19) allow special applications like creation of inert/reactive gas atmosphere or reactions in pre-pressurized vessels. Solid-phase synthesis can be performed as well when PTFE-TFM liners in combination with a special filtration device are employed. For filtration, two liners are attached to the filtration unit that allows further workup of the filtrate (e.g. after cleavage or when polymer-supported reagents, scavengers or catalysts are employed) as well as performing further reaction steps on the solid support that remains in the filter and thus in one of the two liners.

Table 3.3 Anton Paar Synthos 3000 features.

Feature	Description
Cavity size (volume)	$42 \times 57 \times 62$ cm (66 L)
Installed power	1700 W (2 magnetrons)
Max. output power	1400 W
Temperature control	immersed gas balloon thermometer (max. 300 °C); outside IR remote sensor (max. 400 °C)
Pressure control	hydraulic system (max. 86 bar)
Cooling system	190 $m^3 h^{-1}$ (forced air flow through rotor, 4 steps adjustable)
Magnetic stirring	600 rpm (4 steps software adjustable)
Rotor speed	3 rpm
External PC	optional, not required as key panel + keyboard is standard equipment

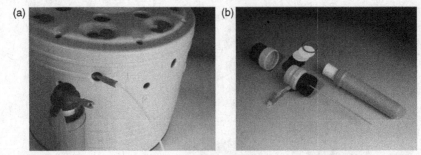

Figure 3.19 Anton Paar accessories for special applications:
(a) gas loading system and (b) filtration unit.

Temperature measurement is achieved by a remote IR sensor from the bottom on the outer surface of the vessels. The operating limit of the IR sensor is 400 °C, but is regulated by the software safety features to 280 °C, as operating limits of the used materials are also at a maximum of 300 °C. For additional control, temperature measurement in one reference vessel by an immersed gas balloon thermometer is available. The operation limit of this temperature probe is 310 °C, suitable for reactions at extreme temperature and pressure conditions.

The pressure is measured by a hydraulic system, either in one reference vessel of the 16- and 48-vessel rotor or simultaneously for all vessels of the 8-vessel rotor. The operation limit is 86 bar, sufficient for most synthetic applications. In addition, a pressure rate limit is set to 3.0 bar s^{-1} by the included control software. Protection against sudden pressure peaks is achieved by metal safety disks, integrated in the vessel caps (safety limit 70 or 120 bar, respectively) and by software settings, dependent on the rotor used and the vessel type.

All parameters are transmitted wirelessly by IR data transfer from the sensors to the system control computer of the instrument to eliminate disturbing cables and hoses from inside the cavity.

3.5.1.1 Rotors and Vessels
Five individual rotor types with different vessels in addition to the combichem rotor accommodating the silicon carbide well plates are available. Depending on the vessel- or the pressure jacket material different temperatures and pressures can be achieved.

- **8-Vessel Rotor (8SXF100, 8SXQ80)** (Figure 3.20a, Table 3.4)
  This rotor system was specially designed for high-temperature and high-pressure reaction conditions. Two different vessel types are provided, PTFE-TFM liners or quartz vessels, which enable reactions to be performed at 260 °C and 60 bar for the former and at 300 °C and 80 bar for the latter. Optional for this rotor system is a gas loading system which allows reactions to be performed under inert or reactive gas atmosphere. Pre-pressurization up to 20 bar is possible.

Figure 3.20 Anton Paar 8- (a) and 16- vessel (b) rotor with PTFE-TFM and quartz liners and pressure jackets.

- **16-Vessel Rotor (16MF100, 16HF100)** (Figure 3.20b, Table 3.4)
  This tool is dedicated for standard synthetic reactions up to 240 °C and offers 100 mL screw-cap vessels with PTFE-TFM liners. Applying different pressure jackets allows continuous operation of these vessels at a maximum of 20 or 40 bar and temperatures up to 200 or 240 °C, respectively.

- **48-Vessel Rotor (48MF50)** (Figure 3.21a, Table 3.4)
  For library generation on the gram scale this rotor with its 50 mL vessels can be used. The vessels are arranged in three circles of 16 vessels each, temperature measurement is performed at the center circle via IR in addition to internal measurement in one reference vessel. For this rotor system, the vessels employ screw caps with conical seals for proper sealing. Operation limits are 200 °C and 20 bar.

Table 3.4 Synthos 3000 – features of rotor systems.

	4×24MG5/ 64MG5	48MF50	16MF100	16HF100	8SXF100	8SXQ80
No. of vessels	96/64	48	16	16	8	8
Volume (mL)	5	50	100	100	100	80
Operating volume (mL)	0.3–3	6–25	6–60	6–60	6–60	6–60
Max. temperature (°C)	200	200	200	240	260	300
Max. pressure (bar)	20	20	20	40	60	80
Liner material	glass	PFA	PTFE-TFM	PTFE-TFM	PTFE-TFM	quartz
Pressure jacket	×	PEEK	PEEK	ceramics	ceramics	×
Pre-pressurizing	×	×	×	10 bar	20 bar	20 bar

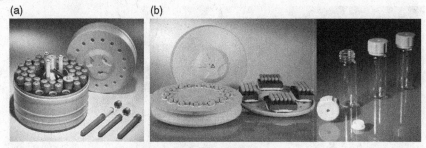

Figure 3.21 Anton Paar 48-vessel rotor (a) and 64- and 96-position rotor (b) with corresponding vials.

- **64-Vial Rotor (64MG5)** (Figure 3.21b, Table 3.4)
  Higher throughput on a smaller scale can be performed with this rotor type featuring 64 disposable 5 mL standard glass vials with screw cap sealing. The vials are arranged in 16 groups of four and allow reaction conditions up to 200 °C and 20 bar.

- **96-Vial Rotor (4 × 24 MG5)** (Figure 3.21b, Table 3.4)
  This high-throughput rotor is specially designed for parallel method optimization in microwave synthesis. The same glass vials as used with the 64-vessel rotor are here arranged in silicon carbide (SiC) blocks within a 6 × 4 matrix. Silicon carbide ensures maximum temperature homogeneity and therefore even allows the use of low-absorbing solvents. Reaction conditions are again 200 °C and 20 bar.

- **Combichem Rotor (4 × 48 MC Well Plate)** (Figure 3.22)
  Here, four SiC microtiter well plates with a standard 6 × 8 matrix and a maximum filling volume of 300 μL per well are arranged on the rotor enabling the performance of 192 reactions in parallel. Due to the advanced sealing mechanism temperatures up to 200 °C and pressures up to 20 bar can be achieved. The plates are covered with an aluminum top plate with corresponding bore holes which allow sample withdrawal.

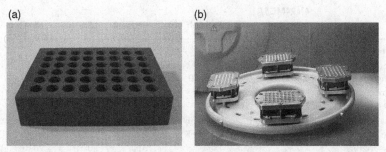

Figure 3.22 Anton Paar SiC 48-well microtiter plate (a) and rotor set-up (b).

(a)   (b)

Figure 3.23 Biotage Emrys Advancer scale-up instrument (a) with automated/tilting cavity lid (Advancer Kilobatch (b)).

3.5.2
Biotage AB

For scale-up applications Biotage offers the Emrys Advancer batch-reactor (see Figure 3.23), with a multimode cavity for operations with one 350 mL Teflon reaction vessel at high-pressure conditions. An operating volume of 50–300 mL at a maximum of 20 bar enables the production of 10 to 100 g product within one run. Homogeneous heating is ensured by a precise field tuning mechanism and vigorous overhead stirring (up to 1000 rpm) of the reaction mixture. Direct scalability allows translation of the optimized reaction conditions from the Initiator/Emrys systems (see Section 3.4.1) to larger scale.

The maximum output power of the Emrys Advancer is 1200 W, generating a heating rate of 0.5–4 °C s^{-1} to reach the maximum temperature of 250 °C for 300 mL reaction volume in comparable times to the monomode experiments. The temperature is controlled via an internal fiber-optic probe. Several connection ports in the chamber head (see Figure 3.23) enable the addition of reagents during irradiation, sample removal for analysis, *in situ* monitoring by real time spectroscopy or creation of inert/reactant gas atmosphere. Cooling is achieved by an effective gas-expansion mechanism (adiabatic flash cooling) to ensure drastically shortened cooling periods (200 mL EtOH within 30 s from 180 to 65 °C).

In 2008, the Advancer Kilobatch was launched. This microwave reactor provides a kilogram scale-up of both homogeneous and heterogeneous reactions in a sequential batch format. With the liquid loading device, liquid components are delivered unattended and homogeneous reaction mixtures can be processed up to 12 h. In addition, by using the automated solid loading carousel, heterogeneous

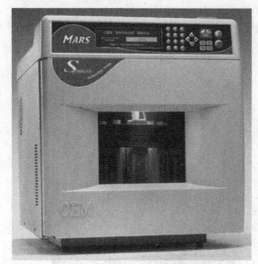

Figure 3.24 CEM MARS S synthesis system.

reactions can be performed in four sequential cycles of 250 mL each to obtain a 1 L batch. For an appropriate mixing, a mechanical overhead stirrer is featured (see Figure 3.23).

Due to the dimensions of the instrument (140 × 65 × 185 cm) extra laboratory space is required to make operations comfortable.

3.5.3
CEM Corporation

The MARS S microwave synthesis system (Figure 3.24, Table 3.5) is based on the related MARS 5 digestion instrument and offers different sets of rotor systems with

Table 3.5 MARS S Synthesis system – general features.

Feature	Description
Cavity size (volume)	48 L
Delivered power	1600 W
Max. output power	1600 W
Temperature control	outside dual IR remote sensor; immersed fiber-optic probe (optional)
Pressure control	pneumatic pressure sensor (optional)
Cooling system	air flow through cavity $210\,m^3\,h^{-1}$
Stirring	magnetic stirring at variable speed overhead stirring (optional for open vessel)
Rotor speed	8.5 rpm
External PC	optional, not required as integrated key panel is standard equipment

several vessel designs and sizes for various synthesis applications under open- and closed-vessel conditions.

Temperature measurement in the rotor systems is accomplished by an immersed fiber-optic probe in one reference vessel or by two IR sensors on the surface of the vessels from the bottom of the cavity. Pressure measurement in HP- and XP-rotors is achieved by an electronic sensor in one reference vessel. Correct temperature and pressure measurement via the sensors is ensured up to 300 °C and 100 bar. The simultaneous use of the fiber-optic probe and dual IR sensors provides a temperature measurement in all vessels of the turntable, for example, for the MARSXpress option, the dual IR sensors allow temperature measurement in up to 40 vessels simultaneously. For reactions at high pressures, the HP- and XP-rotor vessels offer a choice of seals and covers, fully-sealed or self-venting. The temperature and pressure feedback control of the MARS system monitors and regulates the amount of power being applied to the reactions to provide optimum reaction control. The system will automatically shut the microwave power down if the temperature in the control vessel rises too high or if the vessel starts to over-pressurize. In addition, a built-in pressure limit control for all vessels is offered where the magnetron is turned off when the sensor detects excessive venting in the cavity. For the MARSXpress and the GlassChem rotors, no internal pressure measurement is available as the vessels are "self-regulating" to prevent overpressure. The MARSXpress maintains reaction control via the self-regulating pressure vessels and the temperature feedback control. The temperature control sensor monitors the temperature in each vessel and adjusts the power output accordingly to maintain the user-defined temperature set point. The MARSXpress offers real-time display of the temperature in each vessel. All of CEMs high-pressure vessels have an open-architecture design that allows airflow within the cavity to cool the vessels quickly.

The general maximum output power of the instrument is 1600 W, but the MARS control panel offers two additional low-energy levels with unpulsed microwave output power of 400 and 800 W, respectively. This feature avoids overheating of the reaction mixture and unit when small amounts of reagents are used.

The MARS comes with a software package, operated from the integrated spill-proof keypad. The instrument can be connected to an external PC, but this is not required for most common operations. Methods and reaction protocols can be designed as temperature/time profiles or with precise control of constant power during the reaction.

3.5.3.1 MARS Scale-Up System Accessories

For reactions at atmospheric pressure standard laboratory glassware such as round-bottom flasks from 250 mL up to 5 L can be used (see Figure 3.25). An inlet/outlet port on the top of the cavity allows connection of a reflux condenser or distillation equipment as well as addition of reagents, sample withdrawal or overhead stirring. With the Reflux Accessory Kit, all the parts needed, like adapters, a Teflon vessel stand and 3 and 5 L reaction vessels are included.

In addition, reactions can be performed in continuous flow mode with the appropriate flow cells which were recently introduced. The Flow Cell Accessory Kit

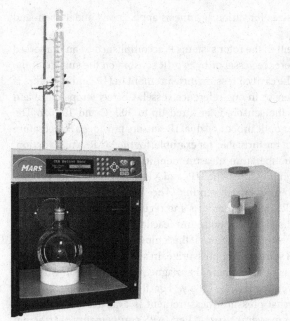

Figure 3.25 CEM MARS scale up systems: open vessel round-bottom flask with reflux condenser and high-pressure 300 mL vessel.

(Figure 3.26) delivers a 2 or a 4 L flow vessel, inlet/outlet tubes, all necessary adapters and the Teflon vessel stand. For this system, the port on the top of the cavity provides the entrance and exit for the inlet and outlet lines.

For both systems, the reflux approach using standard round-bottom flasks and flow processing with flow cells, fiber-optic temperature control is available.

In addition to the open vessel systems, a 300 mL closed vessel for performing larger pressurized reactions up to a volume of 150 mL is also available (Figure 3.25). Operating limits for this vessel are 250 °C and 35 bar.

3.5.3.2 MARS Parallel System Accessories

Four individual rotor systems with different reaction vessels for performing reactions under closed vessel and moderate- to high-pressure conditions are available. Inlet and outlet ports on the side of the cavity allow the introduction of an inert atmosphere into the reaction vial or addition and withdrawal of reagents.

For open vessel high-throughput parallel synthesis a rotor system that holds microtiter plates is also available.

- **Microplate Rotor** (Figure 3.27a)
  96-Well microtiter plates can be used in combination with a turntable for a combinatorial approach under open vessel conditions at up to 150 °C. Three different styles of rotors are available that hold 3, 4 or 8 microtiter trays and allow

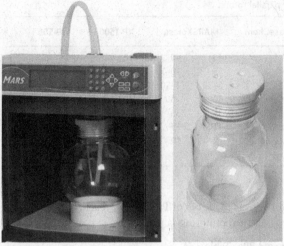

Figure 3.26 CEM MARS Flow Cell Accessory Kit: 4 L flow cell located in the cavity with inlet and outlet tubes and 2 L flow cell.

the performance of up to 768 reactions in parallel. A Teflon stand holds the fiber-optic temperature probe which is immersed in one reference well.

- **GlassChem Rotor** (Figure 3.27b, Table 3.6)
  With this rotor system, reactions in up to 24 glass vessels (Figure 3.27c) can be run at up to 200 °C and 14 bar. The 20 mL vessels use the same screw cap design as the MARSXpress that consists of a liner which holds the reaction mixture, a sealing plug and a cap. Temperature measurement is provided via a fiber-optic probe.

- **MARSXpress Rotor** (Figure 3.28, Table 3.6)
  In this 40-position rotor system, reaction volumes of up to 2 L per run can be accommodated. The vessels also include the screw cap design and are available in four different sizes (10, 25, 55 and 75 mL). Temperatures up to 260 °C for PFA

(a)   (b)   (c)

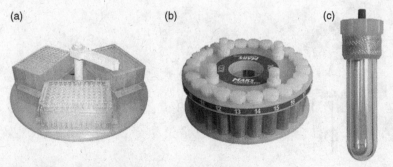

Figure 3.27 CEM MARS parallel rotors: Microplate (a), GlassChem rotor (b) and GlassChem vessel (c).

Table 3.6 MARS S – features of parallel rotors.

	GlassChem	MARSXpress		XP-1500+	HP-500+
No. of vessels	24	40	40	12	14
Vessel volume (mL)	20	55	10–75	100	100
Operating volume (mL)	3–14	6–35	1–50	10–70	10–70
Max. temperature (°C)	200	300	260	300	260
Max. pressure (bar)	14	35	35	100	34
Vessel material	glass	TFM	PFA	Teflon, Pyrex, quartz	
Temp. control	fiber-optic	IR	IR	fiber-optic + optional IR DuoTemp	

vessels and up to 300 °C for vessels made out of TFM can be reached. Here, an all-vessel temperature control via IR monitoring is provided.

- **XP-1500 Plus Rotor** (Figure 3.29a, Table 3.6)
  With this 12-position rotor, reactions at high pressure and temperature (100 bar or 300 °C) can be performed. Temperature and pressure measurement is possible via immersed sensors in one reference vessel. Dual IR sensors in conjunction with a fiber-optic probe provide an all-vessel temperature control, the so called DuoTemp. Teflon, Pyrex and quartz vessels are optional.

- **HP-500 Plus Rotor** (Figure 3.29b, Table 3.6)
  In this rotor type 14 Teflon, Pyrex or quartz vessels with operating limits of 33 bar and 260 °C are accommodated. Temperature and pressure measurement is identical to the XP-1500 Plus system.

3.5.4
Milestone s.r.l.

Milestone offers a large range of multimode instruments with appropriate accessories to perform reactions at volumes up to *circa* 3.5 L under closed-vessel conditions.

Figure 3.28 CEM MARSXpress rotor: Xpress 55 mL and screw-cap vial.

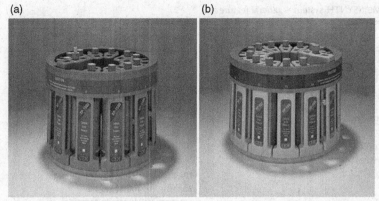

Figure 3.29 CEM MARS high-pressure rotors: XP-1500 Plus (a) and HP-500 Plus (b).

For open-vessel conditions round-bottom flasks up to a size of 4 L can be utilized. For both applications several rotor systems as well as single vessel modules or microtiter plates are available. A continuous flow system is also provided when a larger scale-up is considered.

3.5.4.1 MultiSYNTH System

The MultiSYNTH instrument, a so-called "hybrid" instrument, was introduced in 2006 (see Figure 3.30, Table 3.7). It has the unique feature of being able to merge both single-mode and multimode technologies in a single unit. The benefits of single-mode reactor features like fast heating, full control of single vessels and fast cooling are combined with the main advantage of multimode technologies of performing reactions in parallel using rotor systems. In the single-mode set-up one vessel is located at a defined position where the microwave energy intensity is the highest, whereas in the multimode set-up a classical rotor system is employed (see Figure 3.30).

A single magnetron with homogeneous microwave distribution in the cavity delivers 800 W output power. The full 800 W can be employed for the multimode configuration only, while for the single-mode set-up 400 W is the maximum power. In

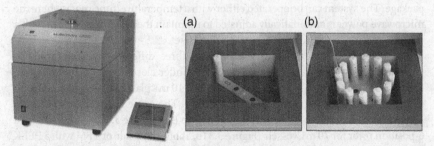

Figure 3.30 Milestone MultiSYNTH with single-mode (a) and multimode (b) set-up.

Table 3.7 MultiSYNTH system – general features.

Feature	Description
Cavity size (volume)	26.5 × 24.5 × 20 cm (13 L)
Delivered power	800 W
Max. output power	800 W (multimode)
	400 W (single-mode)
Temperature control	immersed fiber-optic probe (max. 250 °C);
	outside IR remote sensor (max. 300 °C)
Pressure control	indirect pressure control
Cooling system	compressed air
Agitation	magnetic stirring (0–400 rpm)
	shaking (0–100%)
External PC	external touch-screen terminal

both cases the user can select between continuous and pulsed delivery of the microwave power.

Temperature control is performed by internal measurement via a fiber-optic probe in one reference vessel plus measurement by an IR sensor which is located on the side wall at a defined height that allows temperature recording for all vials. Both sensors are interfaced with a microprocessor-controlled rotor-positioning system which permits vessel recognition and thus profile tracking for all vessels. The MultiSYNTH features an indirect pressure control through a spring-type valve preloaded at 20 bar. This special valve is included in the TFM safety shield in which the standard glass vials are inserted and allows safe pressure release in case of overpressurization, with subsequent resealing of the vessel. An additional built-in sensor is available where the magnetron reduces the power when a set vapor concentration in the cavity is exceeded. In addition to the magnetic stirring a vessel "vibration" system is offered where the vessel is mechanically shaken to prevent hot and cold spot formation, thus ensuring reaction and temperature homogeneity. In the multimode format, the MultiSYNTH labstation oscillates/rotates each vial to achieve a similar mixing effect. Effective cooling is achieved via compressed air.

For the MultiSYNTH system reaction monitoring is achieved via an external control terminal with a touch-screen display utilizing the EasyCONTROL software package. The system can be operated either with a temperature/time mode where the microwave power is automatically adjusted to maintain the set temperature or with a power/time mode where a fixed power value is set.

For this system two different vial types in three different sizes are available permitting a scale-up from 0.25 mL to 300 mL under closed-vessel conditions (see Figure 3.31). For the single-mode format 2.5 and 10 mL glass vials are provided with operating limits of 250 °C and 20 bar (Table 3.8). The same vials can be used for a 12-position rotor system in the multimode set-up (see Figure 3.30b). Additionally, a 6-position rotor for 70 mL vessels (Figure 3.31b, Table 3.8) made of TFM with a PEEK protecting shield and limits of 250 °C and 30 bar is offered. Furthermore, standard

(a)

(b)

Figure 3.31 MultiSYNTH accessories: Teflon/Weflon stand for round-bottom flasks, 70 mL, 2.5 mL, 10 mL vessels with corresponding safety shields (a), and 6-position rotor (b).

round-bottom flasks up to 1 L can be positioned in the cavity and refluxed up to 250 °C under multimode conditions.

Due to the unique feature of the MultiSYNTH in providing single and parallel techniques in a single instrument unit, the user is able to carry out optimization studies via the single-mode set-up and directly scale up the reactions by employing one of the rotor systems in the multimode format.

3.5.4.2 MicroSYNTH Labstation

The MicroSYNTH multimode instrument (also known as ETHOS series, Figure 3.32, Table 3.9) is available with a broad range of accessories. Tools are offered starting from combichem via medchem to parallel synthesis, and large scale synthesis in batch and continuous flow mode. Two magnetrons deliver up to 1000 W microwave output power and a patented pyramid-shaped diffuser ensures homogeneous microwave distribution within the cavity.

In the case of parallel synthesis in multivessel rotors, temperature measurement is achieved by the use of a fiber-optic probe, immersed in one reference vessel. Also available is an IR sensor for monitoring the outside surface temperature of each vessel, mounted in the sidewall of the cavity, about 5 cm above the bottom. The

Table 3.8 MultiSYNTH – features of vessel types.

	Small	Medium	Large
Volume (mL)	2.5	10	70
Operating volume (mL)	0.25–1.2	1–7	10–50
Max. temperature (°C)	250	250	250
Operation pressure (bar)	20	20	30
Vessel material	glass	glass	TFM
Safety shield	TFM	TFM	PEEK
Single-mode setup	✓	✓	×
Rotor system	12-position	12-position	6-position

Figure 3.32 Milestone MicroSYNTH labstation.

reaction pressure is measured by a pneumatic sensor connected to the reference vessel. An additional sensor monitors the vapor concentration in the microwave cavity and thus controls all vessels simultaneously. This technology switches off the applied microwave power until the vapors have been cleared from the cavity by the exhaust module.

Post-reaction cooling of the reaction mixture is achieved by a constant airflow through the cavity and a stream of compressed air (see Table 3.9). Due to the design of the thick polymer/composite segments, the cooling of the high-pressure rotors is not very efficient, external cooling by immersing the rotor in a water bath is recommended. Thus, vessels under pressure have to be handled outside the cavity but a special cooling rack is available for this purpose.

For all MicroSYNTH systems reaction monitoring is achieved via an external control terminal with a touch-screen display utilizing the EasyCONTROL software package. The runs can be controlled either by temperature, pressure or microwave

Table 3.9 MicroSYNTH platform – general features.

Feature	Description
Cavity size (volume)	$35 \times 35 \times 35$ cm (43 L)
Delivered power	1600 W (2 magnetrons)
Max. output power	1000 W
Temperature control	immersed fiber-optic probe (max. 250 °C); outside IR remote sensor (max. 300 °C)
Pressure control	pneumatic pressure sensor (max. 55 bar) indirect pressure control
Cooling system	built-in exhaust air flow 1.8 m^3 min^{-1}, compressed air
Magnetic stirring	0–400 rmp (adjustable by software)
External PC	external touch-screen terminal

Figure 3.33 Milestone CombiCHEM system: rotor with Weflon-made wellplates.

power output. The software enables on-line modifications of any method parameter and the reaction process is monitored by an appropriate graphical interface. An included solvent library and electronic lab journal feature simplifies the experimental documentation.

The MicroSYNTH platform offers a broad range of different rotor and vessel systems, enabling reactions from 0.5 mL up to 4 L under open- and sealed-vessel conditions in batch and parallel manner up to 55 bar. The spectrum ranges from microtiter plates, single reaction vessels operational at different pressures and rotor systems to large volume batch reactors.

- **CombiCHEM System** (Figure 3.33)
  For small-scale combinatorial chemistry applications this barrel-type rotor is available. It can hold up to two 24-, 48- or 96-well microtiter plates utilizing glass vials (0.5–4 mL) up to 4 bar at 150 °C. The plates are made of Weflon (graphite-doped Teflon) to ensure uniform heating and are sealed by an inert membrane sheet. Axial rotation of the rotor tumbles the microwell plates to admix the individual samples. Temperature measurement is achieved by a fiber-optic probe immersed in the center of the rotor.

- **QV50 Module** (Figure 3.34a)
  This tool is designed for single optimization runs at elevated pressure on a comparatively small scale. It includes a quartz vessel, a TFM cover and a safety shield with built-in safety valve. This module is suitable for volumes from 3 to 30 mL at operating limits of 250 °C and 40 bar. The air flow device allows rapid cooling of the reaction mixture using a stream of compressed air. Temperature measurement is conducted via both an internal fiber-optic probe and a surface IR sensor. A special cover that allows working under inert atmosphere or pre-pressurization of the system with a reactive gas is also available.

- **MedCHEM Kit** (Figure 3.34b)
  This accessory is especially designed for medicinal chemistry laboratories covering working volumes from 2 to 160 mL. The package contains a 12 mL glass and a 50 mL quartz vessel, single pressure reactor segments made of TFM (100 mL and 270 mL) as well as Teflon and Weflon stir bars. Reactions up to 250 °C and 55 bar can be performed with this Kit.

(a) (b)

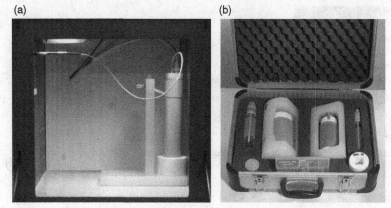

Figure 3.34 Milestone MedCHEM system: QV50 (a) and MedCHEM Kit (b).

- **PRO-16/24 Rotor** (Figure 3.35a, Table 3.10)
  This is an advanced rotor for high-throughput purposes at elevated conditions utilizing 16 or 24 reaction containers. Each 75 mL PTFE-TFM vessel offer up to 50 mL working volume at 250 °C up to 30 bar. The special pressure release valve in the vessels vents in case of overpressure and reseals automatically.

- **Q20 Rotor** (Figure 3.35b, Table 3.10)
  This rotor type is the 20-position parallel version of the QV50 module. Volumes up to 600 mL can be processed at temperatures up to 250 °C and pressures up to 40 bar.

- **High-Pressure Rotor** (Figure 3.36b, Table 3.10)
  This rotor is the 10-position version of the 100 mL pressure reactor included in the MedChem Kit. The TFM reactors are arranged in single segments around the rotor and are suitable for reactions up to 250 °C at 55 bar.

(a) (b)

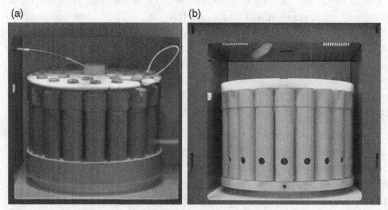

Figure 3.35 Milestone rotor systems: PRO-16/24 (a) and Q20 (b).

Table 3.10 MicroSYNTH – features of parallel rotors.

	PRO-16/24	Q20	High-pressure	Large-volume
No. of vessels	16–24	20	10	6
Vessel volume (mL)	75	45	100	270
Operating volume (mL)	10–50	3–30	10–60	50–180
Vessel material	TFM	quartz	TFM	TFM
Safety shield material	PEEK	PEEK	PEEK	PEEK
Max. pressure (bar)	30	40	55	10
Max. temperature (°C)	250	250	250	250
Internal temp. control	fiber-optic in one reference vessel			
External temp. control	IR for all vessels			

- **Large-Volume Rotor** (Figure 3.36b, Table 3.10)
  Dedicated mainly for parallel scale-up at relatively low pressure, this rotor comes with six segments of the 270 mL TFM vessel that is included in the MedChem Kit. Thus volumes up to 1080 mL can be processed in a single run up to 250 °C and 10 bar.

- **Batch Scale-Up Pressure Reactor** (Figure 3.37a)
  This drum-shaped TFM reactor is available in two sizes (500 mL and 1 L with maximum working volumes of 250 and 500 mL, respectively) and is designed to scale up reactions in a batch mode. The operating limits for this reactor are 200 °C and 14 bar for the 500 mL or 8 bar for the 1 L container, respectively. The same temperature and pressure sensors as described above can be used with this system. An additional high-pressure valve allows pressurization with inert gas.

- **Scale-Up at Normal Pressure** (Figure 3.37b)
  Standard laboratory glassware such as round-bottom flasks or simple beakers from 50 mL to 4 L can be used for reactions at atmospheric pressure. A protective mount in the ceiling of the cavity enables connection of reflux condensers or distillation equipment. Additional mounts in the sidewall allow sample withdrawal or flushing with gas to create inert atmospheres.

(a)

(b)

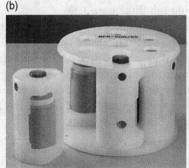

Figure 3.36 Milestone rotor systems: High-Pressure (a) and Large-Volume rotor (b).

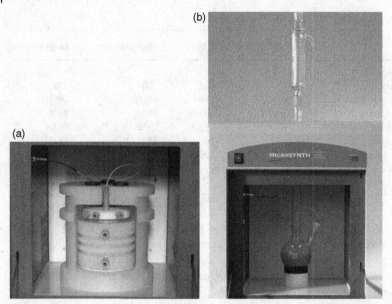

Figure 3.37 Milestone scale-up pressure reactor (a) and scale-up at atmospheric pressure (b).

3.5.4.3 StartSYNTH

This microwave platform is specifically designed for the academic laboratory with proper safety features that incorporate all the safety elements of the MicroSYNTH system. Here, a single magnetron system with a rotating diffuser provides a homogenous microwave distribution in the cavity and delivers an output power up to 1200 W. Magnetic stirring of the reaction mixtures further ensures homogeneous temperature distribution. Temperature and pressure control is achieved via the same sensor systems as described above for the MicroSYNTH (direct temperature and pressure control in one reference vessel and contact-less for all vessels). Reaction monitoring is achieved via an external control terminal with a touch-screen display utilizing the EasyCONTROL software package. The runs can be controlled either by temperature or by microwave power.

- **Teaching Lab Kit** (Figure 3.38)

  This is a basic rotor for standard organic reactions allowing an initial approach toward microwave-mediated chemistry in teaching laboratories. It is designed for 16×20 mL glass vessels and for temperatures up to $150\,°C$. Reactions can either be performed at ambient pressure, employing a special valve that uses a reflux-style pressure venting mechanism, or at pressures up to 1.5 bar using weighted valves that work with a gravity-based venting mechanism that vents when overpressure occurs and reseals afterwards.

- **Research Lab Kit**

  This package contains the same QV50 module that is offered for the MicroSYNTH with the required holder, cover, stir bars and Weflon buttons.

Figure 3.38 Milestone StartSYNTH with Teaching Lab Kit.

For reaction scale-up under normal pressure open-vessel conditions, standard laboratory glassware like round-bottom flasks up to 4 L can be employed. The same assembly as for the MicroSYNTH with a reflux condenser attached through the opening in the ceiling of the cavity is utilized also for the StartSYNTH.

3.5.4.4 Scale-Up Systems

As already mentioned, the scale-up of microwave-assisted reactions from the gram to the kilogram region is of specific interest in many industrial laboratories. For this purpose Milestone offers specific batch and continuous flow reactors that are based on the MicroSYNTH^{plus} unit that incorporates all the features of the MicroSYNTH reactor. Parallel, batch and continuous flow reactions can thus be performed using a single microwave system.

- **BatchSYNTH** (Figure 3.39)

  To address the needs for scaling up reactions in a batch format under high temperatures and pressures the BatchSYNTH was developed. It is equipped with a vertically mounted 300 mL TFM reactor (working volume 250 mL) that is inserted into a HTC (high temperature compound) safety shield and operates up to 230 °C and 30 bar. The temperature is measured internally via a K-type thermocouple and pressure monitoring is also carried out via an internal probe. Efficient cooling is provided by an internal cooling finger. Inlet and outlet ports for creating an inert atmosphere inside the vessel, reagent addition and reaction mixture sampling are included as well (see Figure 3.39c).

- **FlowSYNTH** (Figure 3.40a)

  The FlowSYNTH is a continuous flow system where gram to kilogram scale-up is accomplished. The reagents are pumped through the microwave field from the bottom of the TFM reactor (200 mL) to the top, at maximum operating conditions of 200 °C and 30 bar. The reaction mixture is then cooled by flowing through a water-cooled heat exchanger which is also located on the mobile platform.

(a)

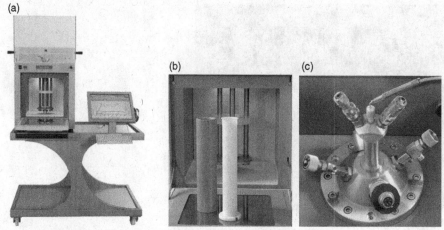

Figure 3.39 Milestone scale-up reactor: BatchSYNTH (a), liner and safety shield (b), and stainless steel flange with sensors and inlet/outlet ports (c).

Temperature and pressure control during the entire course of the process is achieved by built-in sensors. Temperature and reaction homogeneity along the entire length is ensured by a paddle-stirrer homogenizer. Flowrates from 12 to $100\,\text{mL}\,\text{min}^{-1}$, reaction times from 120 s to 16 min and a throughput from 0.8 to $6\,\text{L}\,\text{h}^{-1}$ can be achieved with this system.

(a)

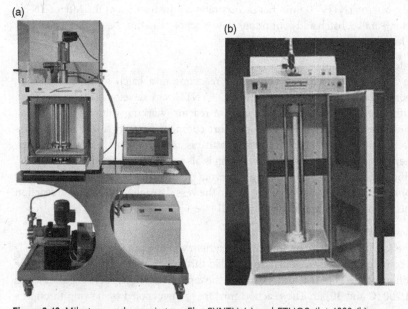

Figure 3.40 Milestone scale-up reactors: FlowSYNTH (a) and ETHOSpilot 4000 (b).

A prototype reactor designed for scale-up on the kilogram scale is the ETHOS pilot 4000 labstation, built from two regular Milestone cavities (see Figure 3.40b). The reaction mixtures can be heated either in continuous flow or batch-type manner. The delivered microwave output power is 2500 W, which can be extended to 5000 W if required [5]. The reaction tubes (quartz or ceramics) are custom built, with several diameters or lengths available, covering a broad range of flow rates and pressure conditions. For reaction control the temperature is monitored over the whole length of the reaction cell, similar to the FlowSYNTH system.

- **UltraCLAVE** (Figure 3.41)
An addition to the spectrum of available reactors for scale-up is the UltraCLAVE batch reactor that is a large scale microwave autoclave for high-temperature/high-pressure batch and parallel reactions. Microwave power settings from 0 to 1000 W may be selected. Continuous unpulsed microwave energy from the system's magnetron is introduced into the vessel through a special microwave-transparent port. The internal geometry of the vessel is optimized for direct microwave coupling with zero reflectance, ensuring maximum sample heating efficiency. A unique feature of this system is the pressurization of the reaction chamber prior to heating so that boiling of the reaction mixture is prevented. In this way, reactions can be conducted at temperatures up to 260 °C and pressures up to 200 bar. The 4.2 L chamber of the UltraCLAVE allows processing of either multiple reactions by employing rotors with as many as 77 positions, or a single large reaction mixture of 3.5 L utilizing the installed PTFE vessel. For the rotor systems, standard glass, quartz or PTFE vessels can be employed. This system is fully automated, only the

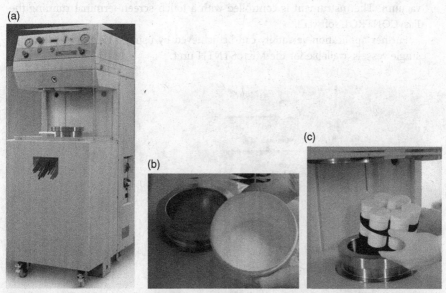

Figure 3.41 Milestone scale-up reactor: UltraCLAVE (a), PTFE vessel (b) and loading of a 6-position rotor system (c).

loading and unloading of reactor vessels has to be done manually. All other functions (raising/lowering the reaction chamber, inert gas pressurization, venting, etc.) are performed automatically under computer control. Reaction monitoring and programming is achieved via a touch-screen control terminal and the EasyCLAVE software.

3.5.4.5 RotoSYNTH

This microwave reactor system is especially designed to perform reactions under solventless conditions, for example where reagents are adsorbed on solid supports like silica or alumina (Figure 3.42). A glass vessel that is fitted in a tilted position in the microwave cavity rotates to obtain proper mixing of the reactants, thus ensuring a homogeneous temperature distribution, even for larger amounts of solids. Furthermore, the glass vessels have special paddles built in for more effective mixing of reaction mixtures. Vessels in different sizes are available, from 300 mL up to 4 L with working volumes of 150 mL to 2 L. However, the RotoSYNTH can also be used for liquid phase reactions. A special application in this case would be product or byproduct distillation out of the reaction mixture which can be realized by the attachment of a vacuum pump to an opening on the outside of the microwave unit on the left-hand side (see Section 4.3).

A single magnetron delivers up to 1200 W microwave output power ensuring a homogeneous microwave distribution in the cavity via a rotating diffuser. Temperature measurement can be conducted either by an IR sensor located in the right-hand wall of the microwave unit controlling the temperature at the bottom of the vessel or by a fiber-optic probe inserted directly into the reaction mixture. Reactions can be performed at temperatures up to 250 °C at atmospheric pressure or under dynamic vacuum. The instrument is controlled with a touch screen terminal running the EasyCONTROL software.

Further application versatility can be achieved by using the complete series of single vessels available for the MicroSYNTH unit.

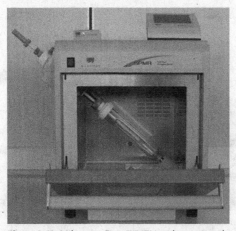

Figure 3.42 Milestone RotoSYNTH with rotating glass vessel.

References

1 For an overview of microwave equipment available prior to 2005, see: Kappe, C.O. and Stadler, A. (2005) *Microwaves in Organic and Medicinal Chemistry*, Wiley-VCH, Weinheim, pp. 29–55 (Chapter 3).

2 Decareau, R.V. and Peterson, R.A. (1986) *Microwave Processing and Engineering*, VCH, Weinheim.

3 Metaxas, A.C. and Meredith, R.J. (1983) *Industrial Microwave Heating*, P. Peregrinus, London, UK.

4 Balanis, C.A. (1989) *Advanced Engineering Electromagnetics*, Wiley, New York, USA.

5 Ondruschka, B., Bonrath, W. and Stuerga, D. (2006) in *Microwaves in Organic Synthesis*, 2nd edn (ed. A. Loupy), Wiley-VCH, Weinheim, pp. 62–103 (Chapter 2).

6 Bogdal, D. (2005) *Microwave-assisted Organic Synthesis. One Hundred Reaction Procedures*, Elsevier, Oxford, UK, pp. 23–32 (Chapter 3).

7 Gedye, R., Smith, F., Westaway, K., Ali, H., Baldisera, L., Laberge, L. and Rousell, J. (1986) *Tetrahedron Letters*, **27**, 279–282.

8 Giguere, R.J., Bray, T.L., Duncan, S.M. and Majetich, G. (1986) *Tetrahedron Letters*, **27**, 4945–4948.

9 Karstädt, D., Möllmann, K.-P. and Vollmer, M. (2004) *Physik in unserer Zeit*, **35**, 90–96.

10 Anton Paar GmbH, Anton-Paar-Str. 20, A-8054 Graz, Austria; phone: (internat.) +43-316-257101; fax (internat.) +43-316-257257; http://www.anton-paar.com.

11 Biotage AB, Kungsgatan 76, SE-753 18 Uppsala Sweden; phone: (internat.) +46-18565900; fax: (internat.) +46-18591922; http://www.biotage.com.

12 CEM Corporation, P.O Box 200, Matthews, NC 28106, USA; Toll-free: (inside US) +1-800-7263331; phone: (internat.) +1-704-8217015; fax: (internat.) +1-704-8217894. http://www.cemsynthesis.com (US website for CEM synthesis equipment); www.cem.com (main website), www.cem.de (German website).

13 Milestone s.r.l, Via Fatebenefratelli, 1/5, I-24010 Sorisole, BG, Italy; phone: (internat.) +39-035-573857; fax: (internat.) +39-035-575498; MLS GmbH, Auenweg 37, D-88299 Leutkirch, Germany; phone: (internat.) +49-7561-9818; fax: (internat.) +49-7561-981812; Milestone Inc., 25 Controls Dr., Shelton, CT 06484, USA; phone: (internat.) +1-866-9955100; http://www.milestonesci.com (US website), www.milestonesrl.com (Italian website), www.mls-mikrowellen.de (German website).

14 Commarmont, R., Didenot, R. and Gardais, J.F. (1986) (Prolabo). French Patent 84/03496.

15 Cleophax, J., Liagre, M., Loupy, A. and Petit, A. (2000) *Organic Process Research & Development*, **4**, 498–504.

16 Chemspeed Technologies AG, Rheinstrasse 32, 4302 Augst, Switzerland; phone: (internat.) +41-61 816 95 00; fax: (internat.) +41-61 816 95 09; http://www.chemspeed.com.

4
Microwave Processing Techniques

In modern microwave synthesis a variety of processing techniques can be utilized, supported by the availability of different types of dedicated single-mode and multi-mode microwave reactors. While in the past much interest has focused on solvent-free reactions under open-vessel conditions [1], today the large majority of the published examples in the area of controlled microwave-assisted organic synthesis (MAOS) involves the use of organic solvents under sealed-vessel conditions [2]. In this chapter, general processing techniques in MAOS as well as processing techniques used in drug discovery and high-throughput synthesis are discussed.

4.1
Solvent-Free Reactions

A frequently used processing technique employed in microwave-assisted organic synthesis since the early 1990s involves solvent-less ("dry-media") procedures [1] where the reagents are reacted neat in the absence of a solvent (see Figure 4.1a). Alternatively, reagents can be pre-adsorbed onto either an essentially microwave transparent (silica, alumina or clay) or a strongly absorbing (graphite) inorganic support, that additionally can be doped with a catalyst or reagent.

When reactions between solids are considered it is important to differentiate between solvent-free, solid-phase, and solid-state synthesis [3]. Solid-phase synthesis indicates a reaction of molecules from a fluid phase with a solid substrate, as in polymer-supported synthesis (see Figure 4.1b and Section 4.7.2.1), whereas in solid-state synthesis two solids interact directly and form a third solid product (Figure 4.1c).

Particularly in the early days of MAOS the solvent-free approach was very popular since it allowed the safe use of domestic household microwave ovens and standard open-vessel technology. Even today, many of the solvent-free chemistries reported in the literature are performed in domestic microwave ovens taking advantage of the benefits described in Box 4.1. The solvent-free technique is claimed to be environmentally benign and contributes to the green chemistry philosophy. In some cases

Practical Microwave Synthesis for Organic Chemists: Strategies, Instruments, and Protocols
C. Oliver Kappe, Doris Dallinger, and S. Shaun Murphree
Copyright © 2009 WILEY-VCH Verlag GmbH & Co. KGaA, Weinheim
ISBN: 978-3-527-32097-4

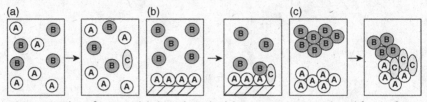

Figure 4.1 Solvent-free (a), solid-phase (b) and solid-state (c) reactions. Adapted from Ref. [3].

work-up procedures are simplified since the products can be obtained by simple extraction, distillation or sublimation techniques [4].

One of the simplest methods involves the mixing of the neat reagents and subsequent irradiation by microwaves. In general, pure, dry solid organic substances do not absorb microwave energy, therefore almost no heating will occur. If none of the reagents is a microwave absorbing liquid, small amounts of a polar solvent (e.g., N,N-dimethylformamide, water) can be added to the reaction mixture in order to allow dielectric heating by microwave irradiation (see also Scheme 4.2). A case study demonstrates this point: according to a literature procedure, the reaction of benzoin with urea yields 4,5-diphenyl-4-imidazolin-2-one within 4 min of microwave irradiation in a domestic oven (Scheme 4.1). The preparation of the reaction mixture is achieved by thoroughly mixing the two solid reactants. Interestingly, this reaction could not be reproduced in a dedicated reactor [5]. As it turned out, in a domestic oven, the utilized glass rotary plate is to some extent microwave absorbing and warms to temperatures of 120–140 °C by microwave irradiation, at

Box 4.1
Characteristics of dry media reactions.

Advantages

- Easy to handle
- No specialized equipment
- High reactivity due to catalysts/reagents on porous supports
- Safe, since no flammable solvent is involved
- Environmentally benign, "Green Chemistry" (no organic solvent?).

Disadvantages

- Temperature measurement difficult
- Localized superheating possible
- Macroscopic hotspots
- Stirring troublesome
- Limited possibilities for scale-up (penetration depth)
- Reproducibility controversial.

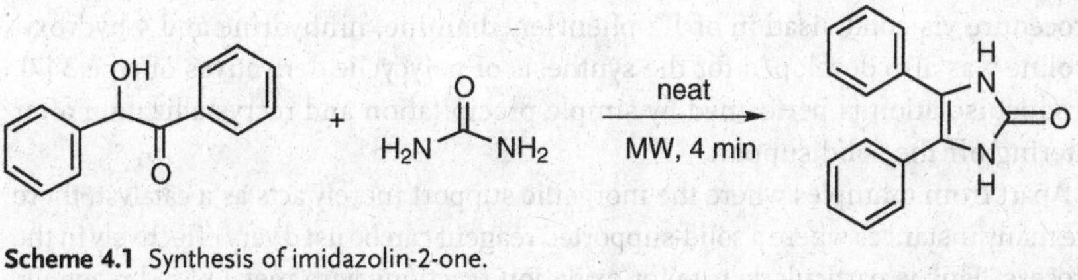

Scheme 4.1 Synthesis of imidazolin-2-one.

which point benzoin melts through convective heating and the "microwave reaction" is started. Only by addition of small amounts of water could this transformation eventually be performed in a Teflon reactor inside a dedicated microwave instrument [5].

An alternative technique utilizes microwave transparent or only weakly absorbing inorganic supports such as silica, alumina, or clay materials [1]. These reactions are effected by the reagents/substrates immobilized on the porous solid supports and have advantages over the conventional solution phase reactions because of their good dispersion of active reagent sites, associated selectivity and easier workup. The recyclability of some of these solid supports also renders these processes eco-friendly "green" protocols. In general, the substrates are pre-adsorbed onto the surface of the solid support and then exposed to microwave irradiation. In the example shown in Scheme 4.2 Montmorillonite K10 clay, an acidic solid support, serves to catalyze the synthesis of pyrrole derivatives [6]. Indeno[1,2-b]quinoxalinones 1 or isatin derivatives 2 are reacted with 4-hydroxyproline in the presence of a few drops of DMSO in only 2 min under microwave irradiation in a domestic microwave oven to give the desired pyrrole products. A one-pot three-component

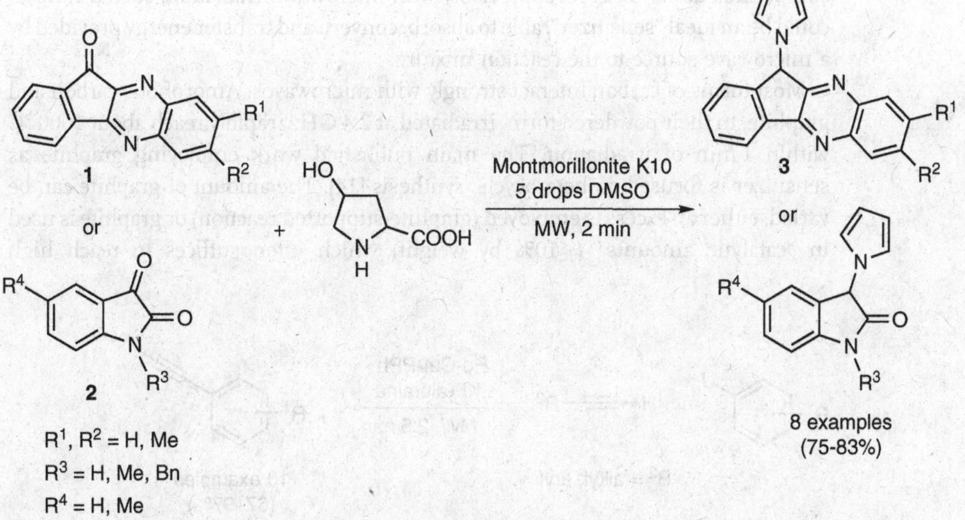

R^1, R^2 = H, Me

R^3 = H, Me, Bn

R^4 = H, Me

8 examples
(75–83%)

Scheme 4.2 Synthesis of pyrrole derivatives on Montmorillonite K10.

procedure via condensation of 1,2-phenylenediamine, ninhydrine and 4-hydroxy-proline was also developed for the synthesis of polycyclic derivatives of type **3** [7]. Product isolation is performed by simple precipitation and recrystallization after filtering off the solid support.

Apart from examples where the inorganic support merely acts as a catalyst, there are many instances where a solid-supported reagent can be used very effectively in the process. This is particularly true for oxidation reactions with metal-based reagents. For example, Varma and Dahiya have developed a method where Montmorillonite K10 clay-supported iron(III) nitrate (so-called clayfen) is used under solvent-free conditions for the oxidation of alcohols to carbonyl compounds in less than 1 min [8]. Related microwave-assisted solvent-free oxidations were also carried out with manganese [9], copper [10], and chromium-based [11] oxidation reagents adsorbed on suitable inorganic supports.

In addition to cases where the inorganic support itself acts as a catalyst, or where a reagent has been impregnated on the solid support, it is also possible to additionally dope the support material with metal catalysts. Scheme 4.3 illustrates a Sonogashira coupling that was performed on a strongly basic potassium fluoride/alumina support, doped with a palladium/copper(I) iodide/triphenylphosphine mixture. The resulting aryl alkynes were synthesized in very high yields (67–97%) [12]. Many more examples of solvent-free microwave-assisted transformations can be found in review articles [1] and recent books on microwave synthesis [4, 13].

In contrast to solvent-free ("dry-media") microwave processing involving (when properly dried) weak microwave absorbing supports such as silica or alumina, an alternative is to use strongly microwave absorbing supports such as graphite. For reactions which require high temperatures, the idea of using a reaction support which takes advantage of both strong microwave coupling and strong adsorption of organic molecules has been considered in recent years [14]. Since many organic compounds do not interact appreciably with microwave irradiation, such a support could be an ideal "sensitizer", able to absorb, convert, and transfer energy provided by a microwave source to the reaction mixture.

Most forms of carbon interact strongly with microwaves. Amorphous carbon and graphite, in their powdered form, irradiated at 2.4 GHz, rapidly reach about 1000 °C within 1 min of irradiation. The main published work employing graphite as sensitizer is focused on heterocyclic synthesis [14]. The amount of graphite can be varied, either an excess is employed (graphite-supported reaction) or graphite is used in "catalytic amounts" ($\leq$10% by weight) which often suffices to reach high

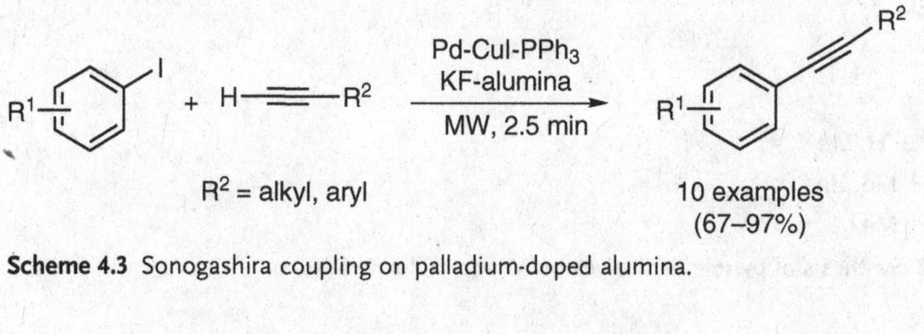

Scheme 4.3 Sonogashira coupling on palladium-doped alumina.

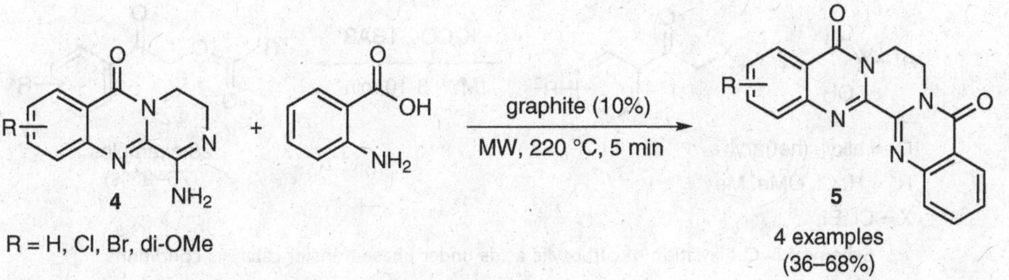

Scheme 4.4 Niementowski-type reaction using graphite as sensitizer.

temperatures in a short time. One example of a solvent-free synthesis of a pentacyclic heterocycle utilizing graphite in catalytic amounts is shown in Scheme 4.4. Here, quinazolinone 4 is reacted with anthranilic acid to give tetraaza-pentaphene-5,8-diones 5 via a Niementowski reaction [15]. When the reaction is performed under the same conditions in the presence of a solvent (1-methyl-2-pyrrolidone or N,N-dimethylformamide), lower product yields are obtained in addition to a large amount of by-product formation. Sealed-vessel conditions turned out to be superior since sublimation of anthranilic acid, that usually occurs at atmospheric pressure, is avoided.

In addition to graphite being used as a "sensitizer" (energy converter), there are several examples in the literature where the catalytic activity of metal inclusions in graphite has been exploited ("graphimets"). The Friedel–Crafts acylation of anisole is a case in point, where the presence of catalytic amounts of iron oxide magnetite crystallites (Fe_3O_4) allowed efficient acylation with benzoyl chloride within 5 min of irradiation [14].

Since graphite is a very strong absorber of microwave heating, the temperature must be carefully controlled to avoid melting of the reactor. The use of a quartz reactor is highly preferable.

4.2
Phase-Transfer Catalysis

In addition to solvent-free processing, phase-transfer catalytic conditions (PTC) have also been widely employed as a processing technique in MAOS [16]. In phase-transfer catalysis the reactants are situated in two separate phases, for example liquid–liquid or solid–liquid. In liquid–liquid PTC the phases are mutually insoluble, ionic reagents are typically dissolved in the aqueous phase, while the substrate remains in the organic phase. In solid–liquid PTC, on the other hand, ionic reagents may be used in their solid state as a suspension in the organic medium. Transport of the anions from the aqueous or solid phase to the organic phase, where the reaction takes place, is facilitated by phase-transfer catalysts, typically quaternary onium salts (e.g. tetrabutylammonium bromide, Aliquat 336) or cation-complexing agents. At least one liquid component in solid–liquid PTC is necessary: usually the electrophilic

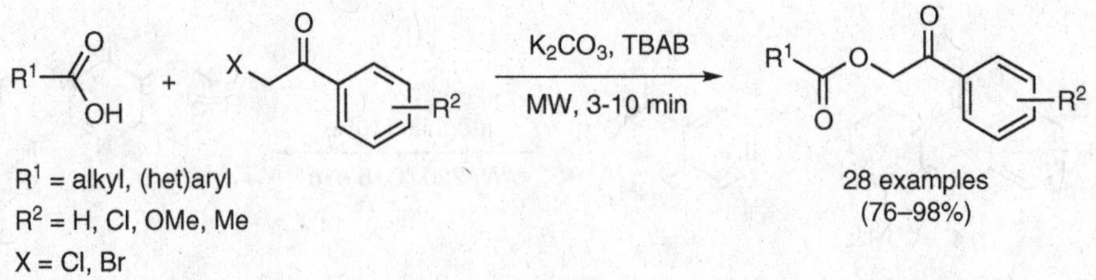

R¹ = alkyl, (het)aryl
R² = H, Cl, OMe, Me
X = Cl, Br

28 examples
(76–98%)

Scheme 4.5 O-Alkylation of carboxylic acids under phase-transfer catalysis conditions.

reagent acts as both substrate and liquid phase. Phase-transfer catalytic reactions are perfectly tailored for microwave activation, and the combination of solid–liquid PTC and microwave irradiation typically gives the best results in this area. Numerous transformations in organic synthesis can be achieved under solid–liquid PTC and microwave irradiation in the absence of solvent, generally under atmospheric pressure in open vessels [16].

In Scheme 4.5 O-alkylations of carboxylic acids under solvent-free solid–liquid phase-transfer catalysis conditions are described [17]. These reactions are performed using ω-haloacetophenones as electrophiles, K_2CO_3 as base and 10 mol% of tetra-butylammonium bromide (TBAB) as phase-transfer catalyst. All reactants are finely ground before microwave irradiation in a domestic microwave oven for 3–10 min. The impact of the phase-transfer catalyst is demonstrated since without the addition of the catalyst the products are obtained in significantly lower yields (45% versus 90%).

Although solid–liquid PTC is best suited for microwave irradiation, in recent years, liquid–liquid PTC has also become a popular technique. It has found widespread application in palladium-catalyzed carbon–carbon cross-coupling reactions (Heck, Suzuki, Sonogashira) [16]. Here, TBAB is again the preferred phase-transfer catalyst and water is used as solvent which renders the process environmentally benign and "green" (for reactions in water as solvent see also Section 4.5.1).

While a large number of interesting transformations using "dry-media" reactions, involving classical solvent-free synthesis as well as phase-transfer catalysis under solvent-free conditions, have been published in the literature [13, 14, 16], technical difficulties relating to non-uniform heating, mixing (stirring), and the precise determination of the reaction temperature remain unsolved [5], in particular when scale-up issues need to be addressed (see Box 4.1).

Recent evidence suggests that a correct temperature measurement in microwave chemistry relies on the use of internal fiber-optic probes (especially when comparing conventional oil-bath with microwave heating). In addition, it now becomes evident that proper stirring during the reaction is also an important factor (see also Section 2.5.1). Particularly in solvent-free synthesis, effective stirring is essential in order to avoid temperature gradients which can lead to irreproducible results.

Bogdal and coworkers have investigated the correlation between stirring and temperature gradients in the solvent-free reaction of chloroacetic ethyl ester and

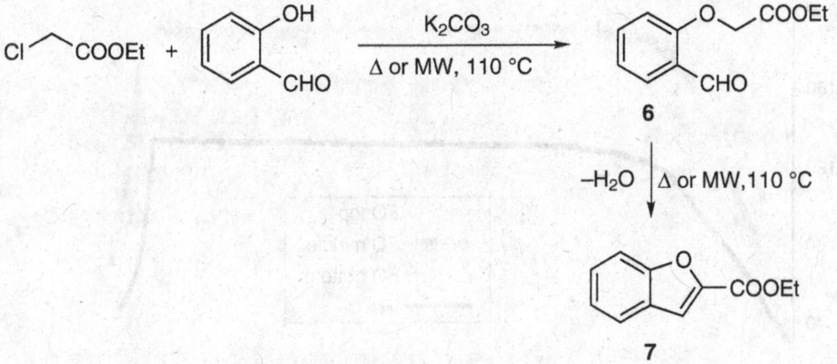

Cl‿COOEt + [2-hydroxybenzaldehyde structure with OH and CHO] →(K₂CO₃, Δ or MW, 110 °C)→ [structure 6: O‿COOEt with CHO]

6

$-H_2O$ | Δ or MW, 110 °C

[benzofuran structure 7] —COOEt

7

Scheme 4.6 Synthesis of benzofuran-2-carboxylic acid ethyl ester.

salicylaldehyde (Scheme 4.6) since different distributions of intermediate **6** and final product **7** were obtained when performing the reaction at identical temperatures under both conventional and microwave heating [18]. Under conventional conditions it was revealed that at temperatures as high as 150 °C the final product **7** could be obtained in about 90% selectivity whereas at lower temperatures intermediate **6** is favored (110 °C, 74 : 26). In the microwave experiment using a single-mode reactor (Synthewave 402, Prolabo), the internal temperature of the reaction mixture was recorded with a fiber-optic probe whereas the surface temperature was recorded with a thermovision camera. If the reaction mixture is not stirred, temperature differences of up to 130 °C (P1: 70 °C, P2: 125 °C, P3: 200 °C, see Figure 4.2a) were observed on the surface and a product distribution of 33 : 67 **6/7** was obtained. The existence of the products could also be detected visually since intermediate **6** is light yellow in color and product **7** is dark brown when adsorbed on K_2CO_3 (Figure 4.2). The rotation of the reaction vessel in the microwave reactor decreases the temperature gradient leading to a product distribution closer to that under conventional conditions at 110 °C. When the reaction mixture was additionally stirred with a quartz spatula, the thermal homogeneity was greatly improved, which was also detected via the surface

(a) P2 — P1 — P3 (b) (c)

Figure 4.2 Surface images of the reaction mixture: no stirring (a), vessel rotation (b), vessel rotation and stirring (c). Reproduced with permission from Ref. [18].

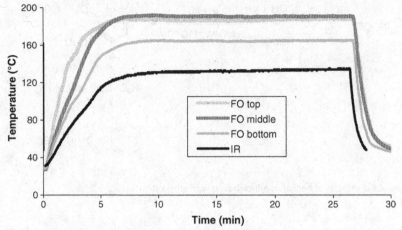

Figure 4.3 Temperature profiles for a sample of Montmorillonite K10 in a 10 mL quartz vessel equipped with three internal fiber-optic sensors (CEM Discover). The sample was heated with 20 W constant power for 27 min. Adapted from Ref. [19].

homogeneity (Figure 4.2c), and the results were now comparable to those obtained under conventional conditions at 110 °C.

As demonstrated in these experiments, in order to maintain good temperature homogeneity effective stirring has to be provided, which is sometimes difficult to achieve for heterogeneous reaction mixtures or under solvent-free conditions. Even temperature monitoring using a fiber-optic probe can lead to incorrect measurements (see also Section 2.5.1). By employing a multiple fiber-optic probe device, temperature inhomogeneities in a sample of Montmorillonite K10 clay were observed when the reaction vessel was heated in a standard single-mode microwave reactor [19]. The temperature was measured with an external IR sensor and additionally with three fiber-optic sensors that were located at different heights inside the reaction vessel. As can be seen in Figure 4.3, the temperature gradients could not be eliminated by magnetic stirring since the stirring proved to be inefficient.

4.3
Open- versus Closed-Vessel Conditions

Microwave-assisted syntheses can be carried out using standard organic solvents under either open- or sealed-vessel conditions. If solvents are heated by microwave irradiation at atmospheric pressure in an open vessel, the boiling point of the solvent (as in an oil-bath experiment) typically limits the reaction temperature that can be achieved. In the absence of any specific- or non-thermal microwave effects (such as the superheating effect at atmospheric pressure, see Section 2.5.3) the expected rate enhancements would be comparatively small. In order nonetheless to achieve high reaction rates, high-boiling microwave-absorbing solvents such as dimethylsulfoxide, 1-methyl-2-pyrrolidone, 1,2-dichlorobenzene, or ethylene glycol (see Table 2.3) have frequently been used in open-vessel microwave synthesis [20]. However, the use of these solvents presents serious challenges during product isolation. Because of the recent availability of modern microwave reactors with on-line monitoring of both temperature and pressure (see Chapter 3), MAOS in sealed vessels is the preferred technique today. This is clearly evident on surveying the recently published literature in the area of MAOS [21]. The main advantage of closed-vessel conditions is that solvents can be heated far above their boiling points, eliminating the need for high-boiling solvents to reach high temperatures, and therefore significant rate enhancements may be achieved compared to reactions that are performed under conventional reflux conditions.

The main advantage of using open-vessel microwave processing is the reduced safety risk, since pressurized reaction vessels can be avoided. This is particularly important for the scale-up of microwave-assisted reactions (see Section 4.8). Apart from that, microwave heating using open vessels can allow a more rapid heating compared to conventional techniques, provided that the reaction mixture is highly microwave absorbing.

An area where high-boiling solvents are frequently used is microwave-assisted solid-phase organic synthesis. In solid-phase synthesis, the synthesized compounds are attached to an insoluble polymer support, which can easily be removed by filtration. Therefore, high boiling solvents do not represent a problem during work-up. An example of solid-phase microwave synthesis where the use of open-vessel technology is essential is shown in Scheme 4.7. The transesterification of β-keto esters with a supported alcohol (Wang resin) is carried out in 1,2-dichlorobenzene (DCB) as a solvent under controlled microwave heating conditions [22]. The temperature is kept constant at 170 °C, about 10 degrees below the boiling point

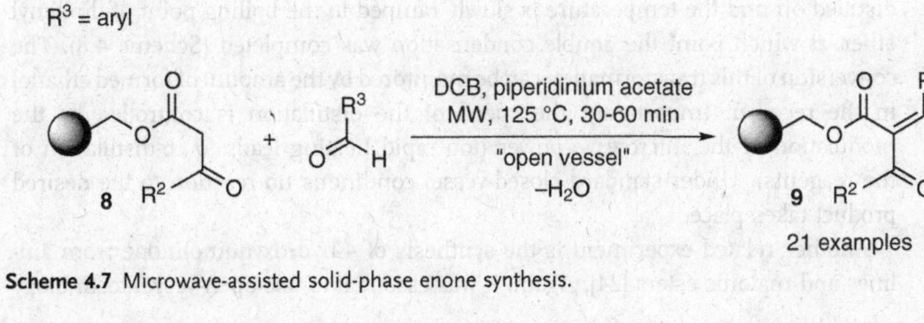

R^1 = Me, Et, t-Bu
R^2 = alkyl, aryl
R^3 = aryl

21 examples

Scheme 4.7 Microwave-assisted solid-phase enone synthesis.

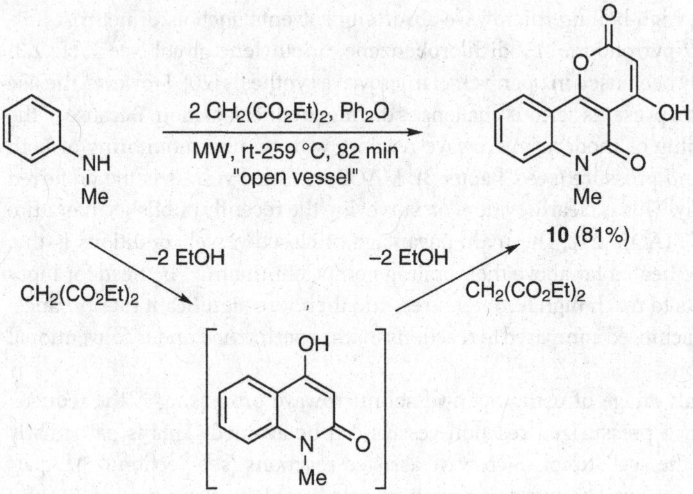

Scheme 4.8 Formation of pyrano[3,2-*c*]quinolone.

of the solvent, thereby allowing safe processing in the microwave cavity. In order to achieve full conversion to the desired resin-bound β-keto esters **8** it is essential that the alcohol formed can be removed from the equilibrium. The second step, a Knoevenagel condensation, is performed under open-vessel conditions as well. Here, a lower temperature of 125 °C is necessary to prevent cleavage from the resin but is still sufficient to remove the water from the reaction mixture and to obtain the resulting polymer-bound enones **9** in quantitative conversion [22].

Another example where open-vessel microwave processing is required is outlined in Scheme 4.8, namely the formation of pyranoquinolone **10** from *N*-methylaniline and diethyl malonate in diphenyl ether as high-boiling solvent (bp 259 °C) (see Section 6.5.1.1) [23]. This double condensation process requires the use of open-vessel technology, since four equivalents of the volatile ethanol by-product are formed that are continuously removed from the reaction mixture by fractional distillation from the reagents. The reaction was performed on a 0.2 mol scale in a 500 mL round-bottom flask in the MicroSYNTH, a multimode microwave reactor (Figure 3.37b) equipped with a Vigreux column through the mount in the ceiling of the instrument and a standard distillation kit. After rapid heating to about 200 °C, the ethanol is distilled off and the temperature is slowly ramped to the boiling point of diphenyl ether, at which point the double condensation was completed (Scheme 4.8). The conversion of this transformation can be monitored by the amount of formed ethanol in the receiver. Importantly, the speed of the distillation is controlled by the modulation of the microwave power (too rapid heating leads to co-distillation of the reagents). Under standard closed-vessel conditions no reaction to the desired product takes place.

Another related experiment is the synthesis of 4-hydroxyquinolinone from anilines and malonic esters [24]. Again, it was essential to use open-vessel technology

Table 4.1 Dependence of the yield of quinolone product on the use of closed- or open-vessel microwave heating[a] (data from Ref. [24]).

Entry	Reagents (mmol)	Solvent (mL)	Yield (%)	Pressure (bar)
1	1	2	76	3.6
2	2	2	67	5.3
3	4	2	60	7.4
4	1	0.5	91	2.0
5	1	neat	90	
6[b]	2	neat	92	
7[b]	4	neat	90	

[a]Microwave heating, 250 °C, 10 min, 1,2-dichlorobenzene or neat.
[b]Open vessel.

here, since the formed 2 equivalents of ethanol need to be removed from the equilibrium (Table 4.1, entries 6–7). Preventing the removal of ethanol from the reaction mixture, for example by using a standard closed-vessel microwave system, results in a pressure build-up in the reaction vial and leads to significantly lower yields (Table 4.1, entries 1–3). Comparable high yields could only be achieved when the reaction under closed-vessel conditions was conducted on a 1 mmol scale with 0.5 mL of solvent (Table 4.1, entry 4). In this case, the pressure build-up was relatively small since enough headspace in the vial is guaranteed due to the low filling volume (see also Section 5.2.2). Scale-up of this synthesis would clearly only be feasible using open-vessel technology [25].

A similar situation was encountered by Leadbeater and Amore when conducting esterification reactions [26]. Typically, either the ester product or the generated water needs to be removed from the equilibrium in order to drive these reactions to completion. Since, under standard sealed-vessel conditions, this was not feasible, a multimode microwave instrument was utilized where the water could be distilled out from the reaction mixture by connecting a vacuum pump to the outside of the reactor (RotoSYNTH, Figure 3.42). Excellent yields (82–94%) were obtained after 15 min heating to 100–120 °C for primary alcohols on a 0.3 or 3 mol scale in a 0.3 or 2 L vessel, respectively [26].

Loupy and coworkers have described a set-up for the removal of water from the reaction medium in an open-vessel microwave system in the synthesis of imines from aromatic ketones [27]. A pyrex beaker was introduced in a cylindrical tube fitted with a drainpipe at the bottom for the water-collection.

Not only liquid by-products can be removed from the reaction mixture, but also gaseous by-products have been successfully purged off to shift the equilibrium to the product. For the ring-closing metathesis (RCM) of diene **11** employing Grubbs II catalyst, it was crucial to remove the developing ethylene during the reaction in order to obtain the tricyclic product **12** in high yields (Scheme 4.9) [28]. This was possible by passing an argon stream through the reaction solution ("gas sparging"). Importantly, under closed-vessel conditions starting material **11** was recovered almost quantitatively.

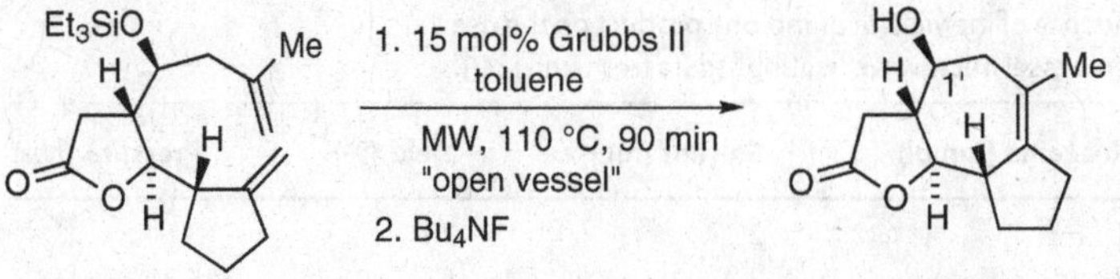

Scheme 4.9 RCM with concurrent removal of ethylene by gas sparging.

Varma and Kim reported the use of a dynamic evaporation method, whereby the formed ammonia that is liberated during the synthesis of cyclic ureas from diamines and urea is removed [29]. A reflux condenser that was attached to a water aspirator was connected to a round-bottom flask in a dedicated single-mode instrument (CEM Discover).

All the experiments discussed in this section highlight the importance of choosing appropriate experimental conditions when using microwave heating technology. In particular, the user should be aware of the consequences of performing reactions in a sealed vessel.

4.4
Pre-Pressurized Reaction Vessels

Relatively little work has been performed with gaseous reagents in sealed-vessel microwave experiments [30]. Although several publications describe this technique in the context of heterogeneous gas-phase catalytic reactions important for industrial processes [31], the use of pre-pressurized reaction vessels in conventional microwave-assisted organic synthesis involving solvents is rare. Several authors have, however, described the use of reactive gases in such experiments and experimental techniques to apply a slight over-pressure (2–3 bar) [32]. Due to the design of most of the scientific single-mode microwave reactors and their reaction vessels, a genuine pre-pressurization is not possible since a pressure limit of about 20 bar exists and no commercial gas-loading accessory typically is available. However, in the last few years multimode- and single-mode reactors have been developed where pre-pressurization up to 20 bar with the appropriate accessories is possible (see Section 3.5.1).

For example, the Diels–Alder cycloaddition reaction of the pyrazinone heterodiene **13** with ethene (Table 4.2) led to the corresponding bicyclic cycloadduct **14** [33]. Under conventional conditions, these cycloaddition reactions have to be carried out in an autoclave applying an ethene pressure of 25 bar before the set-up is heated at 110 °C for 12 h. When the solution of pyrazinone **13** in 1,2-dichlorobenzene (DCB) was saturated with gaseous ethene prior to sealing (1 bar), the cycloaddition was completed (89% yield) after irradiation for 140 min at 190 °C in a single-mode reactor (Biotage). It was, however, not possible to increase the reaction rate further by raising the temperature. At temperatures above 200 °C, an equilibrium between the cycloaddition and the

Table 4.2 Diels–Alder reaction of pyrazinone **13** with ethene under pre-pressurized conditions (data from Ref. [34]).

Entry	Pressure (bar)	Temp. (°C)	Time (min)	Yield (%)
1	1	190	100	12
2	5	190	30	87
3	10	190	20	85
4	10	220	10	85

competing retro-Diels–Alder fragmentation process was observed [33]. Only by using a multimode microwave reactor that allowed pre-pressurization of the reaction vessel with 10 bar of ethene (Anton Paar, Synthos 3000, see Figures 3.18 and 3.19a), could the Diels–Alder addition be carried out much more efficiently at 190 °C within 20 min (Table 4.2, entry 3) [34]. The reaction time could even be reduced to 10 min when the reaction temperature was raised to 220 °C (Table 4.2, entry 4), with the cycloaddition product being still stable at this higher temperature.

Leadbeater and Kormos have performed palladium-catalyzed carbonylation reactions employing gaseous carbon monoxide (CO) using the same multimode instrument with the gas-loading interface as described above. The vessels were pre-pressurized with either 14 bar CO for hydroxy- [35] or 10 bar CO for alkoxycarbonylations [36]. For alkoxycarbonylations (Scheme 4.10b), a lower temperature of 125 °C was sufficient to reach high conversions since carbon monoxide is better soluble in the alcohol than in water which is used as solvent for the hydroxycarbonylations (Scheme 4.10a). In both cases only aryl iodides could be converted to the corresponding acids **15** or esters **16**, aryl bromides were unreactive. For a description of the alkoxycarbonylation of 4-iodoanisole with ethanol and gaseous carbon monoxide, see Section 6.5.2.1.

Furthermore, the same authors have shown that alkoxycarbonylations can be conducted successfully when near-stoichiometric amounts of carbon monoxide are employed [37]. These reactions were performed in a single-mode instrument (CEM Discover, see Figure 3.11) with a gas-loading interface where the exact loading pressure could be monitored since the pressure sensor was directly connected to the reaction vessel. The reactions were run in an 80 mL glass vessel on a 2 mmol scale. First, the reaction mixture was loaded with about 1 bar of carbon monoxide, that corresponds to 2.5 mmol, and subsequently about 9 bar of nitrogen were additionally loaded into the vial to give an initial pressure of about 10 bar. Interestingly, the total pressure had a significant effect on the reaction outcome, and 10 bar proved to give the best results, although the yields of the corresponding esters were somewhat lower than those described in Scheme 4.10b [36].

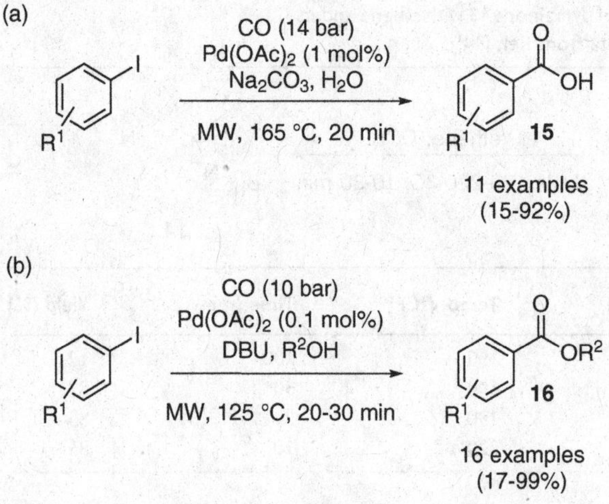

(a)

$$CO \text{ (14 bar)}$$
$$Pd(OAc)_2 \text{ (1 mol\%)}$$
$$Na_2CO_3, H_2O$$

MW, 165 °C, 20 min

15

11 examples
(15-92%)

(b)

$$CO \text{ (10 bar)}$$
$$Pd(OAc)_2 \text{ (0.1 mol\%)}$$
$$DBU, R^2OH$$

MW, 125 °C, 20-30 min

16

16 examples
(17-99%)

Scheme 4.10 Hydroxy- and alkoxycarbonylations using gaseous carbon monoxide and pre-pressurized reaction vessels.

Similarly, hydroformylations of alkenes can be carried out by using a 1 : 1 mixture of carbon monoxide and hydrogen (syngas), a rhodium catalyst and a suitable ligand. Under standard conditions, high pressures (50–80 bar) of syngas and long reaction times, from 1 to 3 days, in an autoclave are required. When a similar set-up to that described above (real-time pressure monitoring) in the CEM Discover is employed, the 80 mL vial including the reaction mixture can be filled with about 3 bar of the syngas [38]. After irradiation for 4–6 min (2 min cycles) at 110 °C, the corresponding aldehydes were obtained in high yields (Scheme 4.11a). Importantly, the ionic liquid is crucial for the reaction, since without bmimBF$_4$ the reaction mixture could not be heated up to 110 °C. Only 60 °C could be reached otherwise, due to the low microwave

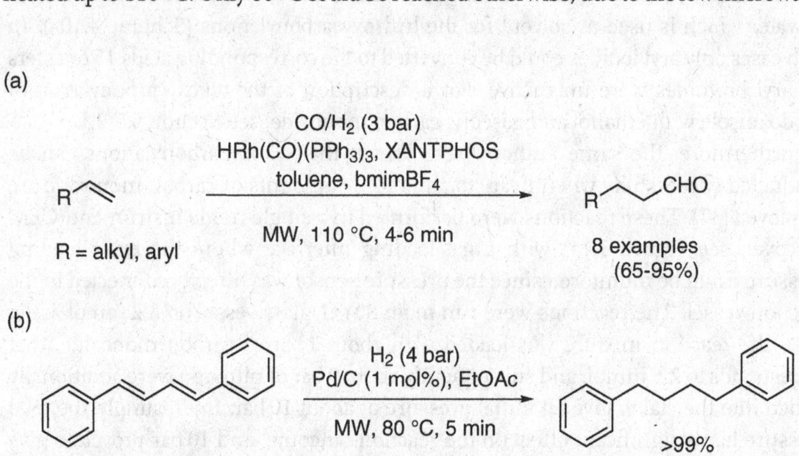

(a)

$$CO/H_2 \text{ (3 bar)}$$
$$HRh(CO)(PPh_3)_3, XANTPHOS$$
$$\text{toluene, bmimBF}_4$$

MW, 110 °C, 4-6 min

R = alkyl, aryl

CHO

8 examples
(65-95%)

(b)

$$H_2 \text{ (4 bar)}$$
$$Pd/C \text{ (1 mol\%), EtOAc}$$

MW, 80 °C, 5 min

>99%

Scheme 4.11 Hydroformylations (a) and hydrogenation (b) reactions performed in a single-mode instrument.

absorbing characteristics of toluene (see also Section 4.5.2) and complete conversion could not be achieved, even after 30 min.

An analogous gas inlet device for the CEM Discover system, where the 10 mL fiber-optic accessory was additionally equipped with a gas inlet, was utilized by Vanier in order to perform hydrogenations with gaseous hydrogen [39]. The reaction vessel, containing the substrate, 1 mol% of palladium-on-carbon (Pd/C) and ethyl acetate as solvent, was charged with about 4 bar of hydrogen (Scheme 4.11b). Different types of substrates were reduced nearly quantitatively after irradiation at 80–100 °C for 3–20 min. It should be noted, that toluene as solvent should be avoided in combination with Pd/C, since it is not able to suspend the catalyst properly and explosions can occur due to deposition of Pd/C on the vessel wall.

Hydrogenation reactions in a specifically designed reactor for a multimode instrument (based on the MicroSYNTH, see Figure 3.32) were reported by Holzgrabe and coworkers [40]. Hydrogen gas of up to 25 bar can be charged to the reaction mixture prior to irradiation. Dearomatizations, debenzylations, azide hydrogenation and double bond reductions are conducted in shorter reaction times than under classical conditions which was rationalized by hotspot formation on the surface of the heterogeneous catalysts due to microwave interaction (see also Section 2.5.3).

Instead of employing gaseous reagents like carbon monoxide or hydrogen gas in MAOS, where special equipment is necessary, a popular and more convenient technique is to utilize solid reagents that liberate, for example, carbon monoxide during the reaction upon heating. One of these solid reagents is $Mo(CO)_6$ that is known to liberate carbon monoxide smoothly at higher temperatures [41]. The Larhed group has extensively employed $Mo(CO)_6$ as a solid carbon monoxide source for diverse carbonylations (such as alkoxy- and aminocarbonylations) under microwave conditions [41]. One example is shown in Scheme 4.12a, where palladium-catalyzed aminocarbonylations were successfully performed with $Mo(CO)_6$ in water as solvent [42]. Aryl iodides, bromides and even the otherwise unreactive aryl chlorides could be reacted with diverse primary and secondary amines to give the aryl amide products **17** in moderate to excellent yields (Scheme 4.12a). The competing hydroxycarbonylation could be inhibited by fine-tuning of the reaction parameters. In particular the stoichiometry of aryl halide to amine was crucial for the successful reaction as was also the choice of the proper catalyst. With this general protocol, aryl iodides could be reacted at 110 °C (see Section 6.5.2.2) whereas the bromides and chlorides needed a higher temperature of 170 °C and sometimes longer reaction times.

Ammonium formate is the most common solid hydrogen donor used for microwave-assisted catalytic transfer hydrogenation reactions. Reduction of the nitro group in the 3-nitrophenyl-dihydropyrimidine derivative **18** to the corresponding amino group was described by Stiasni and Kappe under catalytic transfer hydrogenation conditions, employing excess ammonium formate [43]. After only 2 min of microwave heating at 120 °C the product was obtained in 89% isolated yield (Scheme 4.12b). It should be noted that, when working under sealed-vessel conditions, high pressures are generated (here 12 bar) due to the additional liberation of gaseous carbon dioxide and ammonia.

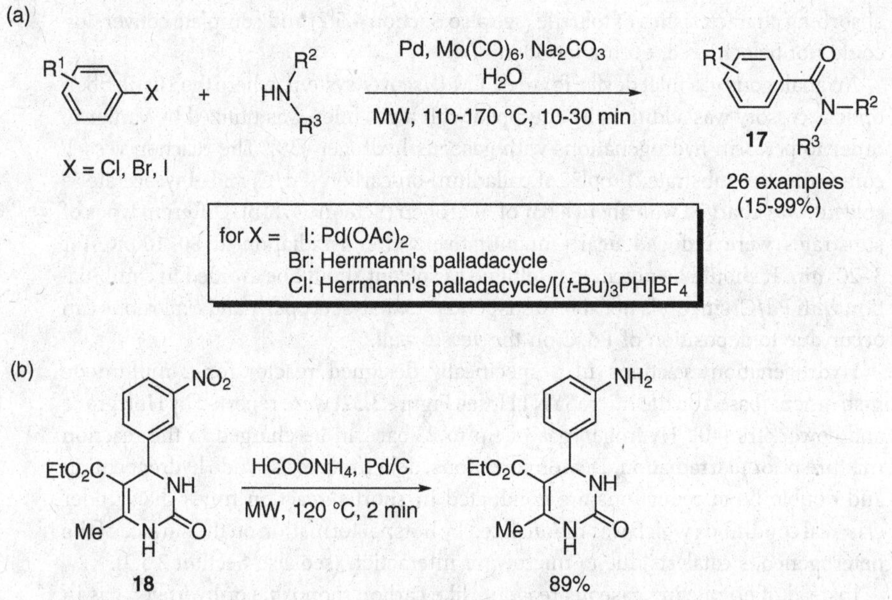
(a)
Pd, Mo(CO)6, Na2CO3
H2O
MW, 110-170 °C, 10-30 min
X = Cl, Br, I
17
26 examples
(15-99%)
for X = I: Pd(OAc)2
Br: Herrmann's palladacycle
Cl: Herrmann's palladacycle/[(t-Bu)3PH]BF4
(b)
HCOONH4, Pd/C
MW, 120 °C, 2 min
18
89%

Scheme 4.12 Aminocarbonylation using a solid carbon monoxide source (a), and catalytic-transfer hydrogenation employing a solid hydrogen donor (b).

4.5 Non-Classical Solvents

Apart from using standard organic solvents in conjunction with microwave synthesis, the use of either water or so-called ionic liquids as alternative reaction media has become increasingly popular in recent years.

4.5.1 Water as Solvent

Synthetic organic reactions in aqueous media at ambient or slightly elevated temperatures have become of great interest as water as a solvent for organic reactions often displays unique reactivity and selectivity [44], exploiting, for example, so-called hydrophobic effects [45]. Apart from performing reactions in aqueous solutions in a moderate temperature range (0 to 100 °C), chemical processing in water is also possible and of considerable interest under "superheated conditions" (>100 °C) in sealed vessels because of the favorable changes that occur in the chemical and physical properties of water at high temperatures and pressures (Table 4.3) [46].

The so-called near-critical (also termed subcritical) region of water at temperatures between 150 and 300 °C is of great importance to organic synthesis [47]. At 250 °C,

Table 4.3 Properties of water under different conditions (data from Ref. [46]).

Fluid	Ordinary water ($T < 150\,°C$) ($p < 0.4\,MPa$)	Near-critical water ($T = 150–350\,°C$) ($p = 0.4–20\,MPa$)	Supercritical water ($T > 374\,°C$) ($p > 22.1\,MPa$)
Temperature (°C)	25	250	400
Pressure (bar)	1	50	250
Density (g cm^{-3})	1	0.8	0.17
Dielectric constant ε'	78.5	27.1	5.9
pK_W	14	11.2	19.4

water exhibits a density and polarity similar to those of acetonitrile at room temperature. The dielectric constant of water (ε') drops rapidly with temperature, and at 250 °C has fallen from 78.5 (at 25 °C) to 27.5 (Table 4.3) [46]. This means that, as the water temperature is increased, it exhibits properties like polar organic solvents and thus organic compounds become soluble in high-temperature water. In addition to the environmental advantages of using water as a so-called pseudo-organic solvent, isolation of products is normally facilitated. Once cooled, most organic products are no longer soluble in ambient temperature water, allowing easy post-synthesis product separation.

Most importantly, the ionic product (dissociation constant) of water is increased by three orders of magnitude on going from room temperature to 250 °C. The pK_W therefore decreases from 14 to 11.2 (Table 4.3). This means that water becomes both a stronger acid and a stronger base as the temperature increases to the near-critical water (NCW) range and can therefore act as an acid, base, or acid–base bi-catalyst [46, 47]. The application of supercritical water (SCW, >374 °C) for preparative organic synthesis is limited due to its degenerative properties.

As with most organic solvents, the loss tangent (tan δ) for water is strongly influenced by temperature (see Section 2.3). Since the dielectric constant ε' for water decreases drastically with temperature (Table 4.3), the loss tangent is also reduced. For that reason it is not a trivial affair to heat pure water to high temperatures under microwave conditions. While water can be heated rather effectively from room temperature to 100 °C, it is more difficult to superheat water in sealed vessels from 100 to 200 °C and very difficult to reach 300 °C by microwave dielectric heating [48]. In fact, SCW is transparent to microwave radiation.

The loss tangent of a solvent such as water can be significantly increased, for example, by addition of small amounts of inorganic salts which improve microwave absorbance by ionic conduction (see Section 2.2). As shown in Figure 4.4, constant microwave irradiation of a 5 mL sample of water with 150 W power leads indeed to a marked difference in heating profiles between pure water and a 0.03 M sodium chloride solution [48]. While the water sample only reaches about 130 °C after 90 s of irradiation and the temperature cannot be further increased, the dilute salt solution can easily be heated to 190 °C (Figure 4.4). For most applications in NCW chemistry (200–300 °C) it can be assumed that the comparatively low salt concentrations

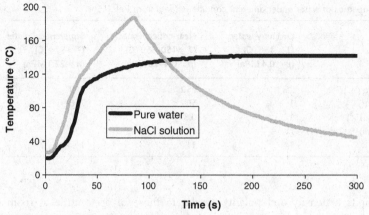

Figure 4.4 Heating profiles for pure water and 0.03 M NaCl
solution at constant 150 W power. CEM Discover, 5 mL sample
volume, quartz vessel and fiber-optic temperature measurement.
Reproduced with permission from Ref. [48].

(0.03 M, 0.18 w/w%) would not significantly influence the reactivity of the water medium [48]. However, pure water can be easily and rapidly heated to temperatures up to 200 °C in the high-temperature water region, if microwave absorbing substrates or a metal catalyst like palladium are present in the reaction mixture [49].

The application of microwave heating in combination with the use of high-temperature water as "green" solvent has received much interest in recent years [49], a field originally pioneered by Strauss and coworkers in the mid 1990s [50–52]. Due to the existing pressure limit of 20 bar for most of today's commercially available single-mode microwave reactors, the majority of published chemistry has generally been restricted to reaction temperatures at or below 200 °C [49]. Only a small number of publications exist where reactions are performed under NCW conditions at temperatures close to 300 °C in one of the few accessible dedicated instruments with higher pressure limits [49]. In Scheme 4.13 examples are highlighted where the Synthos 3000 multimode instrument with a pressure and temperature limit of 80 bar and 300 °C was employed (see Figure 3.18). Kremsner and Kappe have conducted benzamide hydrolysis at 295 °C without the addition of an acid catalyst, taking advantage of the increased ionic product of water at this temperature (Scheme 4.13a and Section 6.5.3.1) [48]. The Diels–Alder reaction shown in Scheme 4.13b produced only trace amounts of product when performed in water at 200 °C, whereas full conversion was achieved at 295 °C within 20 min [48]. For both syntheses, a 0.03 M NaCl solution was employed as solvent because of the improved microwave absorbance (see Figure 4.4). Direct conversion of aryl halides to the corresponding phenols at 300 °C was reported by Leadbeater and Kormos (Scheme 4.13c) [53]. The reactions could be carried out at 200 °C as well, however, higher yields, especially for aryl bromide and chloride substrates, were obtained at 300 °C.

Similar to water, microwave irradiation in the above-mentioned multimode reactor makes possible reactions at near-critical or supercritical conditions of alcohols.

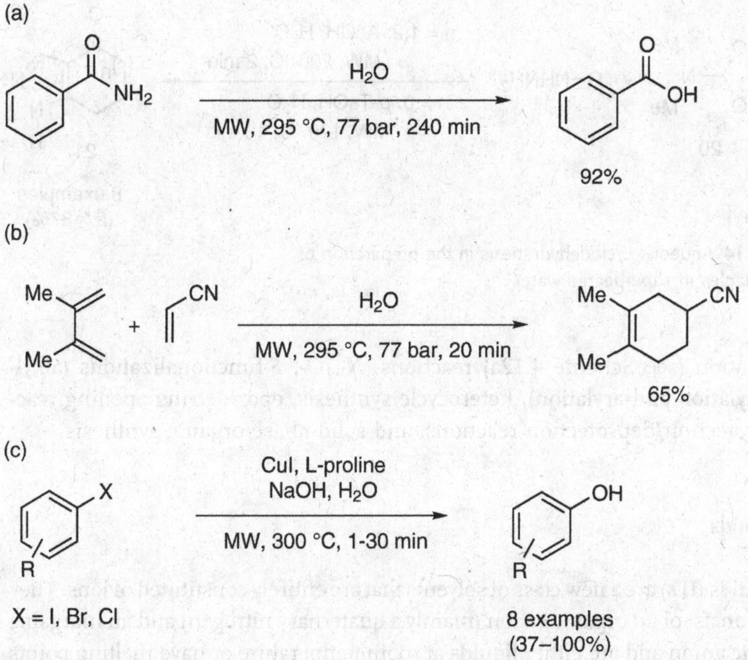

(a)

(b)

(c)

X = I, Br, Cl

8 examples
(37–100%)

Scheme 4.13 Hydrolysis (a), Diels–Alder (b), and phenol synthesis (c) in near-critical water.

Catalyst-free transesterifications of triglycerides (rapeseed oil) with butanol ($T_c =$ 287 °C at 49 bar) under supercritical conditions at 310 °C and 80 bar were performed by Maes and coworkers, resulting in high conversions to the fatty acid butyl ester [54]. To reach such high temperatures, silicon carbide (SiC) passive heating elements had to be used to improve the heating performance due to the low microwave absorbance of butanol at higher temperatures (see Section 4.6).

As already mentioned above, most of the published work in microwave-assisted water chemistry deals with temperatures ≤ 200 °C [49]. For example, Molteni and coworkers have described the three-component, aqueous one-pot synthesis of fused pyrazoles **21** by reacting cyclic 1,3-diketones **19** with *N,N*-dimethylformamide dimethyl acetal **20** (DMFDMA) and a suitable bidentate nucleophile like a hydrazine derivative (Scheme 4.14 and Section 6.1.3.2) [55]. The reaction proceeds via initial formation of an enaminoketone followed by a tandem addition-elimination/cyclo-dehydration step. An amount of 2.6 equivalents of acetic acid is necessary to ensure a clean conversion at 200 °C within 2 min. Upon cooling and stirring, the desired reaction products crystallized and were isolated in high purity by simple filtration. Pyrimidines and isoxazoles could be synthesized in a similar fashion applying the same protocol employing amidines and hydroxylamine as nucleophiles.

Many other examples employing water as solvent, ranging from superheated (>100 °C) to near-critical conditions (200–300 °C), in dedicated microwave instruments can be found in a recent 2007 review [49]. Examples include transition metal-catalyzed transformations such as Suzuki-, Heck-, Sonogashira-, Stille- or

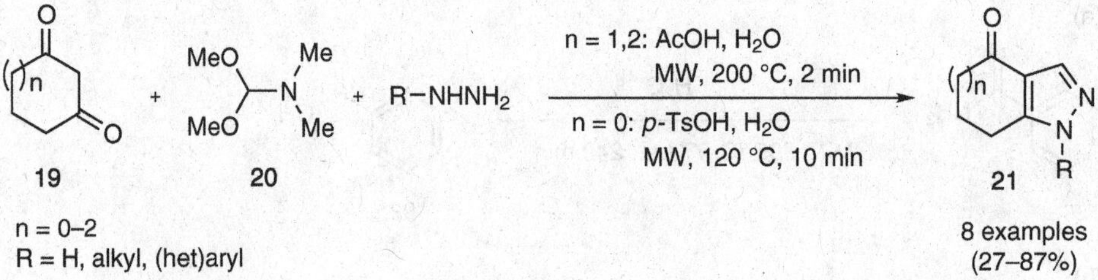

n = 0–2
R = H, alkyl, (het)aryl

8 examples
(27–87%)

Scheme 4.14 Aqueous cyclodehydrations in the preparation of fused pyrazoles in superheated water.

carbonylation (see Scheme 4.12a) reactions, N-, O-, S-functionalizations (alkylation, acylation and arylation), heterocycle synthesis, epoxide ring-opening reactions, protection/deprotection reactions, and solid-phase organic synthesis.

4.5.2
Ionic Liquids

Ionic liquids (ILs) are a new class of solvents that are entirely constituted of ions. They usually consist of an organic cation (mainly a quaternary nitrogen) and an inorganic or organic anion and are either liquids at room temperature or have melting points below 100 °C [56]. They generally have negligible vapor pressures and are immiscible with common nonpolar solvents, meaning that organic products can be easily removed by extraction and the ionic liquid can be recycled. In addition they have a wide accessible temperature range (typically >300 °C), a low toxicity and are non-flammable. Due to those advantages (see Box 4.2) they have attracted much recent attention as environmentally benign solvents [56].

Box 4.2

Characteristics of ionic liquids.

- Organic salts that are liquids at room temperature
- Large liquid temperature range (300 °C)
- Polar, non-volatile
- Dissolve organic and inorganic compounds
- Environmentally benign?
- Couple very effectively with microwaves (ionic conduction).

From the perspective of microwave chemistry one of the points of key importance is their high polarity, which is variable depending on the cation and anion and hence can effectively be tuned to a particular application. Ionic liquids interact very efficiently with microwaves through the ionic conduction mechanism (see Section 2.2) and are rapidly heated at rates easily exceeding $10 °C s^{-1}$ without any significant build up of pressure [57]. Therefore, safety problems arising from over-pressurization of heated sealed reaction vessels can be minimized. The rapid heating of ionic liquids is

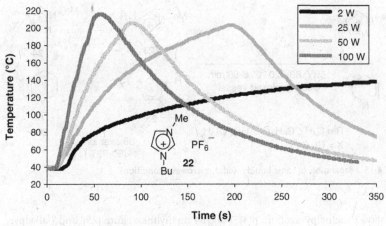

Figure 4.5 Heating profile of neat bmimPF$_6$ at different microwave powers. Single-mode irradiation (CEM Discover), 5 mL sample volume, 10 mL quartz vessel and fiber-optic temperature measurement. Reproduced with permission from Ref. [58].

demonstrated in Figure 4.5, where the heating profile of neat 1-butyl-3-methylimi-dazolium hexafluorophosphate **22** (bmimPF$_6$), that is irradiated at different constant microwave powers, is shown [58]. The very strong coupling characteristics are demonstrated by the fact that only 2 W of microwave energy suffices to heat the IL to 140 °C within about 5 min. Temperatures up to 220 °C can be rapidly reached within less than 1 min when higher power is applied.

The use of ionic liquids in conjunction with MAOS can be classified into three areas [59–61]:

1. Employment of microwave heating in the synthesis of ILs.
2. Use of ILs as solvents, reagents and catalysts for microwave-assisted synthesis.
3. Application of ILs as "doping agents" for poor microwave absorbing solvents.

Although promoted as environmentally benign reaction media, the synthesis of ionic liquids is not. Especially for the purification step, large volumes of organic solvents are typically required under conventional conditions. Halide-based ionic liquids can easily be prepared by reacting nitrogen-containing heterocyclic starting materials with an appropriate alkyl halide. To reduce the large excess of alkyl halides needed under traditional conditions, microwave-assisted methods have been deve-loped. Much of the early work in this field was published by Varma and coworkers, applying open-vessel conditions [62]. Later, Khadlikar and Rebeiro demonstrated that the preparation of ionic liquids was also feasible by applying closed-vessel microwave conditions, eliminating the dangers of working with toxic and volatile alkyl halides in the open atmosphere [63]. A comprehensive study on the microwave-assisted prepa-ration of ionic liquids was published by Deetlefs and Seddon in 2003. These authors synthesized a large number of ionic liquids based on the 1-alkyl-3-methylimidazolium

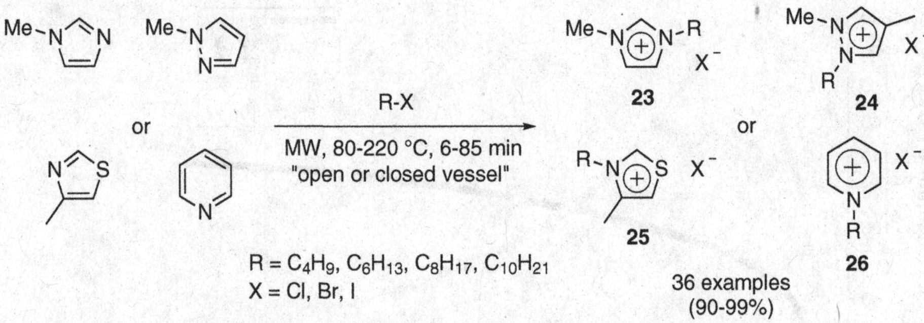

Scheme 4.15 Preparation of ionic liquids under microwave conditions.

(23), 1-alkyl-2-methylpyrazolium (24), 3-alkyl-4-methylthiazolium (25), and 1-alkylpyridinium (26) cations by solvent-free alkylation of the corresponding basic heterocyclic cores (Scheme 4.15) [64]. The published procedures feature dramatically reduced reaction times as compared to conventional methods, minimize the generation of organic waste, and also afford the ionic liquid products in excellent yields and with high purity. The syntheses were performed on flexible reaction scales ranging from 50 mmol to 2 mol in either sealed or open vessels, with a significant reduction of the large molar excess of haloalkane.

Using a similar protocol, the Loupy group has reported the synthesis of chiral ionic liquids based on $(1R,2S)$-$(-)$-ephedrinium salts under microwave irradiation conditions [65]. Importantly, the authors were also able to demonstrate that the desired hexafluorophosphate salts could be prepared in a one-pot protocol by *in situ* anion-exchange metathesis. In 2007, Gaertner and coworkers developed chiral imidazolium ionic liquids bearing a bornyl moiety from $(1S)$-$(+)$-camphorsulfonic acid and $(+)$-camphene precursors with the aid of microwave heating in the quaternization step [66]. These ILs were utilized as solvents in diastereoselective Diels–Alder reactions.

In the last few years, the scope of ionic liquids has been expanded by the introduction of additional functional groups in the ionic liquid structure [67]. In these so-called task-specific ionic liquids (TSILs), the functional group is tethered either to the cation or the anion (or both) and can be utilized as a soluble support. De Kort and coworkers have designed N-methylimidazolium-based ionic support 27 that incorporates an aldehyde functionality and can be seen as the ionic analog of the AMEBA solid support [68]. The ionic AMEBA support is synthesized via simple alkylation of N-methylimidazole with the corresponding aldehyde-functionalized alkyl chloride (Scheme 4.16) and is further converted to supported amines by reductive amination of the aldehyde.

In recent publications both the synthesis of TSILs [69] and further transformations of the introduced functional groups were reported using microwave irradiation [60, 61]. Especially, heterocycle synthesis has been performed on these ionic supports, like the synthesis of dihydropyrimidines and dihydropyridines via Biginelli and Hantzsch reactions, respectively, on PEG-ionic liquid matrices [70].

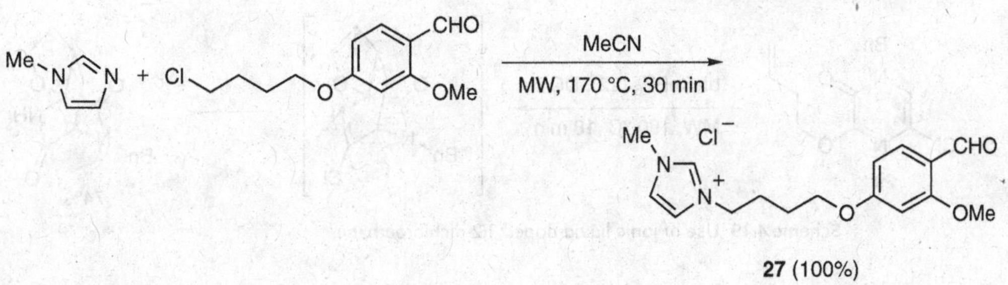

Scheme 4.16 Preperation of task-specific ionic liquids.

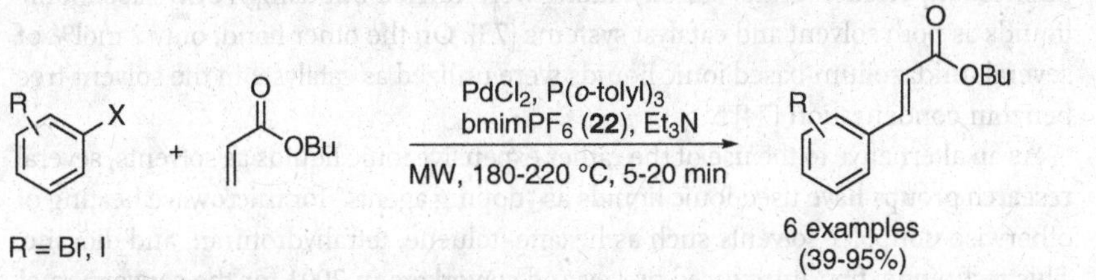

R = Br, I

6 examples
(39–95%)

Scheme 4.17 Heck reactions in ionic liquids.

More established than the synthesis of ionic liquids under microwave irradiation is the employment of ionic liquids as solvents in microwave chemistry because of their efficient coupling characteristics (see Figure 4.5). One example is shown in Scheme 4.17, where the Larhed group performed Heck reactions in bmimPF$_6$ as solvent [71]. Aryl bromides and iodides were reacted with butyl acrylate and PdCl$_2$/P (o-tolyl)$_3$ as catalyst/ligand system. Full conversions were achieved within 5 min at 180 °C (X = I) and 20 min at 220 °C (X = Br). A key feature of this catalytic/ionic liquid system is its recyclability: the phosphine-free ionic catalyst phase PdCl$_2$/bmimPF$_6$ was recyclable at least five times. After each cycle, the volatile product was directly isolated in high yield by rapid distillation under reduced pressure.

Apart from the application of neat ionic liquids as solvents, they can also act as a combined solvent–reagent system in microwave-assisted reactions. Leadbeater and coworkers have disclosed the synthesis of primary alkyl halides from the corresponding alcohols where the 1,3-dialkylimidazolium halide-based ionic liquids 28 serve as both reagents and solvents (Scheme 4.18) [72]. Depending on the employed ionic liquid, reactions could be completed in 30 s to 10 min at 200 °C.

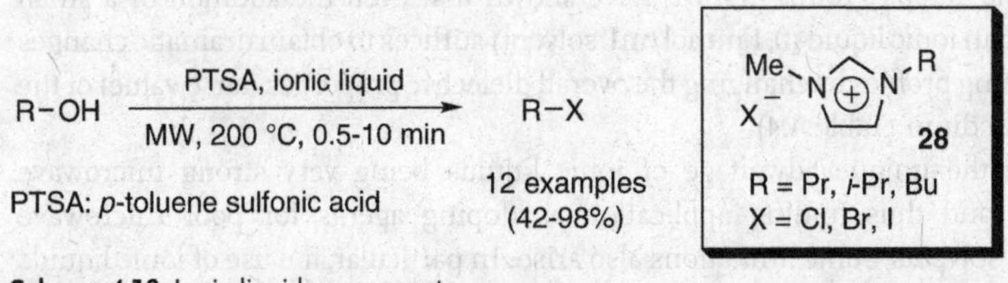

PTSA: p-toluene sulfonic acid

12 examples
(42–98%)

Scheme 4.18 Ionic liquids as reagents.

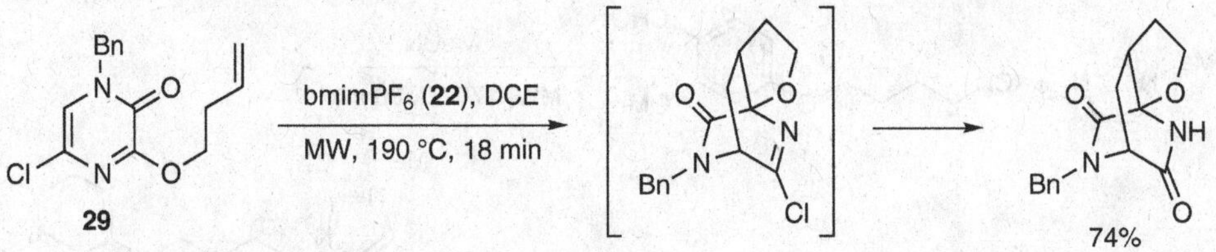

Scheme 4.19 Use of ionic liquid-doped 1,2-dichloroethane.

Ionic liquids can also serve as a solvent–catalyst combination. In a recent publication, Friedel–Crafts sulfonylations were carried out using $FeCl_3$-based ionic liquids as both solvent and catalyst systems [73]. On the other hand, only 2 mol% of several imidazolium-based ionic liquids were utilized as catalysts in the solvent-free benzoin condensation [74].

As an alternative to the use of the rather expensive ionic liquids as solvents, several research groups have used ionic liquids as "doping agents" for microwave heating of otherwise nonpolar solvents such as hexane, toluene, tetrahydrofuran, and dioxane. This technique, first introduced by Ley and coworkers in 2001 for the conversion of amides to thioamides using a polymer-supported thionating reagent in a weakly microwave absorbing solvent such as toluene [75], is becoming increasingly popular, as demonstrated by the many recently published examples [59–61]. One representative example is presented in Scheme 4.19. Under conventional reflux conditions in chlorobenzene at 135 °C, the intramolecular hetero-Diels–Alder reaction of alkenyl-tethered pyrazinone **29** required 1–2 days for completion. When performing the cycloaddition in 1,2-dichloroethane (DCE) as solvent at 170 °C under microwave irradiation, the reaction time could be reduced to about 1 h [33]. To further accelerate the reaction, the authors doped the weakly microwave-absorbing DCE with a small amount of thermally stable bmimPF$_6$ (0.15 mmol for 2 mL DCE) in order to reach a higher temperature in a shorter time. Under these conditions, a temperature of 190 °C could be attained and the reaction was completed within 18 min (Scheme 4.19).

The concept of performing microwave synthesis aided by room temperature ionic liquids as reaction media has been applied to several different organic transformations, such as 1,3-dipolar cycloaddition reactions, ring-closing metathesis, Knoevenagel reactions, multicomponent reactions, and several others. These have been summarized in recently published reviews or book chapters [59–61].

Systematic studies on temperature profiles and the thermal stability of ionic liquids under microwave irradiation conditions by the Leadbeater [76], Ondruschka [60], and Kappe groups [33, 61] have shown that even the addition of a small amount of an ionic liquid (0.1 mmol/mL solvent) suffices to obtain dramatic changes in the heating profiles by changing the overall dielectric properties (tan δ value) of the reaction medium (Table 4.4).

Despite the unique advantage of ionic liquids being very strong microwave absorbers and thus finding application as doping agents for poor microwave absorbing solvents, some limitations also arise. In particular, the use of ionic liquids is sometimes incompatible with certain reaction types and even small amounts of an

Table 4.4 Microwave heating effects of doping organic solvents with ionic liquids **30** and **31** (data from Ref. [76]).[a]

Solvent	IL added	Temperature attained (°C)	Time taken (s)	Temperature attained without IL (°C)[b]
Hexane	30	217	10	46
	31	228	15	
Toluene	30	195	150	109
	31	234	130	
THF	30	268	70	112
	31	242	60	
Dioxane	30	264	90	76
	31	246	60	

[a]Experiments run using a constant 200 W irradiation power (CEM Discover), with 0.2 mmol IL/ 2 mL solvent under sealed vessel conditions.
[b]Temperature attained during the same irradiation time but without any IL added.

ionic liquid may prevent specific reaction pathways. Cavaleiro and Kappe observed the complete decomposition of a specific porphyrin starting material in the presence of small amounts of bmimPF$_6$ as doping reagent in the course of Diels–Alder reactions [78]. Other authors have shown that the thermal stability of alkylimidazolium-based ionic liquids is reduced in the presence of nucleophiles due to reaction of the nucleophile with the alkyl groups of the IL [79]. Similar observations were made by Leadbeater on the alkylation of pyrazoles with alkyl halides and therefore reactions which use or generate nucleophiles are generally not compatible with ionic liquids [76].

Stability studies of a neat IL at high temperatures have been presented in a recent 2007 publication [77]. The authors investigated the recyclability of a bicyclic imidazolium IL (b-3C-imNTf$_2$) in microwave-assisted Claisen rearrangements at 250 °C. In comparison to other ILs, these bicyclic imidazolium structures proved stable at temperatures up to 250 °C and could be reused several times.

In addition to the thermal stability and chemistry incompatibility problem, a more instrument design related issue arises when biphasic mixtures (e.g., ILs that are insoluble in nonpolar organic solvents) are heated under microwave irradiation. The immiscibility creates a severe problem related to the accurate temperature measurement and reproducibility since differential heating will occur. Depending on the position of the IR sensor, either the temperature of the very hot ionic liquid phase (IR from the bottom) or the temperature of the cooler organic layer (IR from the side) will be recorded (see Figure 4.6) [58]. In order to avoid these problems it is recommended

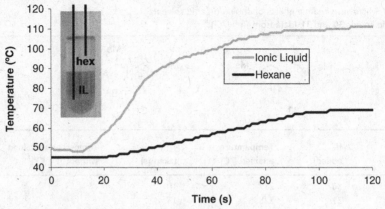

Figure 4.6 Differential heating of a biphasic mixture of bmimPF$_6$ and hexane. CEM Discover, 4 mL total sample volume, 10 mL quartz vessel, fiber-optic temperature measurement at different positions (heights) in the vessel, no stirring. Reproduced with permission from Ref. [58].

to use only ILs that are soluble in the particular solvent in question. In general, differential heating of biphasic mixtures by microwave irradiation is likely to occur when effective stirring is not possible [58], especially in solvent-free chemistry or in the presence of heavy slurries and viscous liquids (see also Section 4.1).

4.6
Passive Heating Elements

As already discussed in Section 4.5, microwave synthesis in low absorbing or microwave transparent solvents such as dioxane, tetrahydrofuran, toluene, hexane or carbon tetrachloride (CCl$_4$) is often not feasible, since the required temperatures for a particular transformation to proceed cannot be reached. For this reason, many nonpolar solvents, that are popular in conventional chemistry, are potentially precluded from the use as solvents in microwave synthesis. To overcome this problem and to avoid the switch to a polar solvent, it is sometimes sufficient to add a small amount of an ionic liquid (see Table 4.4). As an alternative to ionic liquids, a small quantity of a strongly microwave absorbing solvent can be added to an otherwise low absorbing solvent. By adding only 5% of ethanol to 2 mL of CCl$_4$, the temperature can be raised from about 50 to 100 °C at 150 W constant power [58]. Water, which is known to be only a moderate microwave-absorbing solvent, can be doped with sodium chloride (see Figure 4.4) or TBAB [58]. All these above-mentioned methods are so-called invasive methods, having the disadvantage that the polarity of the original solvent system is modified. In particular, severe problems can arise when ionic liquids are employed due to a possible incompatibility of the ionic liquid with certain substrates.

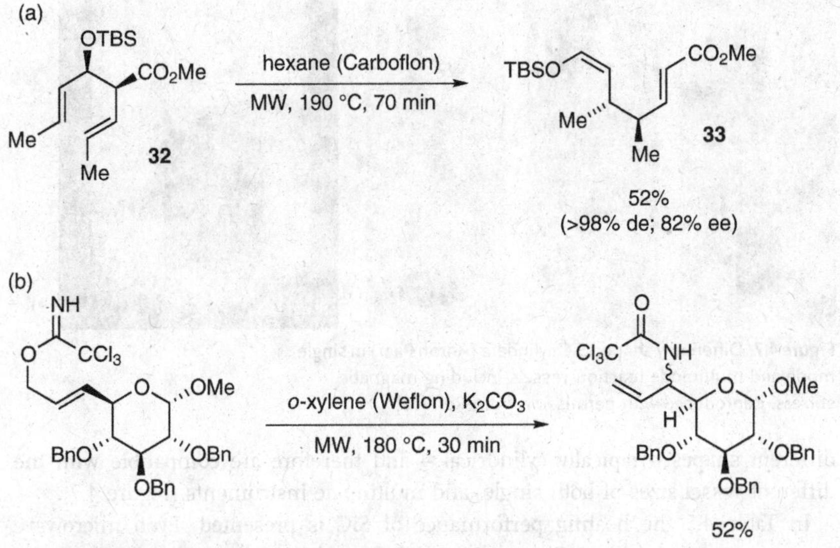

(a)

52%
(>98% de; 82% ee)

(b)

52%

Scheme 4.20 Rearrangement reactions in low-absorbing solvents with the aid of Carboflon (a), and Weflon (b).

Passive heating elements (PHEs) are noninvasive heating aids since they are chemically inert and therefore avoid the above-mentioned difficulties that arise with invasive heating aids. They are strongly microwave-absorbing materials (see Section 2.2) that transfer the generated heat via conduction phenomena to the reaction mixture, similar to conventional oil-bath heating. Furthermore, the use of PHEs is more practical since they can be mechanically removed from the reaction mixture and thus facilitate the purification step.

However, not many publications employing PHEs exist in the field of MAOS. Carboflon (CEM), a fluoropolymer doped with carbon black, was applied in the siloxy-Cope rearrangement of **32** in hexane as solvent (Scheme 4.20a) [81]. Importantly, desilylation of the Cope product **33** could be avoided in contrast to the ionic liquid doped rearrangement. Weflon (Teflon doped with graphite, Milestone), a related PHE, was employed in aza-Claisen rearrangements of allylic imidates to the corresponding amides (Scheme 4.20b) [82]. Due to the higher temperatures that could be achieved, higher product yields in shorter reaction times were obtained. Since both types of PHEs are based on an organic polymer, problems regarding deformation or degradation occur when heated to high temperatures and extended reaction times and their use is limited to about 200 °C bulk temperature [58].

In a 2006 study, silicon carbide (SiC) was introduced as a novel passive heating element for MAOS [58]. SiC is a very strong microwave absorber, has a high thermal conductivity and a low thermal expansion coefficient [83]. Moreover, it is mechanically, thermally and chemically resistant up to 1500 °C and is thus compatible with any solvent or reagent, virtually indestructible, and can be reused for an unlimited number of times without loss of efficiency. The SiC heating elements are available in

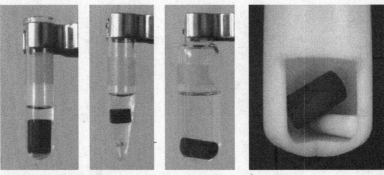

Figure 4.7 Differently shaped SiC cylinders (Anton Paar) in single-mode and multimode reaction vessels including magnetic stirrers. Reproduced with permission from Ref. [58].

different shapes – typically cylindrical – and therefore are compatible with the different vessel sizes of both single- and multimode instruments (Figure 4.7).

In Table 4.5 the heating performance of SiC is presented. Even microwave transparent solvents like CCl_4 can be rapidly heated to temperatures far above the boiling point that, for example, could not be reached without the added SiC heating element.

One application of SiC as a noninvasive heating element was found in the Claisen rearrangement of allyl phenyl ether (Scheme 4.21) [58]. In general, thermal Claisen rearrangements require high temperatures and proceed quite slowly. When a solution of the allyl ether in toluene in a Pyrex vessel was heated under microwave heating, the attained 160 °C proved not high enough to induce the rearrangement (Figure 4.8b). In contrast, by adding a SiC cylinder, a temperature of 250 °C was easily realized within 30 s and full conversion to the desired allyl phenol was accomplished after 105 min (Figure 4.8c and Scheme 4.21). In addition, the alkylation of pyrazole with alkyl halides, previously reported by Leadbeater as unfeasible using ionic liquids as doping agents, was also successfully carried out (see Section 6.1.2.1) [58, 76].

Table 4.5 Temperatures of nonpolar solvents attained by microwave heating in the absence and presence of SiC heating elements (data from Ref. [58]).[a]

Solvent	bp (°C)	T without SiC (°C)[b]	T with SiC (°C)	Time taken (s)[c]
CCl_4	76	40	172	81
Dioxane	101	41	206	114
Hexane	69	42	158	77
Toluene	111	54	231	145
THF	66	93	151	77

[a]CEM Discover, single-mode sealed vessel microwave irradiation, 150 W constant power, 2 mL solvent, 10 mL quartz or Pyrex vessel.
[b]After 77–145 s of microwave irradiation (see footnote c).
[c]Time until the maximum pressure limit of the instrument (20 bar) was reached and the experiment had to be aborted (with SiC).

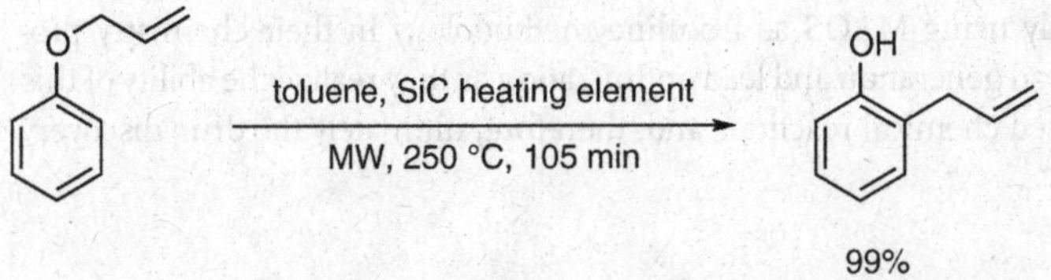

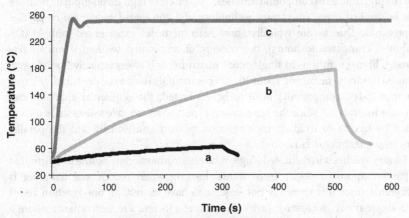

Scheme 4.21 Claisen rearrangement in toluene with the addition of a SiC cylinder.

Figure 4.8 Heating profiles for the Claisen rearrangement shown in Scheme 4.21 at a set temperature of 250 °C in a single-mode instrument. (a) Pure toluene in a quartz vessel. (b) Reaction mixture in a Pyrex vessel. (c) Reaction mixture with SiC cylinder.

4.7
Processing Techniques in Drug Discovery and High-Throughput Synthesis

The generation of diverse compound libraries has been shown to be a valuable tool for lead structure identification in the drug discovery process [84]. However, lead compound optimization and traditional medicinal chemistry remain the bottlenecks in this high-throughput discipline. Microwave-assisted heating under controlled conditions is an invaluable technology for medicinal chemistry and drug discovery applications because it often dramatically reduces reaction times, typically from days or hours to minutes or even seconds [2, 85]. Many reaction parameters such as reaction temperature and time, variations in solvents, additives and catalysts, or the molar ratios of the substrates can be evaluated in a few hours to optimize the desired chemistry.

As speed is a critical factor in the field of drug discovery and medicinal chemistry, the combination of microwave heating with high-throughput techniques for compound library generation is nowadays a popular and convenient application [85]. Therefore it is not surprising that most pharmaceutical and biotechnology companies are already heavily using MAOS as frontline methodology in their chemistry programs, both for lead generation and lead optimization as they realize the ability of this technology to speed chemical reactions and, therefore, ultimately the drug discovery process [85].

4.7.1
Automated Sequential versus Parallel Processing

For the preparation of compound libraries, two different high-throughput techniques can be applied using microwave technology, the automated sequential or parallel approaches. Due to the typically short reaction times experienced with MAOS (minutes compared to hours), the concept of automated sequential microwave-assisted library synthesis in single-mode instruments is a very attractive tool if small focused libraries containing about 20–100 compounds need to be produced. If larger libraries (>200 compounds) need to be generated, the sequential approach can become impractical since the time-saving aspect of microwave synthesis is diminished by having to irradiate each reaction mixture individually and the parallel processing technique is favored.

Library synthesis in dedicated single-mode instruments can become as efficient as a parallel approach under conventional heating when robotic vial handling is integrated since it is currently not feasible to have more than one reaction vessel in a single-mode microwave cavity. Even more efficient are instruments where a liquid handler additionally allows dispensing of reagents into sealed reaction vials, while a gripper moves each sealed vial in and out of the microwave cavity after irradiation (see Section 3.4). Some instruments can process up to 120 reactions per run with a typical throughput of 12–15 reactions per hour in an unattended fashion (see Figure 3.9). In contrast to the parallel synthesis application in multimode cavities, this approach allows the user to perform a series of optimization or library production reactions with each reaction separately programmed.

In a case study, a 48-member library of dihydropyrimidines (DHPMs) via the Biginelli reaction (Scheme 4.22 and Section 6.3.1.2) – a one-pot, three-component condensation of a CH-acidic building block (34), aldehyde (35) and (thio)urea (36) – was performed employing automated sequential processing in a fully automated microwave unit specifically designed for library production (Biotage Emrys Liberator,

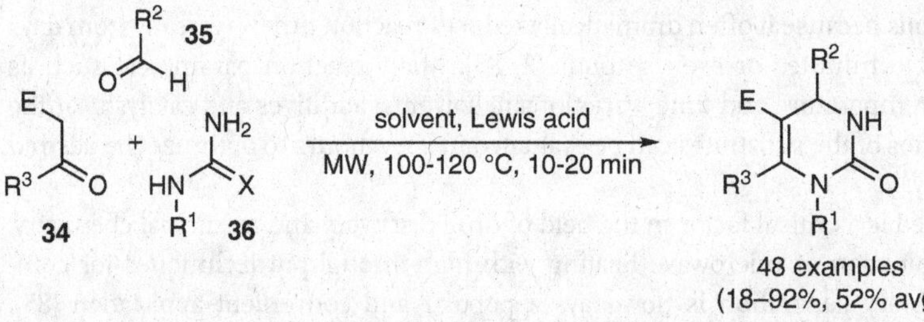

48 examples
(18–92%, 52% avg)

Scheme 4.22 DHPM-library generation employing the automated sequential process technique.

(a)    (b)

Figure 4.9 Abbott Laboratories (Illinois, USA) robotic microwave facility (a) and Novartis (Basel, Switzerland) high-throughput microwave synthesis factory (b). Courtesy of D. Sauer (Abbott) and S. Chamoin (Novartis).

see Figure 3.9) [86]. With the incorporated software, liquid dispensing of stock solutions or liquid reagents was possible and each experiment was generated and carried out individually. In summary, out of the 3400 possible DHPM derivatives, a subset of 48 compounds was prepared within 12 h on a 0.2–1 g scale. Compared to a conventional protocol the reaction times were reduced from 3–12 h to 10–20 min, with initial reaction optimization being accomplished within a few hours. Importantly, the sequential treatment allowed the use of optimized conditions for specific building block combinations, not possible in parallel processing. For a more detailed description of this case study, see Section 6.3.1.2.

As an alternative to the instrument employed above, fully automated workstations that integrate either a CEM-Discover (Navigator) or a Biotage Initiator (SWAVE, see Figure 3.10) single-mode instrument can be utilized for compound library synthesis. Due to the advantage of microwave heating for library synthesis, it is not surprising that pharmaceutical companies have established microwave facilities in their high-throughput synthesis divisions. As shown in Figure 4.9a, at Abbott Laboratories (Illinois, USA) a microwave station combining two single-mode instruments with additionally incorporated liquid reagent addition, automated capping and solid-phase extraction (SPE) purification tools has been employed for high-throughput microwave synthesis. With this set-up, a 480-member library of DHPM-5-carbox-amides was generated by reacting 10 different DHPM-acid cores with 48 diverse amines (Scheme 4.23) [87]. A 76% success rate (365 compounds isolated) could be achieved with a 55% average isolated yield and >95% purity after SPE. The whole processing time per compound took 30–45 min. In the Novartis high-throughput microwave synthesis factory (Basel, Switzerland) four single-mode instruments are integrated in the robotic station where several hundred reactions can be performed within 24 h (Figure 4.9b).

Despite the current trend in the pharmaceutical industry to synthesize smaller, focused libraries, medicinal chemists in a high-throughput synthesis environment often still need to generate large compound libraries using a parallel synthesis approach. Parallel microwave synthesis can be performed in multimode instruments using either dedicated multivessel rotor systems or deep-well microtiter plates that

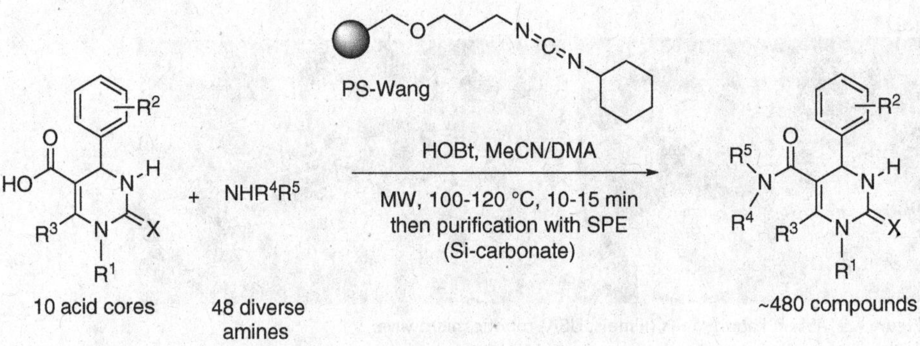

Scheme 4.23 Fully automated DHPM amide synthesis performed with the set-up shown in Figure 4.9a.

allow higher throughput (see Section 3.5). The first published example of parallel reactions carried out under microwave irradiation conditions involved the nucleophilic substitution of an alkyl iodide with 60 diverse piperidine or piperazine derivatives [88]. Reactions were carried out in a multimode microwave reactor in individually sealed polypropylene vials using acetonitrile as solvent.

In the early days of microwave synthesis, domestic microwave ovens were used as heating devices, without utilizing specialized reactor equipment for parallel processing. Reactions in household multimode ovens are notoriously difficult to reproduce due to the lack of temperature and pressure control, pulsed irradiation, uneven electromagnetic field distribution, and the unpredictable formation of hotspots (Section 3.2). Most of the parallel reactions performed in domestic ovens were conducted under solvent-free conditions (see also Section 4.1), for example involving a Biginelli synthesis to form DHPMs [89], Ugi-type multicomponent reactions for the synthesis of imidazoles, pyrazines, and pyrimidines [90], thioamide synthesis via amides and Lawesson's reagent [91], or the preparation of pyrido[2,3-d]pyrimidinones [92]. Today, in most of the published protocols, dedicated commercially available multimode reactor systems for parallel processing are used. The large selection of commercially available multivessel rotor systems for use in different microwave reactors is described in detail in Section 3.5.

An important issue in parallel microwave processing is the homogeneity of the electromagnetic field in the microwave cavity. Inhomogeneities in the field distribution may lead to the formation of so-called hot and cold spots, resulting in different reaction temperatures in individual vessels or wells and thus different product conversions. Conducting microwave-assisted parallel synthesis in a reproducible manner is, therefore, often a nontrivial affair. In this context, investigations of the reaction homogeneity in a 36 sealed-vessel rotor system (MicroSYNTH) were conducted [93]. For that purpose, 36 Biginelli condensations using six different aldehydes, ethyl acetoacetate, and urea as building blocks (see Scheme 4.22) were performed employing ethanol as solvent and hydrochloric acid as catalyst [93, 94]. Importantly, the yields of isolated products did not vary significantly depending on

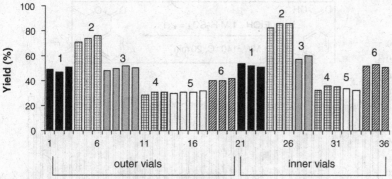

Figure 4.10 Isolated yields of Biginelli DHPM products in different reaction vessels of a 36 vessel rotor (MicroSYNTH). Outer ring, vessels 1–20; inner ring, vessels 21–36. Aldehydes: **1**: benzaldehyde, **2**: 2-hydroxybenzaldehyde, **3**: 3,4-dimethoxybenzaldehyde, **4**: 3-nitrobenzaldehyde, **5**: 2-chlorobenzaldehyde, **6**: 4-(N,N-dimethylamino)benzaldehyde. Adapted from Ref. [2].

the position in the rotor, although slightly increased yields were obtained for mixtures that were placed in the inner circle, which would indicate a somewhat higher temperature in those reaction vessels (Figure 4.10).

Similar results were achieved when Biginelli reactions in acetic acid/ethanol (3 : 1) as solvent (120 °C, 20 min) were run in parallel in an eight-vessel rotor system (Anton Paar, Synthos 3000, see Figure 3.20a and Section 6.4.2.1) on an 8×80 mmol scale [95]. Here, the temperature in one reference vessel was monitored with the aid of a suitable internal probe, while the surface temperatures of all eight quartz reaction vessels were also monitored (deviation less than 10 °C). The product yield in all eight vessels was nearly identical and the same set-up was also used to perform a variety of different chemistries in parallel mode [95].

The two-step synthesis of a 21-member library of polymer-bound enones depicted in Scheme 4.7 under open-vessel conditions was conducted in a parallel fashion employing PFA (perfluoroalkoxyethylene) vessels in a 50-position rotor (Micro-SYNTH) [22]. Here, the temperature was monitored with the aid of a fiber-optic probe inserted into one of the reaction vessels. It was confirmed, by standard temperature measurements performed immediately after the reaction period, that the resulting end temperature in each vial was the same to within ±2 °C.

The same group reported on temperature and reaction homogeneity studies for a 48-vessel rotor system introduced in 2006 (Synthos 3000, Figure 3.21a) by performing the acid-catalyzed esterification of benzoic acid with ethanol [96]. As can be seen in Figure 4.11, good homogeneity for this temperature sensitive reaction between the individual vessels was guaranteed with only 1.7% standard deviation (60–67% conversion). Similar results were also obtained for a 64-vessel rotor system (Synthos 3000, Figure 4.21b). The utility of the 48-vessel rotor system for library generation was demonstrated by synthesizing a set of 16 5-aroyl-DHPM derivatives by the Liebeskind–Srogl coupling of thiol esters with boronic acids at 130 °C within 1 h

EtOH / 1 M H_2SO_4 = 2:1
MW, 140 °C, 20 min
Conversion (%)
Vessel Number

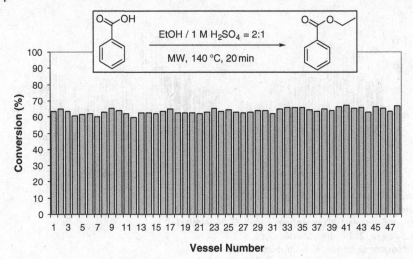

Figure 4.11 Reaction homogeneity for an esterification reaction in a 48-vessel rotor. Adapted from Ref. [96].

(Scheme 4.24) [96]. Only a minimal yield deviation (3–4%) compared to an automated sequential single-mode protocol was reported. Importantly, when performed in a parallel format, a significant time saving is possible for the library synthesis, since it requires 16 h to produce the 16-member library under automated sequential conditions.

The construction of a custom-built parallel reactor with expandable reaction vessels that accommodate the pressure build-up during a microwave irradiation experiment has also been reported [97]. The system was used for the parallel synthesis of a 24-member library of substituted 4-sulfanyl-1*H*-imidazoles that was subsequently converted into a second library of bicyclic imidazothiazoles and imidazothiazines, respectively.

Since, in parallel microwave synthesis, all reaction vessels are exposed to the same irradiation conditions, problems regarding the individual conversions in specific vessels can occur. Investigations of this phenomenon have been conducted by Leadbeater in a combined multi- and single-mode instrument (Milestone MultiSYNTH, Figure 3.30) [98]. In particular, when low microwave-absorbing solvents are

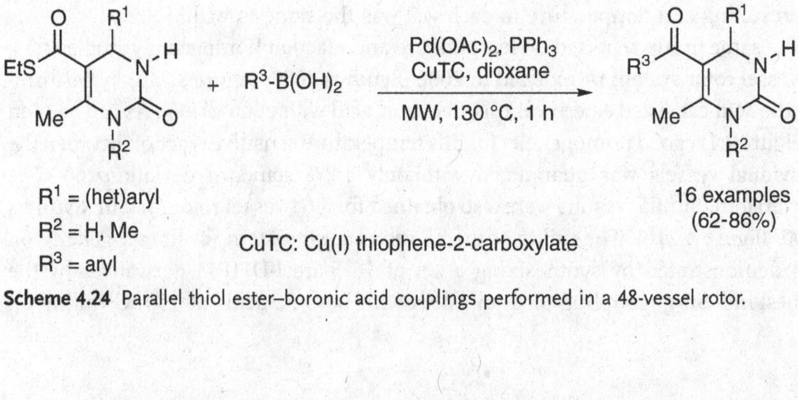

EtS
Me
R^3-$B(OH)_2$
$Pd(OAc)_2$, PPh_3
CuTC, dioxane
MW, 130 °C, 1 h
R^1 = (het)aryl
R^2 = H, Me
R^3 = aryl
16 examples
(62-86%)
CuTC: Cu(I) thiophene-2-carboxylate

Scheme 4.24 Parallel thiol ester–boronic acid couplings performed in a 48-vessel rotor.

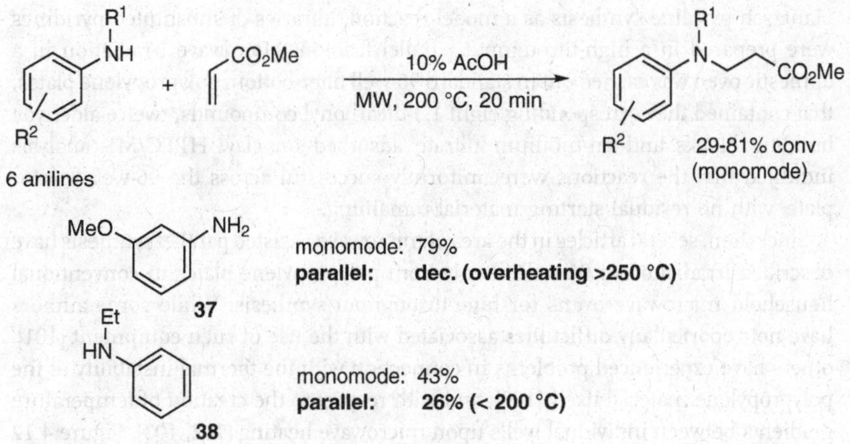

Scheme 4.25 Different microwave absorptivity of anilines in Michael additions performed in parallel.

used (or no solvent at all), the heating characteristics of the reaction mixture are strongly dependent on the microwave absorptivity of the building blocks, thus leading to different temperatures in the reaction vessels, and different conversions compared to a single-mode sequential method can arise. A case in point is the Michael addition of anilines with methyl acrylate, which is best performed under neat conditions at 200 °C, as demonstrated by a single-mode experiment (Scheme 4.25) [99]. Since this reaction is highly temperature dependent, lower temperatures in general give poorer yields whereas higher temperatures may lead to side-product formation and decomposition. When different anilines were heated in parallel, different product yields were observed as compared to the corresponding single-mode experiments. For example, decomposition of the reaction involving *m*-anisidine 37 was observed that indicated a temperature higher than 200 °C, whereas for *N*-ethylaniline 38 a lower conversion was obtained corresponding to a temperature lower than 200 °C (Scheme 4.25) [99]. These observations were additionally confirmed by the evaluation of individual heating profiles recorded with a fiber-optic probe in one reference vessel and an IR sensor that recorded the temperatures for all the rotor vessels. Moreover, the positioning of the fiber-optic sensor is crucial since the vessel with the fiber-optic probe is used as the control. For example, if the fiber-optic probe is placed in a vessel with high absorbing reagents, a lower microwave power will be used to reach and keep the temperature. As a consequence, the set temperature will not be attained for lower absorbing substrates, leading to lower yields. These effects could be minimized either by using a high absorbing solvent or by adding polar additives such as TBAB [99].

In order to achieve an even higher throughput and to address the needs of the combinatorial chemistry community to produce several hundreds of compounds per day, microwave chemistry performed in microtiter plates in multimode instruments has emerged [100].

In a key 1998 publication, the concept of microwave-assisted parallel synthesis in microtiter plates was introduced for the first time [101]. Using the three-component

Hantzsch pyridine synthesis as a model reaction, libraries of substituted pyridines were prepared in a high-throughput parallel fashion. Microwave irradiation in a domestic oven was carried out in standard 96-well filter-bottom polypropylene plates, that contained the corresponding eight 1,3-dicarbonyl compounds, twelve aldehyde building blocks and ammonium nitrate adsorbed on clay. HPLC/MS analysis indicated that the reactions were uniformly successful across the 96-well reactor plate, with no residual starting material remaining.

Since then, several articles in the area of microwave-assisted parallel synthesis have described irradiation of 96-well filter-bottom polypropylene plates in conventional household microwave ovens for high-throughput synthesis. While some authors have not reported any difficulties associated with the use of such equipment [101], others have experienced problems in connection with the thermal instability of the polypropylene material itself [102], and with respect to the creation of temperature gradients between individual wells upon microwave heating [102, 103]. Figure 4.12 shows the temperature gradients after irradiation of a conventional 96-well plate for 1 min in a domestic microwave oven. For the particular chemistry involved, the 20 °C difference between inner and outer wells was, however, not critical. Furthermore, conducting pressurized reactions is troublesome in conventional microtiter plates due to inappropriate sealing devices.

While issues of temperature stability with standard polypropylene microtiter plates can be overcome to some extent by utilizing PTFE (Teflon) or HTPE (high-temperature polyethylene) as plate materials [104, 105], dealing with transient and static temperature gradients across a microtiter plate is a nontrivial affair. In particular, the significantly lower temperature attained in wells located on the outside region of the plate due to both a radiative heat loss from the plate to the ambient air and a lower microwave coupling (due to the absence of neighboring wells) constitutes a problem regarding lower conversions or product purities compared to reactions performed in wells near to the center of the plate. To address these problems, custom-built variations of PTFE microtiter plates were developed by scientists from Sanofi–Aventis [104]. In order to eliminate the temperature gradients across the plate, microtiter plates with heat reservoirs in the form of microwave-absorbing fluid filled

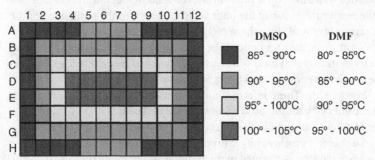

Figure 4.12 Temperature gradients within a microwave-heated microtiter plate; 1 mL per well, heated continuously for 1 min at full power in a conventional microwave oven. Reproduced with permission from Ref. [2].

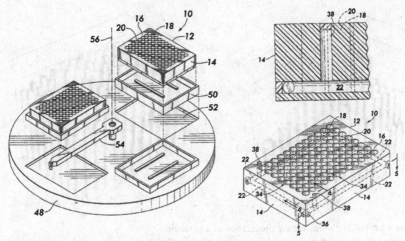

Figure 4.13 Perspective view of the multi-well plates in use (a) and a partial cross-section of the plates showing the reservoirs/channels (b) [104].

channels or elongated solid bodies positioned lengthwise along the outside region of the plate were designed (Figure 4.13).

A similar solution was proposed by a group from Boehringer Ingelheim [105]. By placing graphite pellets in the exterior wells, an improved microwave coupling in these otherwise problematic regions was ensured (see also Section 4.1). This improvement in temperature homogeneity was demonstrated by a temperature sensitive reaction (Scheme 4.26). Without the addition of graphite pellets, the temperature in the outer wells was up to 40 °C lower than in the center ones, leading to lower purities (Figure 4.14a), whereas the purity obtained with added graphite pellets proved to be good across the plate (Figure 4.14b) [105]. In addition, a prototype sealing mechanism has been described that allowed reactions to be run at up to 10 bar pressure.

To overcome the problems associated with using conventional polypropylene deep well plates in a microwave reactor, specifically designed well plates in combination with rotor systems for the combinatorial chemistry approach are available from the instrument vendors. For a detailed description of these rotor systems including microtiter plates, see Section 3.5.

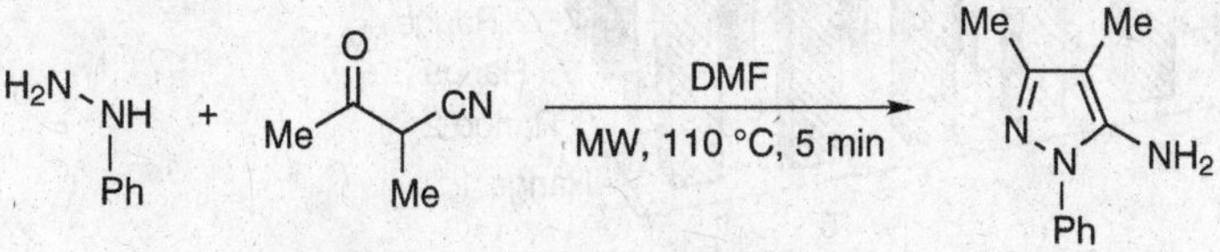

Reaction Temperature <80 °C: no product
Reaction Temperature >120 °C: dec of starting material

Scheme 4.26 Test reaction for evaluation reaction homogeneity across a 96-well plate.

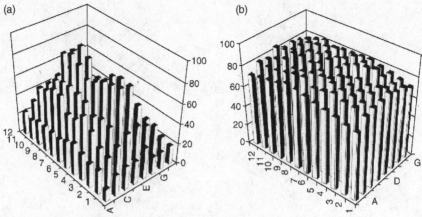

Figure 4.14 Product purity for the preparation of a pyrazole (Scheme 4.26) in sealed microtiter plates without (a) and with (b) the use of graphite pellets. Reproduced with permission from Ref. [105].

The reaction homogeneity in a commercially available 24-well plate that consists of a basis of carbon-doped Teflon (Weflon) for better heat distribution with glass inserts as reaction vessels (CombiCHEM system for the MicroSYNTH, Figure 3.33) was investigated by monitoring the esterification of octanoic acid with 1-octanol at 120 °C for 30 min [94]. The difference in conversion between the individual vessels was 3% with a standard deviation of 2.7% (Figure 4.15). It appears therefore that all individual

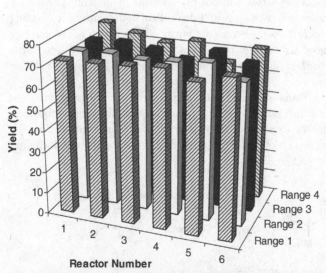

Figure 4.15 Reaction homogeneity expressed as conversion for the synthesis of octyloctanoate in a 24-well reaction plate (Milestone CombiCHEM system). Adapted from Ref. [94].

reactions were irradiated homogeneously in the applied microwave field. It is important to note that with this system, the material used for the preparation of the plates (Weflon) absorbs microwave energy, which means that the sealed glass vials will be heated by microwave irradiation regardless of the dielectric properties of the reactants/solvents. In addition, since up to two of the available 24-, 48- or 96-well plates can be mounted on top of each other (see Figure 3.33), several hundred reactions may potentially be performed in one irradiation cycle.

This Weflon-based microtiter plate system has been used for a variety of library applications [106–108]. Alcazar has recently reported on the synthesis of a 24-member library of tertiary amines via the alkylation of four amines with six alkylation agents, after having confirmed the reaction homogeneity with a model reaction (yields from 80 to 87%) [106]. Importantly, optimized reaction protocols from single-mode microwave synthesis performed in the Emrys Optimizer can be successfully transferred to this microtiter plate system when high boiling solvents are used in order to prevent a significant pressure build-up (2.3% average yield difference).

The preparation of a 96-member hexa-β-peptide library on a solid-phase employing a standard 96-well polypropylene bottom filtration plate in a dedicated rotor system (Microplate system for MARS reactor, Figure 3.27a) was reported by Gellman and Murray [109]. Prior to the library generation, a temperature homogeneity study was conducted by synthesizing model peptide **39** in 26 different wells distributed across the plate. As can be seen in Figure 4.16, lower purities were obtained in the outer wells. The authors have ascribed this to poor stirring that occurred especially in the outer regions of the plate but could be overcome by switching to a smaller stir bar.

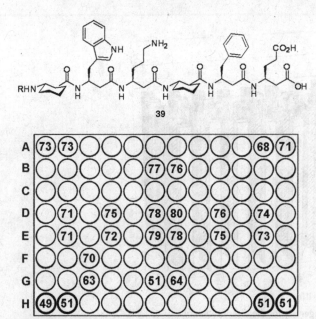

Figure 4.16 Model hexa-β-peptide **39** and its purity distribution across the plate. Reproduced with permission from Ref. [109].

Unfortunately, for the hexa-β-peptide library synthesis it was necessary to couple one β-amino acid at a time due to the different microwave absorption characteristics that would otherwise lead to different temperatures in individual wells.

The deep-well plate systems described so far are limited to the use of high-boiling solvents under atmospheric pressure or to sealed vessels at low pressures up to 4 bar. With these set-ups no direct translation from reactions performed under standard single-mode sealed-vessel conditions (20 bar pressure) is possible. In a 2007 publication, a novel 48-well microtiter plate was introduced that is constructed for performing reactions under sealed-vessel conditions at elevated pressure, similar to what can be achieved with a single-mode reactor (4×48 well plate system for the Synthos 3000, Figure 3.22) [110]. These special plates are made out of strongly microwave absorbing SiC (see Section 4.6) and can hold pressures up to 20 bar. On investigating the reaction homogeneity, it was found that by performing the esterification of benzoic acid with ethanol at 145 °C for 20 min the conversions in all 48 wells were virtually identical with only 0.6% standard deviation (Figure 4.17). The successful application of the SiC microtiter plate set-up was demonstrated by synthesizing a 30-member library of 2-aminopyrimidines by reacting a set of five 2-sulfonyl-pyrimidines with six diverse amines (Scheme 4.27) [110]. The generated 7 bar pressure was tolerated well by the system and in the majority of cases high conversions to the corresponding 2-aminopyrimidines were achieved.

The issue of parallel versus sequential synthesis using multimode or single-mode cavities, respectively, deserves special comment (see also Box 4.3). While the parallel set-up allows a considerably higher throughput in the relatively short timeframe of a

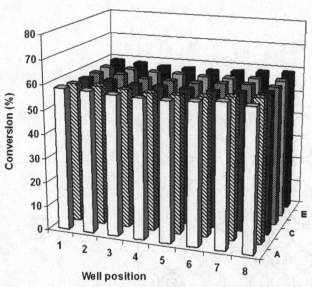

Figure 4.17 Reaction homogeneity for the esterification of benzoic acid in a 48-well SiC microtiter plate (Anton Paar Synthos 3000).

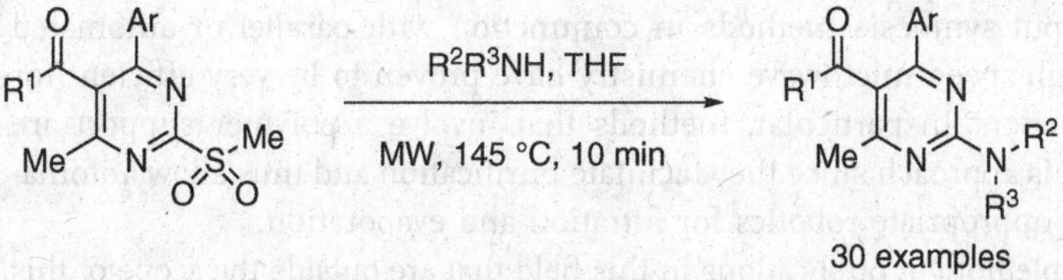

30 examples

Scheme 4.27 2-Aminopyrimidine library generation in a 48-well SiC microtiter plate.

microwave-enhanced chemical reaction, the individual control over each reaction vessel in terms of reaction temperature/pressure is limited. In the parallel mode, all reaction vessels are exposed to the same irradiation conditions. In order to ensure similar temperatures in each vessel, the same amount of the identical solvent should be used in each reaction vessel because of the dielectric properties involved [94, 98]. As an alternative to parallel processing, the automated sequential synthesis of libraries can be a viable strategy if small focused libraries (20–200 compounds) need to be prepared. Irradiating each individual reaction vessel separately not only gives better control over the reaction parameters but also fast iterations in protocol development and individual rapid optimization of reaction conditions are ensured. For the preparation of relatively small libraries, where delicate chemistries are to be performed, the sequential format may be preferable.

Box 4.3
Parallel versus sequential processing: Advantages and disadvantages.

Parallel synthesis (Multimode cavity)	Automated sequential synthesis (Single-mode cavity)
☺ Higher throughput	☹ Lower throughput
☺ Larger scale	☹ Smaller scale
☹ Less control over T and p	☺ Better control over T and p
☹ Same solvent in all vessels (tan δ !)	☺ Different solvents can be used
☹ Similar reagents	☺ Different reagents, catalysts, and so on
☹ All vessels exposed to identical time–temperature profile	☺ Specific time–temperature needs for individual vessels can be addressed

---

4.7.2
High-Throughput Synthesis Methods

Modern drug discovery relies on high-speed organic synthesis and high-throughput chemistry techniques for the rapid generation of compound libraries. Several high-throughput synthesis methods in conjunction with parallel or automated sequential high-speed microwave chemistry have proven to be very efficient for library production. In particular, methods that involve a polymer support are suitable for this approach since they facilitate purification and thus allow automation by using appropriate robotics for filtration and evaporation.

Due to the plethora of publications in this field that are outside the scope of this book, only selected examples that highlight the usefulness of these methods are presented in the following sections. For an in-depth overview of microwave-assisted transformations employing high-throughput synthesis methods that are described in the subsequent sections, the reader is referred to several recent reviews and book chapters [111–115].

4.7.2.1 Solid-Phase Synthesis

In solid-phase organic synthesis (SPOS), a molecule is attached to a solid support and subsequent chemistry is then performed on the molecule until, at the end of the multistep synthesis, the desired product scaffold is released from the support. To accelerate reactions and to drive them to completion, a large excess of reagents can be used, since this can easily be removed by filtration. Thus, even final purification of the desired product is simplified, as by-products formed in solution do not affect the outcome of the target.

Several articles reporting rate-enhancements of this otherwise time-consuming technique by applying microwave irradiation have been published in the literature [111]. An example that involves a prototype microwave reaction vessel that takes advantage of bottom-filtration was presented in 2003 [116]. The authors described the use of a modified reaction vessel for single-mode processing (Biotage) including a polypropylene frit, suitable for the filtration/cleavage steps in solid-phase Sonogashira couplings (Scheme 4.28).

Surprisingly, although solid-phase synthesis was originally introduced for peptide couplings, there are comparatively few published reports on the use of microwave irradiation in solid-phase peptide synthesis (SPPS) [117]. The first example of microwave-assisted SPPS of a decamer was published in 1992 [118] where a significant improvement in the coupling efficiency (two to fourfold) was obtained. However, the procedure is not easily reproducible due to the use of a domestic microwave oven. With the availability of dedicated microwave instruments, the

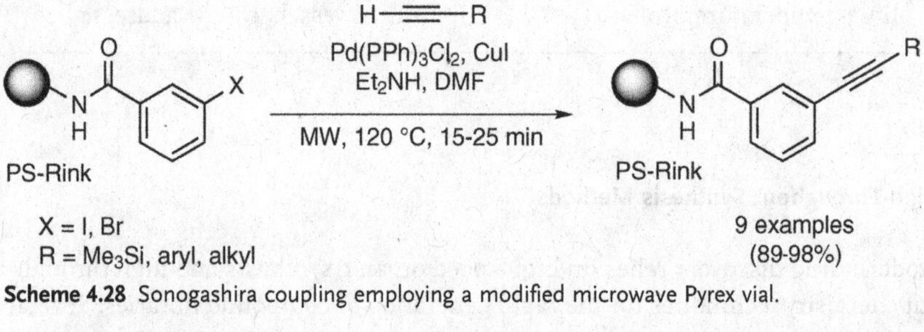

$$H \!-\!\!\equiv\!\!- R$$

Pd(PPh)$_3$Cl$_2$, CuI
Et$_2$NH, DMF

MW, 120 °C, 15–25 min

PS-Rink   PS-Rink

X = I, Br   9 examples
R = Me$_3$Si, aryl, alkyl   (89–98%)

Scheme 4.28 Sonogashira coupling employing a modified microwave Pyrex vial.

synthesis of a small tripeptide containing three of the most hindered natural amino acids was reported in 2002 by Erdelyi and Gogoll [119]. The authors observed enhanced couplings when employing a microwave protocol compared to standard conditions without racemization [119].

Since a standard protocol for one SPPS cycle consists of four steps – deprotection, washing, coupling, washing – the effort for the synthesis of longer peptide sequences in traditional dedicated microwave instruments using the standard glass vials is rather cumbersome due to the transfer of the resin suspension out of the microwave vial for each washing step. To overcome this problem, a MicroKan reactor that contains the resin beads can be introduced into standard microwave process vials [120]. MicroKans are made of a porous Teflon derivative that is fully penetrable by small molecules in solution but not by the resin particles. By using the MicroKan reactor, physical loss of resin particles is avoided [120]. Alternatively, microwave-assisted solid-phase reactions can be performed on SynPhase Lanterns that are rigid polymeric supports onto which reagents can be attached [121].

In order to cope with the labor intensive handling of resin beads in SPPS, an automated microwave peptide synthesizer was introduced in 2003 (CEM Liberty, Figure 3.16). This microwave instrument is able to perform all the necessary SPPS cycles in a fully automated fashion. In addition to the entirely automated microwave peptide synthesizer, a manual version is also available (CEM Discover SPS, Figure 3.17). The reaction vessel in this instrument is designed for solid-phase synthesis, allowing bottom filtration and therefore mimicking the workflow of conventional peptide synthesizers. The preparation of a nonapeptide (Figure 4.18) using conventional Fmoc/Bu^t orthogonal protecting strategy has been described by employing this manual instrument [122]. The coupling steps were performed within 5 min at 60 °C and the Fmoc-deprotection steps were completed within 3 min at 60 °C. The authors demonstrated that the model nonapeptide could be synthesized in a shorter time (about 3.5 h) and with higher purity (>95%) under microwave conditions compared to standard room temperature methods (11 h).

In a 2007 publication, the group of Papini reported comparison studies for the synthesis of difficult α-peptide sequences performed under either conventional conditions in a standard peptide synthesizer or microwave conditions in the automated peptide synthesizer (Liberty, Figure 3.16) [123]. For the hydrophobic

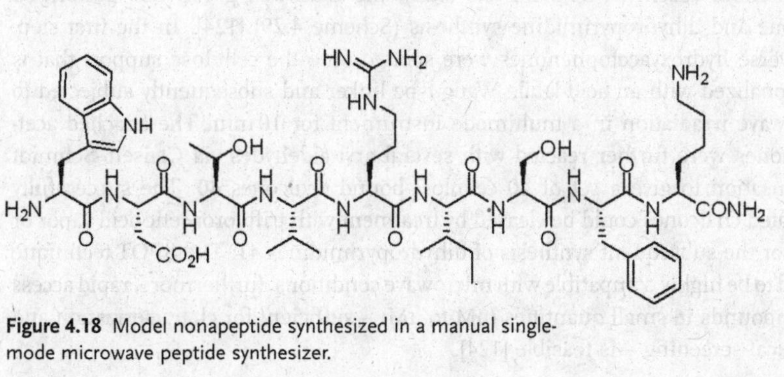

Figure 4.18 Model nonapeptide synthesized in a manual single-mode microwave peptide synthesizer.

Table 4.6 Comparison of peptide yield and purity between conventional and MW-assisted SPPS strategies (data from Ref. [123]).

Peptide	SPPS strategy	Yield (%)[a]	HPLC purity (%)
Gramicidin A (15mer)	rt	11	<20
	MW	59	72
CSF114(Glc) (21mer)	rt	10	<20
	MW	46	74

[a]Yield of crude peptide, desalted.

antibiotic peptide Gramicidin A (15mer) and the glycopeptide CSF114(Glc) (21mer) the microwave SPPS approach was more effective in terms of yield and purity since deletion sequences occurred by the conventional strategy (Table 4.6). In addition, the reaction time for each coupling cycle could be reduced from 2 h to 30 min employing the Liberty system.

The positive impact of microwave heating on the coupling and deprotection steps in peptide synthesis in terms of higher purities and enhanced reaction times could potentially be ascribed to a reduction in chain aggregation [117]. It is proposed that the polar N-terminal amine group and polar backbone constantly try to align with the oscillating field and that this movement could lead to a de-aggregation of the peptide backbones thus allowing reagents to reach the reaction sites at the end of the growing chains more easily.

An interesting alternative technique for conducting solid-phase synthesis is the so-called SPOT synthesis on planar supports. This method involves a spatially addressed synthesis on derivatized cellulose membranes (e.g., standard filter paper) to generate arrays of single compounds (1–10 000 spots per array) [124]. The membrane sheets are mechanically robust and, moreover, are compatible with various "on support" biological screening methods. This technique was applied for the preparation of an 8000-member library of 1,3,5-triazines on a $18 \times 26\,cm^2$ cellulose membrane via microwave-assisted nucleophilic substitution of the corresponding monochlorotriazines [125].

In a more recent SPOT-synthesis study, the Blackwell group has performed chalcone and dihydropyrimidine synthesis (Scheme 4.29) [124]. In the first step, six diverse hydroxyacetophenones were spotted onto the cellulose support that is functionalized with an acid-labile Wang-type linker and subsequently subjected to microwave irradiation in a multimode instrument for 10 min. The attached acetophenones were further reacted with several aryl aldehydes via Claisen–Schmidt condensation to give a set of 40 cellulose-bound chalcones **40**. The successfully generated chalcones could be cleaved by treatment with trifluoroacetic acid vapor or used for the subsequent synthesis of dihydropyrimidines **41**. The SPOT technique proved to be highly compatible with microwave conditions, furthermore, rapid access to compounds in small quantities (nM to μM) – sufficient for characterization and biological screening – is feasible [124].

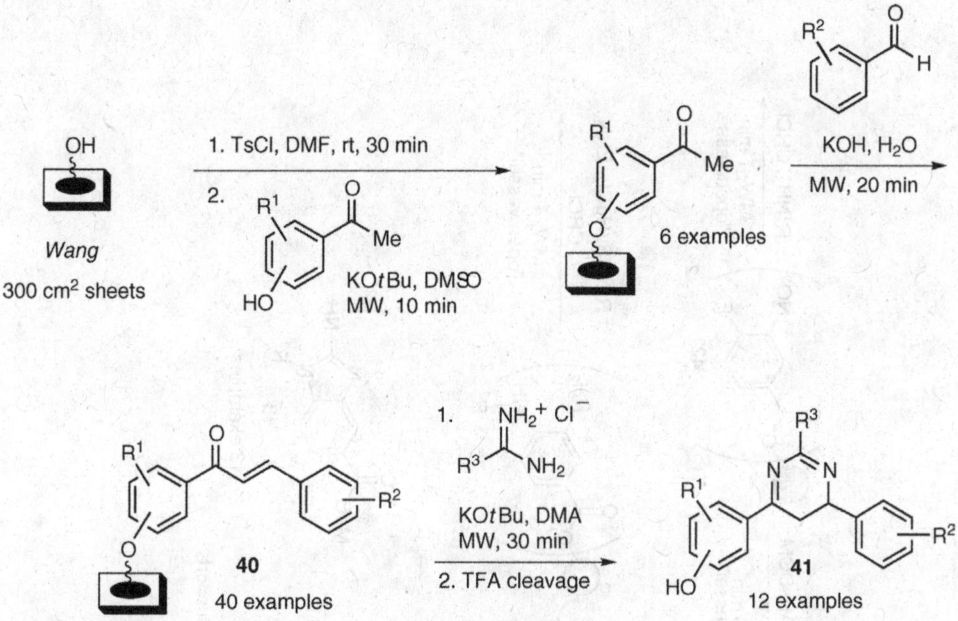

Scheme 4.29 SPOT synthesis of chalcones and dihydropyrimidines.

4.7.2.2 Soluble Polymer-Supported Synthesis

A viable alternative to SPOS is the use of a soluble polymer as support [126]. Soluble polymer-supported synthesis offers several advantages over SPOS: the polymer support is soluble in several organic solvents so that reactions can be carried out under homogeneous conditions, thus allowing standard spectroscopic characterization techniques to be used. Product isolation is performed by precipitation of the soluble support by addition of an appropriate solvent in which the support is insoluble (e.g. diethyl ether or hexane) with subsequent cleavage of the product from the support. The most common soluble supports are polyethylene glycol (PEG) and monomethoxypolyethylene glycol (MeOPEG-OH). The synthesis of a set of diversely substituted benzimidazoles **43** was performed by Sun and coworkers, starting from polyethylene glycol (PEG 6000) and 4-fluoro-3-nitrobenzoic acid to give the polymer-bound intermediate **42** (Scheme 4.30) [127]. In the first reaction sequence, ipso-fluoro displacement with various amines was conducted, followed by nitro-group reduction. For the ring closure step in one pot, the primary amines were converted to thioureas with a variety of isothiocyanates and subsequent intramolecular cyclization was possible under HgCl$_2$ mediation. All reactions were performed under open-vessel conditions in a CEM Discover and after each reaction step, the PEG-bound products were precipitated from a suitable solvent combination of dichloromethane and diethyl ether. The final benzimidazoles were cleaved from the support by using methanolic sodium methoxide and PEG 6000 was removed by precipitation with diethyl ether and filtration. Furthermore, the synthesis of

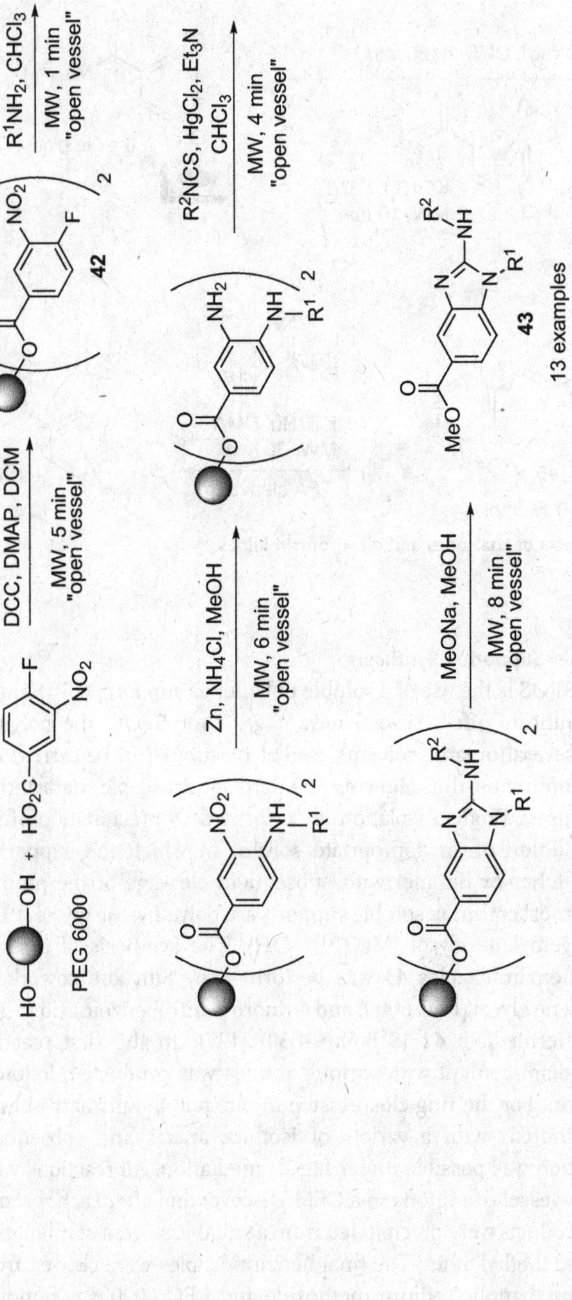

Scheme 4.30 Multistep benzimidazole synthesis on a soluble PEG support.

bis-benzimidazoles using a similar protocol was performed by the same group [128], as well as the traceless synthesis of thiohydantoins [129] and hydantoin-fused β-carboline scaffolds [130].

More literature on the use of soluble polymers in MAOS can be found in reviews and book chapters [111, 113].

4.7.2.3 Fluorous-Phase Organic Synthesis

Fluorous-phase organic synthesis is a separation and purification technique for organic synthesis and process development that combines the advantages of solution-phase reaction conditions with the convenient purification of solid-phase synthesis [114]. Perfluorinated (fluorous) chains such as C_6F_{13} and C_8F_{17} instead of resin beads are employed as phase tags that facilitate product separation. Molecules on which a fluorous tag is attached can be easily isolated from the reaction mixture by fluorous separation techniques such as fluorous solid-phase extraction (F-SPE) on fluorous silica gel [131]. Compounds with a fluorous tag are soluble in a range of organic solvents allowing reactions to be conducted under homogeneous conditions and monitored with standard analytical methods such as TLC, HPLC and NMR. In contrast to SPOS, more than one fluorous tagged compound can be employed in a single reaction.

Zhang and coworkers have reported on fluorous Suzuki couplings applying fluorous aryl sulfonates **44**, obtained by the reaction of phenols with perfluorooctylsulfonylfluoride (Scheme 4.31), as precursors [132]. After the reaction with diverse boronic acids, the biaryl products were isolated via F-SPE by elution with $MeOH/H_2O$ 80 : 20 in high yields and purities >90% while the cleaved fluorous tag remained on the cartridge. Under similar reaction conditions, the perfluorooctylsulfonyl group in **44** serves as a traceless tag for palladium-catalyzed deoxygenations with formic acid [133].

In addition to fluorous-tagged substrates that can be compared to polymer-bound substrates in SPOS, fluorous reagents, catalysts and scavengers are also available. Whereas fluorous substrates are more suitable for multistep syntheses, in particular for combinatorial chemistry approaches, fluorous reagents are commonly used for single-step syntheses [114].

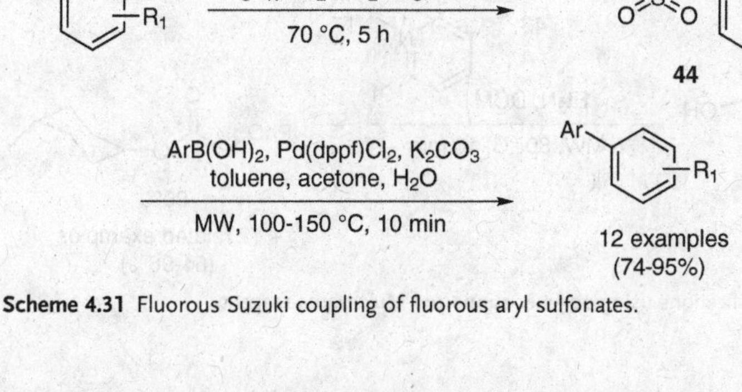

Scheme 4.31 Fluorous Suzuki coupling of fluorous aryl sulfonates.

Numerous applications of microwave-assisted fluorous transformations have been discussed in the recent literature [111, 114, 134].

4.7.2.4 Polymer-Supported Reagents, Catalysts and Scavengers

The use of polymer-supported reagents for solution-phase chemistry has attracted increasing attention in high-throughput organic synthesis [115, 135]. The polymer-assisted solution-phase (PASP) synthesis technique combines the benefits of SPOS in terms of work-up with the advantages of solution-phase synthesis. Excess amounts of polymer-supported reagents can be used to drive reactions to completion without affecting the purification step. Importantly, reactions can be easily monitored in real time by conventional methods such as TLC, HPLC or NMR. In addition, this technique is highly suitable for automation.

For example, the rapid synthesis of esters – starting from carboxylic acids and alcohols – in the solution phase employing polymer-supported Mukaiyama-type reagent **45** was developed by the group of Taddei [136]. The 2-iodo-1-methylpyridinium salt attached to a PS-DVB resin (**45**) was efficient in activating carboxylic acids and thus various esters could be synthesized at 80 °C within 6–18 min (Scheme 4.32). Hindered alcohols such as cyclohexyl alcohol or long chain acids and alcohols reacted in good yields to the corresponding esters which were obtained in high purity after a simple filtration step.

Catalysts immobilized on solid-supports (heterogeneous catalysts) have an important advantage over conventional homogeneous catalysts since work-up is facilitated – the catalyst is simply filtered upon completion of the reaction [137]. In addition, the catalyst system can often be regenerated and recycled several times without significant loss of activity [137].

Immobilized palladium catalysts for microwave-assisted Suzuki couplings were introduced by Sauer and Wang (Scheme 4.33) [138]. In the so-called FibreCats the palladium is coordinated to a phosphine ligand that is covalently bound to a polyethylene support. Compared to reactions that were performed under homogeneous catalysis with $PdCl_2(PPh_3)_2$, the supported palladium reactions were cleaner and, most importantly, no phosphine by-products, that are usually difficult to remove,

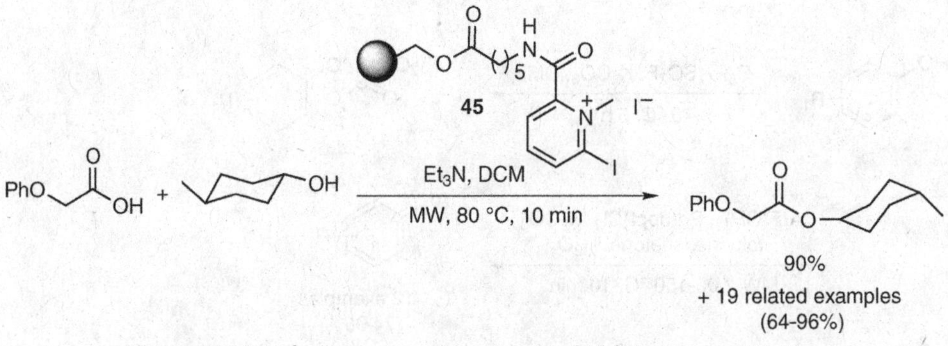

Scheme 4.32 Esterifications using polymer-supported Mukaiyama reagent.

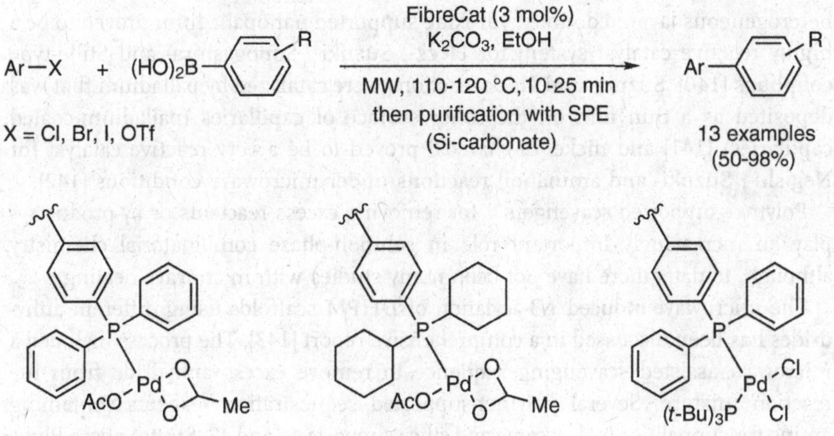

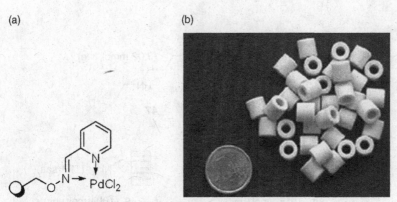

Scheme 4.33 Immobilized palladium catalysts for Suzuki couplings.

were detected. Another feature of this protocol in terms of purification simplification is the use of carbonate functionalized silica (Si-carbonate) for an SPE step after the reaction in order to remove excess boronic acid.

A palladium catalyst that is anchored to a glass/polymer composite material (Raschig rings, Figure 4.19) has been successfully applied for Suzuki couplings of aryl bromides and iodides with boronic acids in water as solvent by Kirschning and Dawood [139]. In a recyclability study, the authors demonstrated that under microwave conditions at 160 °C for 3 min this palladium catalyst can be reused with almost full conversion (97%) up to the seventh run.

Several other supports for transition metals have been introduced in the recent literature for a range of microwave-assisted coupling reactions. Ligand-free

(a)   (b)

Figure 4.19 Palladium (II) complex (a) that is anchored to a glass/polymer composite material – Raschig rings (b). Reproduced with permission from Ref. [139].

heterogeneous layered double hydroxide supported nanopalladium proved to be a highly reactive catalyst system for Heck-, Suzuki-, Sonogashira- and Stille-type couplings [140]. Suzuki- and Heck reactions were catalyzed by palladium that was deposited as a thin film on the inner surface of capillaries (palladium-coated capillaries) [141] and nickel-in-charcoal proved to be a very reactive catalyst for Negishi-, Suzuki- and amination reactions under microwave conditions [142].

Polymer-supported scavengers – for removing excess reactants or by-products – play an increasingly important role in solution-phase combinatorial chemistry although, to date, there have not been many studies with microwave heating.

The microwave-induced N3-acylation of DHPM scaffolds using different anhydrides has been discussed in a comprehensive report [143]. The process included a microwave-assisted scavenging sequence to remove excess anhydride from the reaction mixture. Several polymer-supported sequestration reagents containing amino functionalities (polystyrene and silica supports **46** and **47**, StratoSphere Plugs **48** and SynPhase Lanterns **49**) were employed for scavenging excess benzoic anhydride (Scheme 4.34). In both synthesis and purification, applying microwave heating reduced reaction times from several hours to minutes.

A very efficient method for the scavenging of metal contaminants from reaction mixtures was presented by Pitts and coworkers [144]. With the strict guidelines limiting metal levels in pharmaceuticals there is a growing need for practical techniques for the removal of trace metals from reaction products, in particular, since homogeneous metal-catalyzed reactions are very popular reactions in

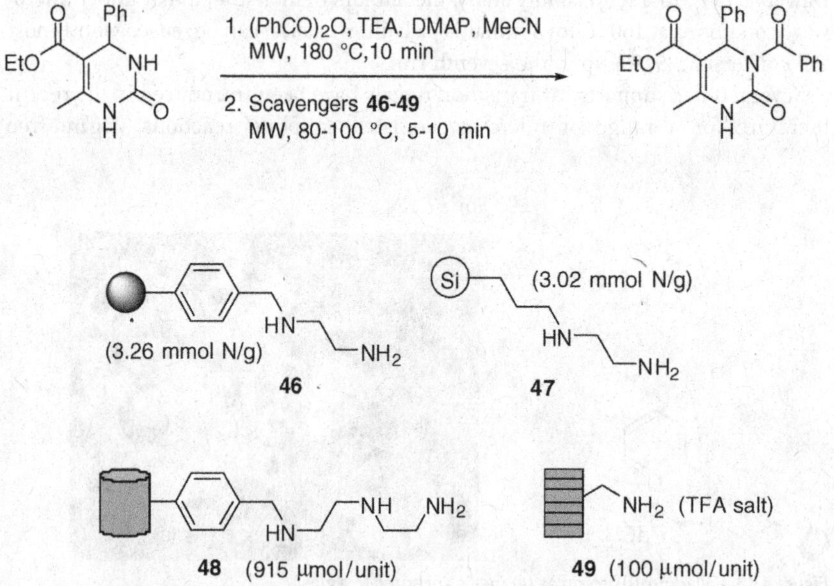

Scheme 4.34 N3-Acylation of DHPM with subsequent anhydride scavenging employing different scavengers **46–49**.

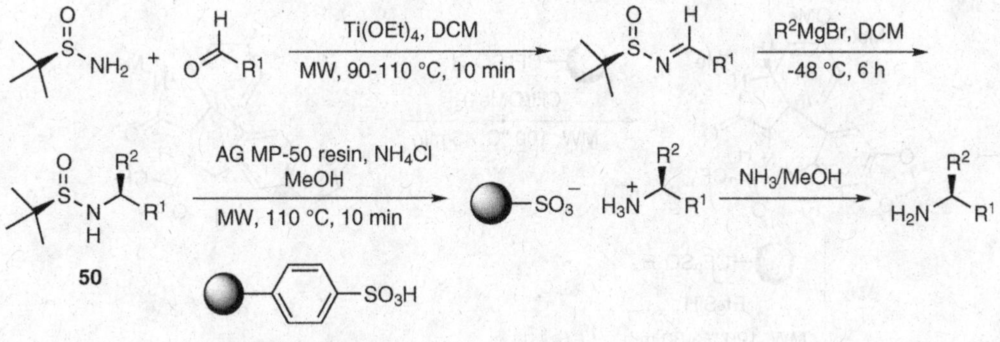

Scheme 4.35 Microwave-assisted resin capture.

medicinal chemistry. Packed cartridges of QuadraPure metal scavengers have been used to reduce levels of metal contaminants in flow. QuadraPure scavengers are functionalized macroporous polystyrene-based resins, highly cross-linked and with low swelling properties in organic solvents. In this study, copper by-products from a Rosemund-von Braun cyanation performed under microwave heating at 250 °C for 30 min were removed by passing the reaction mixture through an imino diacetate-functionalized QuadraPure cartridge. The copper content could be reduced from 345 ppm after the reaction to <1 ppm after the scavenging step. In the same way, but with different functionalized scavengers, palladium, copper, iron and rhodium were efficiently removed up to 99%.

Instead of scavenging excess reagents or by-products out of the reaction mixture an alternative approach enables the selective capture of the product. In the course of the parallel synthesis of α-branched amines by microwave-assisted imine formation, the Ellman group applied resin capture of the amine product when the sulfinyl group of **50** was removed via acidic alcoholysis (Scheme 4.35) [145]. Cleavage of the sulfinyl group and concomitant amine capture was induced by macroporous sulfonic acid **51**. The final product was subsequently released with methanolic ammonia.

An impressive example of applying PASP synthesis was demonstrated by the Ley group (Scheme 4.36). In the course of their investigations toward the application of polymer-supported reagents and scavengers in multistep synthesis of small compound libraries and more advanced natural products, they reported on the synthesis of the alkaloid (+)-plicamine where immobilized reagents and scavengers are employed in every single step [146]. No conventional purification such as chromatography or crystallization of the intermediates was necessary, only filtration was required and the precursors obtained could be used for further synthesis. In Scheme 4.36 the microwave-assisted sequences toward (+)-plicamine are highlighted.

More information on microwave-assisted PASP syntheses can be found in reviews and book chapters [111, 115].

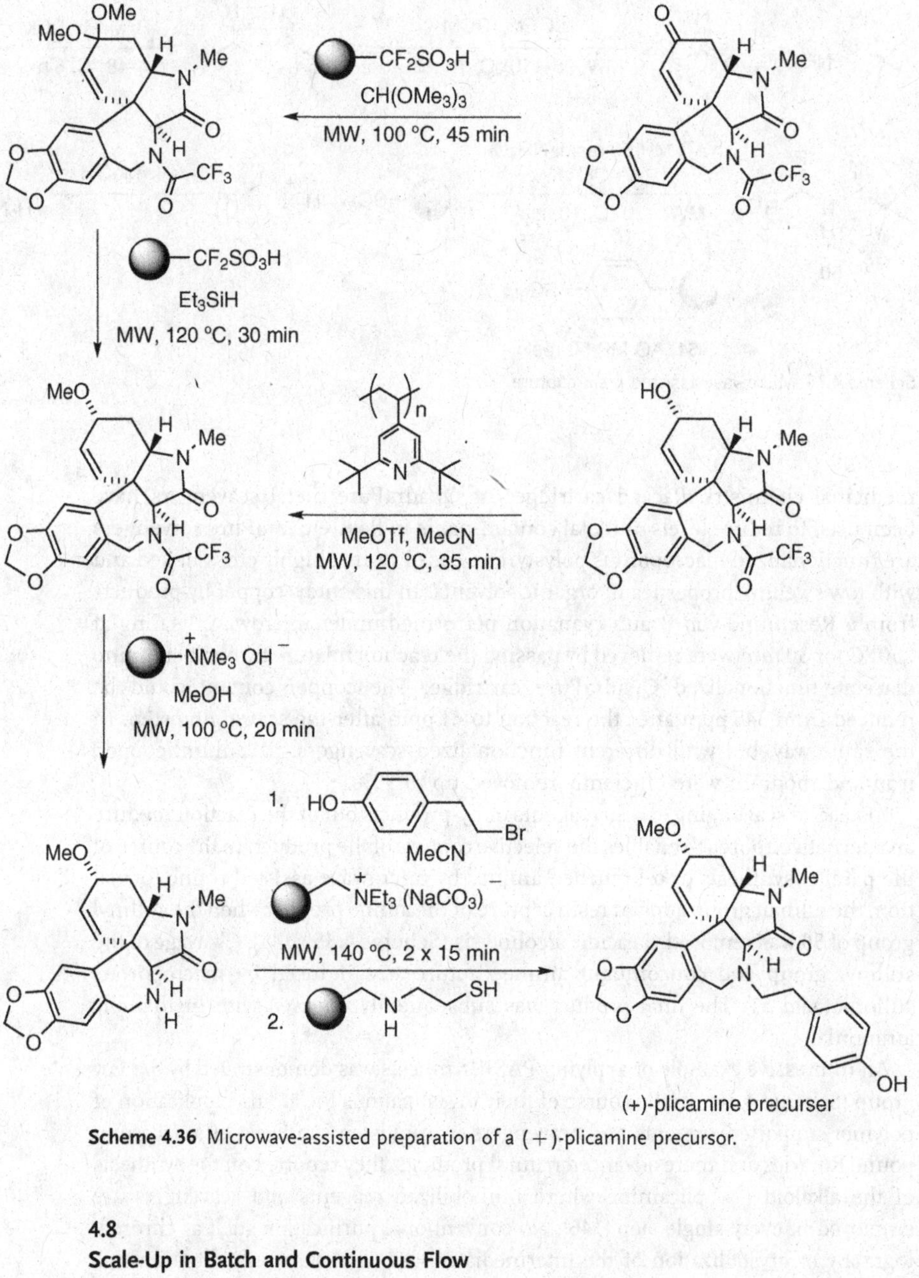

Scheme 4.36 Microwave-assisted preparation of a (+)-plicamine precursor.

4.8
Scale-Up in Batch and Continuous Flow

Most examples of microwave-assisted chemistry published to date have been performed on a less than 1 g scale with a typical reaction volume of 1–5 mL. This is in part a consequence of the availability and popularity of single-mode microwave reactors that allow the safe processing of small reaction volumes under sealed-vessel conditions by microwave irradiation (see Section 3.4). Due to limitations in the vessel and microwave cavity size of these single-mode instruments, microwave-assisted synthesis so far has focused predominantly on reaction optimization and method development on a small scale (<10 mmol). While these instruments have been very successful in this field, it is clear that for microwave-assisted synthesis to become a fully accepted technology in the future there is a need to develop larger scale MAOS techniques, that can ultimately routinely provide products on a multi-kg (or even higher) scale.

Bearing in mind some of the physical limitations of microwave heating technology, such as magnetron power or penetration depth (see Section 2.3), two different approaches for microwave synthesis on a larger scale (>100 mL volume) have emerged. While some groups have employed larger batch-type multimode reactors (≤5 L processing volume), others have used continuous flow (CF) or stop-flow (SF) techniques (multi- and single-mode) to overcome the inherent problems associated with MAOS scale-up. An additional key point in processing comparatively large volumes under pressure in a microwave field is the safety aspect, as any malfunction or rupture of a large pressurized reaction vessel may have significant consequences. In addition, solvents, reagents and products should be stable at temperatures higher than 200 °C, since instability and degradation of the reaction mixture may lead to safety problems as well [147]. The issues that have to be considered when going from small scale in single-mode instruments to a larger scale (>100 mL) in multimode reactors under closed vessels conditions are highlighted in Box 4.4.

Box 4.4
Scale-up of microwave synthesis.

Issues

- Safety
- Penetration depth
- Field homogeneity
- High microwave power required
- Heat and mass transfer – agitation
- Heating/cooling profiles
- Cavity design.

Options

- Batch reactors (multimode)
- Continuous or stop-flow (multimode or single-mode).

When heating rates under microwave conditions of small volumes that are processed in single-mode instruments are compared with those of larger volumes in multimode reactors, it becomes evident that in multimode instruments a higher microwave power has to be employed to reach identical reaction temperatures in the same time-frame. Otherwise, lower heating rates are achieved or, in some cases, the set temperatures cannot be reached when performing large scale experiments. This is true in particular when low absorbing solvents, such as toluene, are utilized. Maes and coworkers experienced this problem in the scale-up of a Buchwald–Hartwig reaction using toluene as solvent [148]. When the reaction is performed on a 1 mmol scale in a 10 mL vessel in a CEM Discover instrument with 300 W maximum power the reaction temperature of 150 °C could be reached within 2 min. However, when performing the reaction on a 20 mmol scale in an 80 mL vessel in the same instrument the set temperature of 150 °C could not be achieved within the 10 min reaction time. With the final maximum temperature being only 128 °C, incomplete conversion to product was observed (38% versus 76% isolated yield). Direct scalability, and thus full conversion, could only be achieved by switching from toluene to higher absorbing benzotrifluoride (BTF) as solvent, allowing the temperature of 150 °C to be reached within the specified ramp time. In Figure 4.20, the heating profiles of pure toluene and

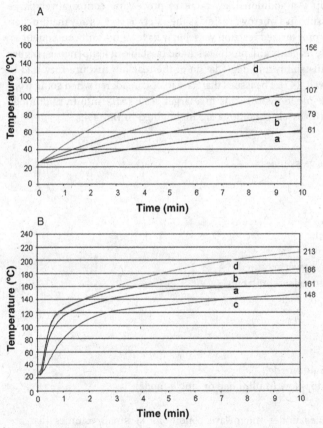

Figure 4.20 (A) Heating profile of 20 mL toluene at a constant power output for 10 min. (B) Heating profile of 20 mL BTF at 300 W power output for 10 min. The following microwave instruments were used: (a) Discover, 80 mL vessel (300 W). (b) MARS, HP-500 vessel (1200 W). (c) MicroSYNTH, high pressure vessel (1000 W). (d) MARS, Greenchem vessel (1200 W). The constant power outputs for the heating of toluene shown in the heating profile (A) are given in parentheses. Adapted from Ref. [148].

pure BTF in different microwave instruments are shown [148]. As can be seen, much higher temperatures can be reached when BTF is used as solvent.

In addition to the heating profile, a significantly different cooling profile also has to be considered as when going from small-scale single-mode to large-scale multimode instruments. In particular, when batch reactors are employed for synthesis, one has to bear in mind that a longer cooling period is necessary, thus resulting in a longer total processing time.

With today's commercially available single-mode cavities having different vessel types available, scale-up in a linear fashion is feasible from 0.05 to 50 mL (CEM Discover platform) or from 0.2 to 20 mL (Biotage Initiator EXP series, see Section 3.4). The scale-up of microwave-assisted reactions can be defined in different ranges, depending on the discipline in which the user is involved. In the case of medicinal chemistry, a scale-up to 50 mL reaction volume – corresponding to a 10- to 100-fold scale-up performed in standard single-mode microwave vials and to multigram quantities of product – is a significant amount. On the other hand, in a preparative laboratory, the synthesis of >100 g compound quantity or the use of at least 1 L reaction volume is required. Ideally, reactions should be directly scalable from small to large scale, heterogeneous mixtures should be processable and the possibility for automation (sequential or continuous) should be feasible [147].

A possibility for further scale-up using the above-mentioned single-mode instruments would be using the "numbering up" approach, where repetitive cycles of small scale runs are performed employing the automated sequential processing technique (Section 4.7.1). Alternatively, those reactions can also be conducted by parallel synthesis in multivessel rotor systems switching to multimode instruments.

Scale-up as defined for this section covers batch reactions in open vessels up to 3 L and closed vessels at the $\geq$50 mL scale, flow systems employing flow-cells $\geq$5 mL and stop-flow vessels of $\geq$50 mL volume. For an overview and more detailed information on the microwave instruments and accessories described in this section or on scale-up in MAOS, see Chapter 3 or Ref. [149], respectively. In general, one should note that published examples of MAOS scale-up experiments are rare, in particular those involving complex organic reactions.

4.8.1
Scale-Up in Batch and Parallel

An important issue for the process chemist is the potential of direct scalability of microwave reactions, allowing rapid translation of previously optimized small-scale conditions to a larger scale. Several authors have reported independently the feasibility of directly scaling reaction conditions from small-scale single-mode (typically 0.5–5 mL) to larger scale multimode batch microwave reactors (20–500 mL) without reoptimization of the reaction conditions [24, 95, 150–152].

The successful scale-up of Suzuki couplings under open-vessel conditions using low palladium concentrations (Scheme 4.37) was demonstrated by Leadbeater and coworkers [153]. In order to prepare multigram quantities of biaryls but at the same time keep a high level of safety, a switch from sealed-vessel to open-vessel conditions

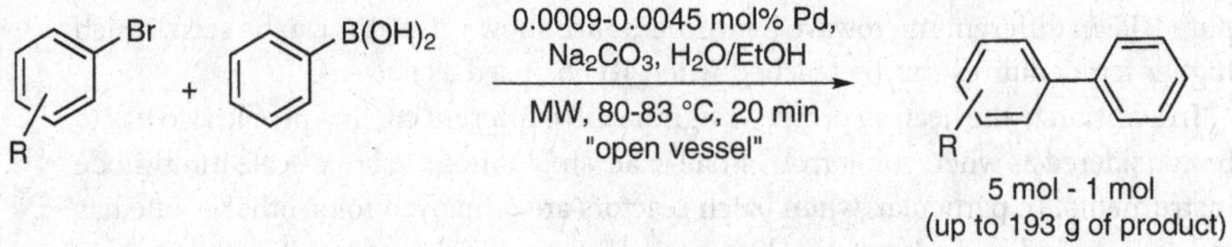

Scheme 4.37 Scale-up of Suzuki couplings under open-vessel conditions.

employing standard round-bottom flasks was performed. Direct scalability without changing any of the reaction condition parameters was possible when going from a 5 mmol (see Section 6.4.1.2) to a 1 mol scale. The small scale reactions were performed in a 100 mL round-bottom flask in a single-mode instrument (CEM Discover) whereas for the scale-up approach a 3 L reaction vessel in a multimode reactor (MARS, Figure 3.25) was employed. As discussed above, a higher microwave power (600 W initial power) is necessary to reach the reflux temperature in a similar time when larger amounts need to be heated. In addition, efficient stirring is a problem on such large scales but could be overcome by using an overhead paddle stirrer. The same authors have also reported on the up to 1000-fold scale-up from the mmol to mol region for Heck couplings [154] and *N*-heterocyclization reactions [155] under open-vessel conditions with the same instrument set-up as described above.

A comprehensive study on the scalability of optimized small-scale microwave protocols in single-mode reactors to large-scale experiments in a dedicated multimode instrument utilizing multivessel rotors (Synthos 3000, Figures 3.18 and 3.20) has been presented by the Kappe group [95]. In Scheme 4.38, the Biginelli reaction leading to dihydropyrimidines is depicted as an example (see also Section 6.4.2.1). By using the eight-position rotor, a scale-up from 4 mmol to 640 mmol (8 × 80 mmol) involving about 400 mL reaction volume was possible. Additional examples performed included Kindler thioamide synthesis, Heck and Negishi couplings, solid-phase amination and Diels–Alder cycloaddition reactions. In all cases, the yields obtained in the small scale single-mode experiments (1–4 mmol) could be reproduced on a larger scale (40–640 mmol) without the need for reoptimizing the reaction conditions. It has to be noted, however, that in many cases the rapid heating and

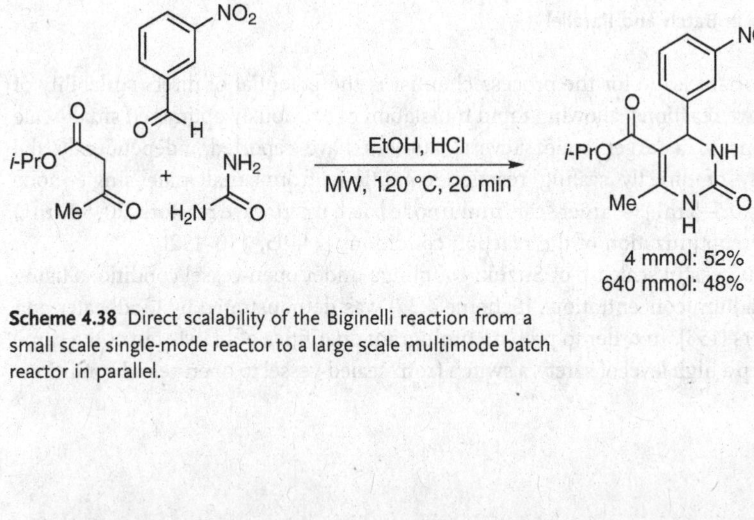

Scheme 4.38 Direct scalability of the Biginelli reaction from a small scale single-mode reactor to a large scale multimode batch reactor in parallel.

cooling profiles seen in a small scale single-mode reactor with high power density cannot be reproduced on a larger scale. Despite the somewhat longer heating and cooling period, no appreciable difference was found in the outcome of the reactions studied.

Similar scale-up results were obtained using a different multimode batch reactor in a single reaction vessel (Emrys Advancer, Figure 3.23). Mannich reactions (2 mmol → 40 mmol) [151], oxidative Heck processes (1 mmol → 10 mmol) [32], 2,5-diketopiperazine synthesis (7 mmol → 42 mmol) [156] and ketone reductions (0.5 mmol → 81 mmol) [157] could be successfully scaled to larger product quantities. Again, yields were comparable on going from a small scale single-mode reactor to a larger multimode reactor. Here, rapid cooling after the microwave heating step is possible by a patented expansion cooling process.

The fact that direct scalability is not always achievable, in particular when heterogeneous reaction mixtures are present, has been recently reported by researchers from Merck [158]. During the synthesis of Rasta resins via living free radical polymerization (LFRP) it was discovered that, when applying the same reaction conditions, previously optimized in a single-mode instrument (180 °C, 10 min), in the multimode reactor (Biotage Emrys Advancer), after 2 min the temperature spiked to above 250 °C and instead of resin beads a polymeric mass was obtained. One explanation for the spiking could be that the resin polymerized around the fiber-optic temperature probe that is used in the multimode instrument for internal temperature measurement whereas the single-mode instrument is equipped with an external IR sensor. However, when the reaction conditions are changed to 30 min at 160 °C and NMP is added as a spectator co-solvent, a 150-fold larger scale compared to single-mode experiments could be prepared with the same loading level.

4.8.2
Scale-Up Using Continuous Flow Techniques

Mainly because of safety concerns and issues related to the penetration depth of microwaves into absorbing materials such as organic solvents, the preferable option for processing volumes of >1 L under sealed vessel microwave conditions is a CF technique, although here the number of published examples using dedicated microwave reactors is limited [159, 160]. In such a system, the reaction mixture is passed through a microwave-transparent coil that is positioned in the cavity of a single- or multimode microwave reactor. The previously optimized reaction time under batch microwave conditions now needs to be related to a "residence time" (the time for which the sample stays in the microwave-heated coil) at a specific flow rate. While the early pioneering work in this area stems from the group of Strauss [161], others have since made notable contributions to this field, often utilizing custom-built microwave reactors or modified domestic microwave units. More information on these systems and on MAOS employing CF techniques in general can be found in a 2007 review [159].

Recently published examples of CF organic microwave synthesis involve, for example, Biginelli reactions and Dimroth rearrangements in the CEM Voyager$_{CF}$

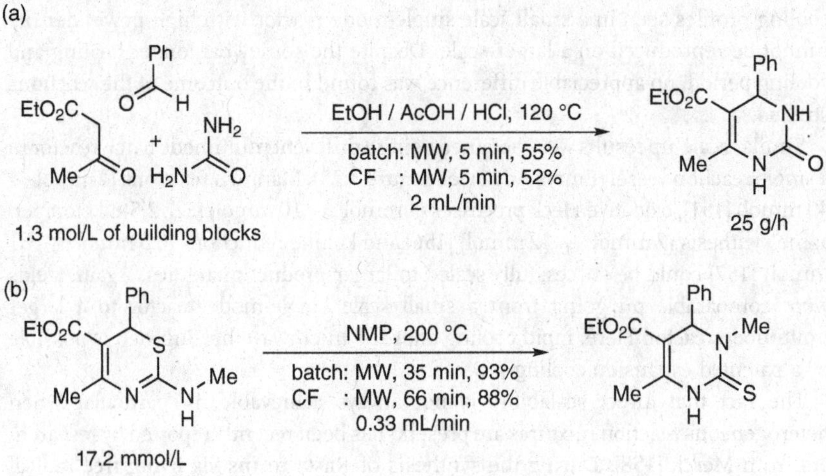

(a)

1.3 mol/L of building blocks

EtOH / AcOH / HCl, 120 °C

batch: MW, 5 min, 55%
CF : MW, 5 min, 52%
2 mL/min

25 g/h

(b)

17.2 mmol/L

NMP, 200 °C

batch: MW, 35 min, 93%
CF : MW, 66 min, 88%
0.33 mL/min

Scheme 4.39 (a) Biginelli reaction and (b) Dimroth rearrangement under CF conditions.

system (see Figure 3.15) [162]. For these flow experiments the 10 mL flow cell was charged with 2 mm-sized glass beads (Figure 3.15b) in order to create microchannels that increase the residence time of the reaction mixture in the microwave heating zone. The reaction mixture was introduced into the flow cell at the bottom of the vial via a Teflon tube using standard HPLC pumps, and the reaction pressure was controlled by a back-pressure regulator connected at the end of the outlet tube. The described set-up was first evaluated with the well-known Biginelli reaction that was carried out at an adjusted flow-rate of $2 \, mL \, min^{-1}$, resulting in an identical residence time compared with the batch experiment (Scheme 4.39a). In contrast, the Dimroth rearrangement was performed with a flow-rate leading to an almost doubled residence time of the substrate within the microwave heating zone compared to the batch attempt (Scheme 4.39b). Importantly, for both reactions the isolated yields were nicely comparable with the corresponding batch results.

In a related study, the group of Bagley has developed a CF microwave flow cell – which works on the same principle as the reactor described above – for use in the CEM Voyager$_{CF}$ unit [163]. The flow cell consists of a standard 10 mL glass tube fitted with a custom-built steel head filled with sand (about 12 g) between two drilled frits to minimize dispersion and create a lattice of microchannels (Figure 4.21). The same set-up as described above was utilized to perform microwave-heated reactions under flow. To evaluate the performance of the custom-built flow cell, two reactions, well-known under microwave conditions, were conducted: the hydrolysis of chloromethyl thiazole and the Fischer indole synthesis of **53** from hydrazine **52** and cyclohexanone in acetic acid (Scheme 4.40). Both syntheses were performed on a gram scale at 150 °C. In addition, comparison studies of the newly designed CF cell for the Bohlmann–Rahtz synthesis of pyridines with batch experiments in a sealed-vessel and with a Teflon heating coil were carried out. The study revealed, that under conditions that produced the same yields, a higher processing rate ($mmol \, min^{-1}$) is

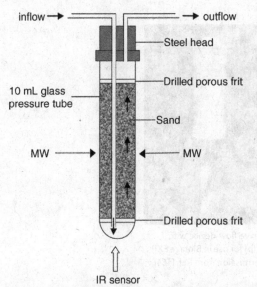

Figure 4.21 Schematic diagram of a sand-filled flow cell employed in the Voyager_CF. Reproduced with permission from Ref. [163].

possible using the sand-filled reactor since a faster flow can be applied. In principle, by immobilizing a catalyst on the support in the glass tube, transformation involving heterogeneous catalysis would be feasible.

The Fischer indole synthesis shown in Scheme 4.40 can also be performed in the commercially available 10 mL flow cell filled with glass beads (Figure 3.15b), as is demonstrated in Section 6.4.3.1.

Another type of custom-built flow microwave device was introduced by the Ley group in 2007 [164]. The design consists of fluorinated polymer tubings wound around a Teflon core that is fitted with a dummy pressure cap. This flow device exhibits the basic shape of a 20 mL vial (Figure 3.8) suitable for the Biotage EXP single-mode instruments with the input and exit tubes, which are connected to HPLC pumps and a 7 bar back-pressure regulator, on the bottom of the microwave unit (Figure 4.22). In addition, with this system purification can be facilitated by passing the exiting flow stream through columns packed with polymer-supported reagents or scavengers. The synthesis of 5-amino-4-cyanopyrazoles **54** via reaction of a set of hydrazines with ethoxymethylene malononitrile in methanol was performed at

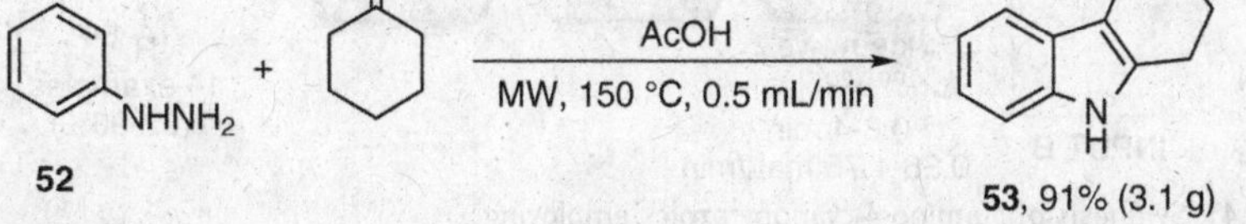

52

AcOH

MW, 150 °C, 0.5 mL/min

53, 91% (3.1 g)

Scheme 4.40 Fischer indole synthesis performed in a CF cell filled with sand (see Figure 4.21).

(a)   (b)

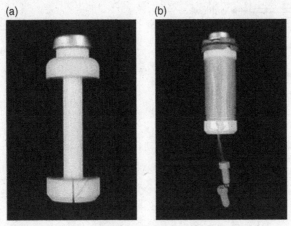

Figure 4.22 Teflon core (a) and microwave flow device with tubings wrapped around the Teflon core (b) for use in Biotage EXP microwave reactors. Reproduced with permission from Ref. [164].

100–120 °C with a residence time of 0.8–4 min (Scheme 4.41). Subsequent passing of the reaction mixture through a column with supported benzylamine to scavenge unreacted malononitrile, followed by a column with activated carbon to remove colored impurities furnished products **54** in high purities, good to excellent yields and in quantities up to 250 g (Scheme 4.41). A benefit of this flow device is its versatility, since different tubing lengths can be wrapped providing reactors with different internal volumes. The application of multiple tubings in fact allows different reactions within one single reactor.

Current single-mode continuous flow microwave reactors only allow processing of comparatively small volumes. Much larger volumes can be processed in CF reactors that are housed inside a multimode microwave system. In a 2001 publication, Shieh and coworkers described the methylation of phenols, indoles and benzimidazoles with dimethyl carbonate under CF microwave conditions, using a Milestone ETHOS-CFR reactor [165]. The same authors also reported the usefulness of this general method for the esterification of carboxylic acids (up to 100 g product within 20 min) [166]. Similar results were also achieved for benzylations employing dibenzyl carbonate using the same CF instrument [167].

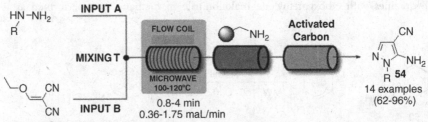

Scheme 4.41 Synthesis of 5-amino-4-cyanopyrazoles employing the microwave flow device shown in Figure 4.22.

In a recent study published in 2007 [168], Moseley and Lawton have reported preliminary results using the FlowSYNTH continuous flow reactor (an improved version of the Milestone ETHOS-CFR reactor, see Figure 3.40a). For their investigations, the Newman–Kwart rearrangement was chosen as model reaction since the authors have previously performed extensive studies under microwave batch conditions [80, 169]. Moreover, this reaction is completely homogeneous, an important issue in CF synthesis regarding line clogging by solids or slurries. Newman–Kwart rearrangements were performed at 200 °C, which is the limit of the instrument, at a flow rate of 2.1 L h^{-1} that would correspond to a reaction time of 10 min under batch conditions (Scheme 4.42). However, the actual residence time was slightly shorter – about 6 min – resulting in reduced conversions, but this problem could be solved by simply passing partially converted reaction mixtures several times through the reactor, thus enabling longer reaction times. Employing the above mentioned flow-rate, 200 g of S-thiocarbamate product per hour could be generated [168]. In addition, with the incorporated product cooler, efficient cooling was provided, hence a heating and cooling profile similar to small-scale runs in single-mode microwave instruments could be achieved. The reported results using microwave flow processing were in good agreement in terms of reaction times, yield and purity with previously reported batch microwave data for the Newman–Kwart rearrangement [168].

The scale-up of polymerization reactions under CF conditions is not a trivial affair since, for the synthesis of well-defined polymers, not only a homogeneous heating profile is required, but also a homogeneous concentration profile through the entire polymerization mixture to ensure narrow molecular weight distribution. In addition, the higher viscosities that are present might lead to undesired concentration profiles under microwave CF conditions. In order to investigate these issues, the Schubert group performed the previously microwave batch tested cationic ring-opening polymerization of 2-ethyl-2-oxazoline in the CEM Voyager$_{CF}$ – using a 10 mL flow cell, a 5 mL Teflon coil and a 10 mL glass coil (see Figure 3.15) – and in the Milestone FlowSYNTH [170]. With the same optimized conditions as in the batch experiment – 4 M monomer concentration in MeCN at 140 °C, a monomer-to-initiator ratio of about 100 and a flow rate that corresponds to 1000 s residence time – the polymerization efficiency was examined. The reactions in batch and the two coil reactors reached full conversion, whereas the flow cell and the FlowSYNTH only showed 80 and 60% conversion, respectively. The authors found that the polymerization results regarding molecular weight distributions and PDI (polydispersity indices) were strongly dependent on the flow profile (laminar versus tubular flow). In general, all the CF polymerizations resulted in broader molecular weight distributions, the coils

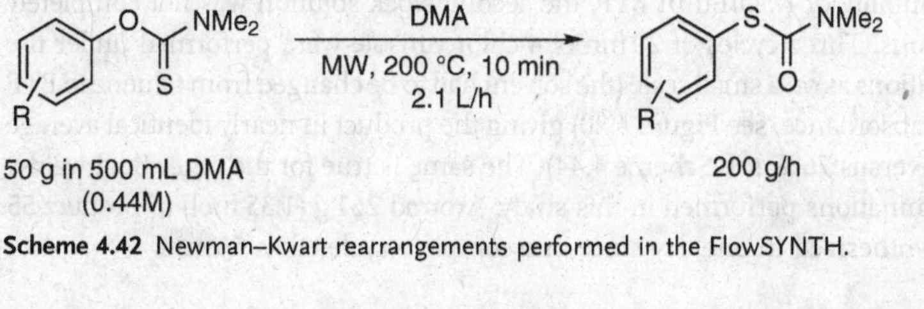

50 g in 500 mL DMA (0.44M)	DMA / MW, 200 °C, 10 min / 2.1 L/h	200 g/h

Scheme 4.42 Newman–Kwart rearrangements performed in the FlowSYNTH.

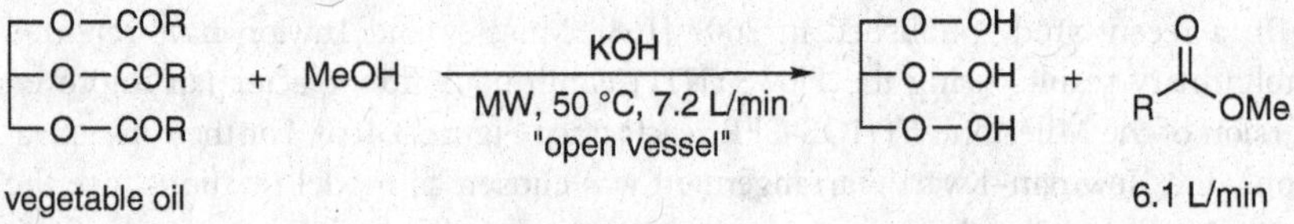

$$\left[\begin{array}{l} O-COR \\ O-COR \\ O-COR \end{array}\right. + MeOH \xrightarrow[\substack{MW,\ 50\ ^\circ C,\ 7.2\ L/min \\ \text{"open vessel"}}]{KOH} \left[\begin{array}{l} O-OH \\ O-OH \\ O-OH \end{array}\right. + \underset{R}{\overset{O}{\|}}OMe$$

vegetable oil   6.1 L/min

Scheme 4.43 Open-vessel CF preparation of biodiesel using a 4 L flow cell in the CEM MARS.

giving better results than the other two systems, which is probably a result of the residence time distributions in the flow reactors.

The reactions reported so far under continuous flow conditions have been conducted in closed systems under pressure using back-pressure regulators. In contrast, the continuous flow synthesis of biodiesel under open-vessel conditions employing the CEM MARS in conjunction with the flow cell accessories (Figure 3.26) was demonstrated by Leadbeater and coworkers [171]. By using the 4 L flow vessel, a maximum flow rate of 7.2 L min^{-1} could be achieved, producing 6.1 L of biodiesel per minute (99% conversion, see Scheme 4.43). The same reaction conditions as for batch synthesis in the same microwave unit could be applied – a 1 : 6 molar ratio of oil/methanol and 1 wt% KOH were heated to 50 °C and held there for 1 min, followed by pumping new material into the flow cell.

4.8.3
Scale-Up Using Stop Flow Techniques

As mentioned above in Section 4.8.2, a serious problem with continuous flow reactors is the clogging of the lines and the difficulties in processing heterogeneous mixtures. Since many organic transformations involve some form of insoluble reagent or catalyst, single-mode so-called stop-flow microwave reactors have been developed (CEM Voyager$_{SF}$, Figure 3.14), in which peristaltic pumps – capable of pumping slurries and even solid reagents – are used to fill a batch reaction vessel (80 mL) with the reaction mixture. After microwave processing in batch, the product mixture is pumped out of the system which is then ready to receive the next batch of the reaction mixture.

In the course of investigations toward the scale-up of palladium-catalyzed Buchwald–Hartwig aminations of aryl chlorides using dedicated multimode and single-mode instruments, Maes and coworkers have employed the CEM Voyager$_{SF}$ for the batchwise scale-up in the coupling of 4-chloroanisole with morpholine [148]. Two different stock solutions were prepared, one containing the palladium catalyst/ligand system in BTF and the other one the aryl chloride, amine and base, also in BTF. Due to the low solubility of NaOtBu in BTF, the second stock solution was not completely homogeneous. Three cycles of 20 mmol 4-chloroanisole were performed under the same conditions as on a small scale (the solvent had to be changed from toluene to BTF due to low absorbance, see Figure 4.20) giving the product in nearly identical average yield (78% versus 76%, see Scheme 4.44). The same is true for the other Buchwald–Hartwig aminations performed in this study. Around 261 g (1.35 mol) of product **55** could be synthesized in one day since one complete cycle takes 16 min.

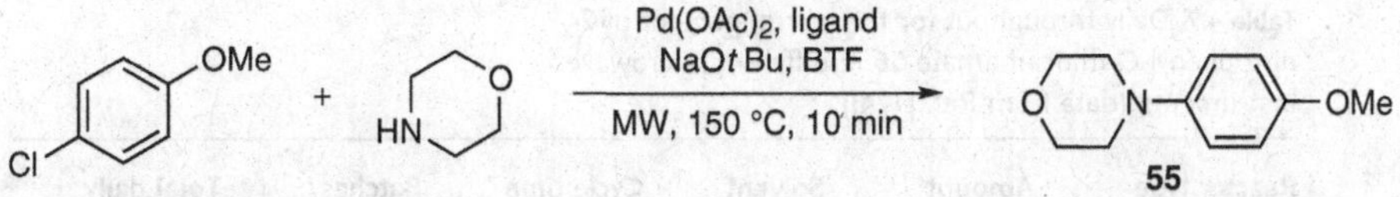

Scheme 4.44 Buchwald–Hartwig aminations under stop-flow conditions.

The same instrument was employed by Leadbeater and coworkers for the scale-up of Suzuki and Heck couplings in water using ultra-low palladium concentrations [172]. When going from a 1 to a 10 mmol scale in the Suzuki reaction of 4-bromoacetophenone and phenylboronic acid using 250 pbb palladium at 150 °C, pumping the reaction mixture out of the 80 mL vessel proved to be problematic since the biaryl product precipitated below 90 °C and blocked the exit tube. This problem could be solved by programming an additional step in the protocol: after cooling to 110 °C, the vessel is vented and ethyl acetate is added to dissolve the biaryl product. A 95% average yield corresponding to 18.6 g of biaryl product over 10 cycles of 10 mmol each – with 15 min processing time per cycle – was obtained. The reaction conditions for the Heck coupling of 4-bromoanisole and styrene at 170 °C had to be slightly modified as well. Here, a small amount of DMF had to be added to the reaction mixture in water to guarantee efficient pumping in and out of the reaction vessel.

The palladium-catalyzed ($Pd_2(dba)_3$/Xantphos) cyanation with $Zn(CN)_2$ as final step in the synthesis of citalopram, an antidepressant drug, could be scaled up under SF conditions, as was demonstrated by Pitts and coworkers [173]. The reaction was performed at 160 °C and a cycle time of about 10 min per batch (200 s hold time under microwave irradiation). After 4 cycles and a total run time of 40 min, 47 g of citalopram was obtained.

A comparison study of seven commercially available microwave reactors for scale-up in terms of process chemistry has been conducted by Moseley and coworkers [174]. The scale-up of Newman–Kwart rearrangements (see Scheme 4.42) was evaluated in single batch (Biotage Advancer, Figure 3.23; Milestone UltraCLAVE, Figure 3.41; CEM MARS, Figure 3.25; Milestone MicroSYNTH, Figure 3.37), multibatch (Anton Paar Synthos 3000, Figure 3.20), stop-flow (CEM Voyager$_{SF}$, Figure 3.14) and continuous flow (Milestone FlowSYNTH, Figure 3.40a) reactors under both open- and closed-vessel conditions. The authors reported a linear and reliable scalability from small scale experiments to larger scales (from 2 mL to >1 L) accomplished by each large-scale microwave reactor. In addition, a comparison of daily throughputs for the rearrangement of 2-nitrophenyl-*O*-thiocarbamate performed in the large-scale microwave instruments indicated in Table 4.7 was conducted. As highlighted in the table, up to 4 kg of product per day could be produced in the FlowSYNTH, but one should bear in mind that those data refer to a highly concentrated, homogeneous reaction. At present, no commercially available scale-up microwave reactor is capable of performing the majority of reactions important for the pharmaceutical industry on a >1 kg scale. For those kinds of reactions, scale-up to several 100 g of product is feasible, useful for the

Table 4.7 Daily throughput for the rearrangement of 2-nitrophenyl-O-thiocarbamate **56** in different microwave instruments (data from Ref. [174]).

Reactor type	Amount of 56 (g)	Solvent (mL)	Cycle time (min)	Batches/ day	Total daily throughput (kg)
Advancer	60	240	30[b]	16	0.96[a]
UltraCLAVE	400	800	100[b]	5	1.92
MARS	500	2000	96[c]	5	2.5
Synthos 3000	200	800	45[c]	10	2.1
Voyager$_{SF}$	10	40	16	30	0.3
FlowSYNTH	500 g h^{-1}	2000	2.1 L h^{-1}	Cont.	4.0

[a]Assumes automated.
[b]Good estimate.
[c]Extrapolated from similar reaction conditions.

top end of medicinal chemistry scale-up, but for genuine process chemistry further improvements with regard to reaction volume and automation are desirable [147, 174].

Similar comparison studies have also been reported by Leadbeater [175] and Lehmann [147] for evaluation of the scale-up of diverse reactions under both batch and continuous flow processing conditions.

4.8.4
Flow-Through Techniques using Microreactors

The concept of performing reactions in a flow format using micro-channels (microreactors, lab-on-a-chip) is a recent addition to the powerful toolbox of high-throughput synthesis technologies [176]. In general, flow-through micro-scale reaction systems have used a sequential-flow approach for the preparation of libraries, whereby compounds are produced, one after another, through the same reactor channel [176]. Virtually all libraries prepared by this approach have been synthesized at room temperature using glass microchips, because existing microfluidic technology does not fully address the issue of applying, controlling and monitoring heating in these reaction systems, which limits its application [176]. Notably, there have been some attempts to apply microwave heating to microreactor technology.

One example of employing microwave heating in combination with the use of microreactors has been described by the group of Haswell [177]. They used micro-wave energy to deliver heat locally to a heterogeneous palladium-supported catalyst (Pd/Al$_2$O$_3$, catalyst channel: $1.5 \times 0.08 \times 15$ mm) located within a microreactor device. A 10–15 nm gold film patch, located on the outside surface of the base of a glass microreactor, was found to efficiently assist in the heating of the catalyst, allowing Suzuki cross-coupling reactions to proceed very effectively [177].

Another custom-built microreactor device was developed by the Organ group [178]. They described the design of a capillary-based (internal diameter: 200–1200 μm) flow system for performing microscale organic synthesis under microwave irradiation

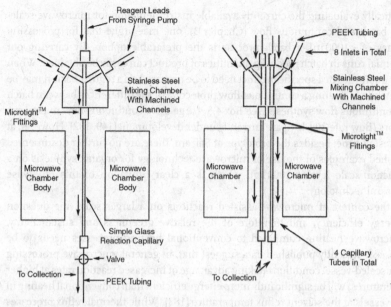

Figure 4.23 Single capillary reactor (a) and multicapillary reactor (b) for microwave-assisted CF organic synthesis. Reproduced with permission from Refs. [178, 179].

(Biotage unit). By using this microreactor device the reactants can be co-injected from separate syringes, mix and react without showing problems of poor kinetics due to laminar flow (Figure 4.23). In addition to the initially developed single capillary reactor [178], an extension to a multireactor where eight reactions can be conducted in parallel was subsequently reported (Figure 4.23b) [179]. The reagents flow in sequence into the multicapillary reactor device, allowing library preparation and combinatorial syntheses where the number of products is not limited to the number of reaction vessels. The capillaries can be additionally coated with a thin film of palladium allowing Suzuki and Heck reactions to be performed, without the addition of any extra catalyst, in very short reaction times – the residence time being less than 1 min [141]. Several metal catalyst-employing reactions such as Suzuki, Heck or ring-closing metathesis, nucleophilic aromatic and Wittig reactions, as well as multicomponent reactions [180] showed excellent conversions in extremely short reaction times (down to several seconds).

In contrast to the combined microreactor/microwave technology for performing reactions in the mg-range, the largest microwave reactor for organic synthesis to date is a pilot plant scale prototype installed at Sairem in France, developed and designed in collaboration with BioEurope and De Dietrich. This custom-built 1 m³ reactor with a powerful 6 kW microwave generator was used for the production of Laury-done [181]. Running in a batch-type recycling process, the equipment accomplished a 40% power reduction compared to the classical thermal approach. Moreover, the overall processing time could be reduced by 80%.

Critically evaluating the currently available instrumentation for microwave scale-up in batch and continuous flow (Chapter 3), one may argue that for processing volumes of <1000 mL a batch process is the preferable option. By carrying out sequential runs in batch mode, kg quantities of product can easily be obtained. When larger quantities of a specific product need to be prepared on a regular basis it may be worthwhile evaluating a continuous-flow protocol. For the differences between batch and continuous flow synthesis, see Box 4.5. Large-scale continuous-flow microwave reactors (flow rate 20 L h^{-1}) are currently under development [149, 182]. However, at the present time, besides the prototype at Sairem, there are no further documented published examples of the use of microwave technology for organic synthesis on a production-scale level (>1000 kg), which is a clear limitation of this otherwise successful technology.

In the context of microwave-assisted reactions on a larger scale, the question of energy efficiency, and therefore of the relative greenness and sustainability, of microwave heating compared to conventional heating processes needs to be addressed. Recently published data suggest that, in general, microwave processing under sealed-vessel conditions (taking advantage of increased reaction rates at higher temperatures) will be significantly more energy efficient than conventional heating in open vessels at the solvent reflux temperature [183]. While thermal reflux processes often require several hours to reach completion, the same transformations can sometimes be completed within a few minutes using sealed-vessel microwave heating. It is important to note that the energy savings in these cases are mainly

---

Box 4.5

Comparison of the advantages and disadvantages of batch and continuous flow synthesis.

Batch synthesis	Continuous flow
♀ Safety issues	♂ Comparatively safe smaller volumes in hot zone
♀ Limited in scale-up power constraints penetration depth	♂ Higher energy transmission
♂ Same processing time as on small scale	♀ Longer processing times
♂ Heterogeneous mixtures	♀ Heterogeneous mixtures or precipitating products troublesome

---

the result of the significantly shortened reaction times and are not directly connected to the heating mode. This has been made evident by comparing the consumed energy from open-vessel microwave heating experiments with the corresponding data from conventional thermal reflux heating runs, lasting for the same period of time [183]. In all the investigated cases the thermal runs were more energy efficient than the microwave experiments, independent of the scale, the absorbance characteristics of the medium or the particular microwave instrument used (single- or multimode). This can be rationalized by considering the moderate energy efficiency (50–65%) of a magnetron (see Section 3.3), the central component of any microwave reactor transforming electrical energy into electromagnetic irradiation [149].

A full energy balance between microwave and conventional heating would also have to take into account losses through the glassware, the oil bath and the condenser. In order to evaluate the overall ecobalance of a particular process, the consumed energy for pre- and post-treatment of the reaction mixture should also be considered. However, the widespread general opinion on the relative "greenness" of microwave heating in chemical processing – at least in terms of energy efficiency – needs to be critically questioned.

References

1 Loupy, A., Petit, A., Hamelin, J., Texier-Boullet, F., Jacquault, P. and Mathé, D. (1998) *Synthesis*, 1213–1234; Varma, R.S. (1999) *Green Chemistry*, **1**, 43–55; Kidawi, M. (2001) *Pure and Applied Chemistry*, **73**, 147–151; Varma, R.S. (2001) *Pure and Applied Chemistry*, **73**, 193–198; Varma, R.S. (2002) *Tetrahedron*, **58**, 1235–1255; Varma, R.S. (2002) *Advances in Green Chemistry: Chemical Syntheses Using Microwave Irradiation*, Kavitha Printers, Bangalore.

2 Kappe, C.O. (2004) *Angewandte Chemie-International Edition*, **43**, 6250–6284; Kappe, C.O. and Stadler, A. (2005) *Microwaves in Organic and Medicinal Chemistry*, Wiley-VCH, Weinheim.

3 Cave, G.W.V., Raston, C.L. and Scott, J.L. (2001) *Chemical Communications*, 2159–2169.

4 Strauss, C.R. and Varma, R.S. (2006) *Topics in Current Chemistry*, **266**, 199–231.

5 Nüchter, M., Müller, U., Ondruschka, B., Tied, A. and Lautenschläger, W. (2003) *Chemical Engineering & Technology*, **26**, 1207–1216.

6 Azizian, J., Karimi, A.R., Kazemizadeh, Z., Mohammadi, A.A. and Mohammadizadeh, M.R. (2005) *The Journal of Organic Chemistry*, **70**, 1471–1473.

7 Azizian, J., Karimi, A.R., Kazemizadeh, Z., Mohammadi, A.A. and Mohammadizadeh, M.R. (2005) *Tetrahedron Letters*, **46**, 6155–6157.

8 Varma, R.S. and Dahiya, R. (1997) *Tetrahedron Letters*, **38**, 2043–2044.

9 Varma, R.S., Saini, R.K. and Dahiya, R. (1997) *Tetrahedron Letters*, **38**, 7823–7824.

10 Varma, R.S. and Dahiya, R. (1998) *Tetrahedron Letters*, **39**, 1307–1308.

11 Varma, R.S. and Saini, R.K. (1998) *Tetrahedron Letters*, **39**, 1481–1482.

12 Kabalka, G.W., Wang, L., Namboodiri, V.N. and Pagni, R.M. (2000) *Tetrahedron Letters*, **41**, 5151–5154.

13 Varma, R.S. and Ju, Y. (2006) in *Microwaves in Organic Synthesis*, 2nd edn (ed. A. Loupy), Wiley-VCH, Weinheim, pp. 362–415 (Chapter 8).

14 Besson, T., Thiéry, V. and Dubac, J. (2006) in *Microwaves in Organic Synthesis*, 2nd edn (ed. A. Loupy), Wiley-VCH, Weinheim, pp. 416–455 (Chapter 9).

15 de Fatima Pereira, M., Picot, L., Guillon, J., Léger, J.-M., Jarry, C., Thiéry, V. and Besson, T. (2005) *Tetrahedron Letters*, **46**, 3445–3447.

16 Deshayes, S., Liagre, M., Loupy, A., Luche, J.-L. and Petit, A. (1999) *Tetrahedron*, **55**, 10851–10870; Loupy, A., Petit, A. and

Bogdal, D. (2006) in *Microwaves in Organic Synthesis*, 2nd edn (ed. A. Loupy), Wiley-VCH, Weinheim, pp. 278–326 (Chapter 6).

17 Li, Z., Quan, Z.J. and Wang, X.C. (2004) *Chemical Papers*, 58, 256–259.

18 Bogdal, D., Bednarz, S. and Lukasiewicz, M. (2006) *Tetrahedron*, 62, 9440–9445.

19 Herrero, M.A., Kremsner, J.M. and Kappe, C.O. (2008) *The Journal of Organic Chemistry*, 73, 36–47.

20 Bose, A.K., Banik, B.K., Lavlinskaia, N., Jayaraman, M. and Manhas, M.S. (1997) *Chemtech*, 27, 18–24; Bose, A.K., Manhas, M.S., Ganguly, S.N., Sharma, A.H. and Banik, B.K. (2002) *Synthesis*, 1578–1591.

21 Kappe, C.O. and Stadler, A. (2005) *Microwaves in Organic and Medicinal Chemistry*, Wiley-VCH, Weinheim, pp. 107–391 (Chapters 6 and 7).

22 Strohmeier, G.A. and Kappe, C.O. (2002) *Journal of Combinatorial Chemistry*, 4, 154–161.

23 Razzaq, T. and Kappe, C.O. (2007) *Tetrahedron Letters*, 48, 2513–2517.

24 Stadler, A., Pichler, S., Horeis, G. and Kappe, C.O. (2002) *Tetrahedron*, 58, 3177–3183.

25 Lange, J.H.M., Verveer, P.C., Osnabrug, S.J.M. and Visser, G.M. (2001) *Tetrahedron Letters*, 42, 1367–1369.

26 Amore, K.M. and Leadbeater, N.E. (2007) *Macromolecular Rapid Communications*, 28, 473–477.

27 Vo-Tanh, G., Lahrache, H., Loupy, A., Kim, I.-J., Chang, D.-H. and Jun, C.-H. (2004) *Tetrahedron*, 60, 5539–5543.

28 Nosse, B., Schall, A., Jeong, W.B. and Reiser, O. (2005) *Advances in Synthesis and Catalysis*, 347, 1869–1874.

29 Kim, Y.J. and Varma, R.S. (2004) *Tetrahedron Letters*, 45, 7205–7208.

30 Petricci, E. and Taddei, M. (2007) *Chimica Oggi-Chemistry Today*, 25, 40–45.

31 Will, H., Scholz, P. and Ondruschka, B. (2004) *Topics in Catalysis*, 29, 175–182; Will, H., Scholz, P. and Ondruschka, B. (2004) *Chemical Engineering & Technology*, 27, 113–122; Will, H., Scholz, P., Ondruschka, B. and Burckhardt, W.

(2003) *Chemical Engineering & Technology*, 26, 1146–1149; Will, H., Scholz, P. and Ondruschka, B. (2002) *Chemie Ingenieur Technik*, 74, 1057–1067; Zhang, X., Lee, C.S.-M., Mingos, D.M.P. and Hayward, D.O. (2003) *Catalysis Letters*, 88, 129–139; Zhang, X., Hayward, D.O. and Mingos, D.M.P. (2003) *Catalysis Letters*, 88, 33–38.

32 Miljanić, O.Š., Vollhardt, K.P.C. and Whitener, G.D. (2003) *Synlett*, 29–34; Andappan, M.M.S., Nilsson, P., von Schenck, H. and Larhed, M. (2004) *The Journal of Organic Chemistry*, 69, 5212–5218.

33 Van der Eycken, E., Appukkuttan, P., De Borggraeve, W., Dehaen, W., Dallinger, D. and Kappe, C.O. (2002) *The Journal of Organic Chemistry*, 67, 7904–7909.

34 Kaval, N., Dehaen, W., Kappe, C.O. and Van der Eycken, E. (2004) *Organic and Biomolecular Chemistry*, 2, 154–156.

35 Kormos, C.M. and Leadbeater, N.E. (2006) *Synlett*, 1663–1666.

36 Kormos, C.M. and Leadbeater, N.E. (2007) *Organic and Biomolecular Chemistry*, 5, 65–68.

37 Kormos, C.M. and Leadbeater, N.E. (2007) *Synlett*, 2006–2010.

38 Petricci, E., Mann, A., Schoenfelder, A., Rota, A. and Taddei, M. (2006) *Organic Letters*, 8 3725–3727; Piras, L., Genesio, E., Ghiron, C. and Taddei, M. (2008) *Synlett*, 1125–1128.

39 Vanier, G.S. (2007) *Synlett*, 131–135.

40 Heller, E., Lautenschläger, W. and Holzgrabe, U. (2005) *Tetrahedron Letters*, 46, 1247–1249.

41 Kaiser, N.-F.K., Hallberg, A. and Larhed, M. (2002) *Journal of Combinatorial Chemistry*, 4, 109–111; Georgsson, J., Hallberg, A. and Larhed, M. (2003) *Journal of Combinatorial Chemistry*, 5, 350–352; Wannberg, J. and Larhed, M. (2003) *The Journal of Organic Chemistry*, 68, 5750–5753; Herrero, M.A., Wannberg, J. and Larhed, M. (2004) *Synlett*, 2335–2338; Wu, X., Roenn, R., Gossas, T. and Larhed, M. (2005) *The*

Journal of Organic Chemistry, **70**, 3094–3098; Wannberg, J., Dallinger, D., Kappe, C.O. and Larhed, M. (2005) *Journal of Combinatorial Chemistry*, **7**, 574–583; Wannberg, J., Kaiser, N.-F.K., Vrang, L., Samuelsson, B., Larhed, M. and Hallberg, A. (2005) *Journal of Combinatorial Chemistry*, **7**, 611–617; Wu, X., Wannberg, J. and Larhed, M. (2006) *Tetrahedron*, **62**, 4665–4670.

42 Wu, X., Ekegren, J.K. and Larhed, M. (2006) *Organometallics*, **25**, 1434–1439.

43 Stiasni, N. and Kappe, C.O. (2002) *ARKIVOC*, **viii**, 71–79.

44 Li, C.-J. and Chan, T.-H. (1997) *Organic Reactions in Aqueous Media*, Wiley, New York; Lindström, U.M.(ed.) (2007) *Organic Reactions in Water*, Blackwell Publishing, Oxford; Lindström, U.M. (2002) *Chemical Reviews*, **102**, 2751–2772; Li, C.-J. (2005) *Chemical Reviews*, **105**, 3095–3166; Li, C.-J. and Chen, L. (2006) *Chemical Society Reviews*, **35**, 68–82; Breslow, R. (2004) *Accounts of Chemical Research*, **37**, 471–478; Pirrung, M.C. (2006) *Chemistry – A European Journal*, **12**, 1312–1317.

45 Blokzijl, W. and Engberts, J.B.F.N. (1993) *Angewandte Chemie (International Edition in English)*, **32**, 1545–1579; Widom, B., Bhimalapuram, P. and Koga, K. (2003) *Physical Chemistry Chemical Physics*, **5**, 3085–3093.

46 Krammer, P. and Vogel, H. (2000) *Journal of Supercritical Fluids*, **16**, 189–206; Krammer, P., Mittelstädt, S. and Vogel, H. (1999) *Chemical Engineering & Technology*, **22**, 126–130.

47 Bröll, D., Kaul, C., Krämer, A., Krammer, P., Richter, T., Jung, M., Vogel, H. and Zehner, P. (1999) *Angewandte Chemie*, **111**, 3180–3196; Savage, P.E. (1999) *Chemical Reviews*, **99**, 603–621; Siskin, M. and Katritzky, A.R. (2001) *Chemical Reviews*, **101**, 825–836; Katritzky, A.R., Nichols, D.A., Siskin, M., Murugan, R. and Balasubramanian, M. (2001) *Chemical Reviews*, **101**, 837–892; Akiya, N. and Savage, P.E. (2002) *Chemical Reviews*, **102**, 2725–2750; Watanabe, M., Sato, T., Inomata, H., Smith, R.L., Jr, Arai, K., Kruse, A. and Dinjus, E. (2004) *Chemical Reviews*, **104**, 5803–5822; Weingärtner, H. and Franck, E.U. (2005) *Angewandte Chemie-International Edition*, **44**, 2672–2692.

48 Kremsner, J.M. and Kappe, C.O. (2005) *European Journal of Organic Chemistry*, 3672–3679.

49 Dallinger, D. and Kappe, C.O. (2007) *Chemical Reviews*, **107**, 2563–2591.

50 Strauss, C.R. and Trainor, R.W. (1995) *Australian Journal of Chemistry*, **48**, 1665–1692.

51 Strauss, C.R. (1999) *Australian Journal of Chemistry*, **52**, 83–96.

52 Roberts, B.A. and Strauss, C.R. (2005) *Accounts of Chemical Research*, **38**, 653–661.

53 Kormos, C.M. and Leadbeater, N.E. (2006) *Tetrahedron*, **62**, 4728–4732.

54 Geuens, J., Kremsner, J.M., Nebel, B.A., Schober, S., Dommisse, R.A., Mittelbach, M., Tavernier, S., Kappe, C.O. and Maes, B.U.W. (2008) *Energy Fuels*, **22**, 643–645.

55 Molteni, V., Hamilton, M.M., Mao, L., Crane, C.M., Termin, A.P. and Wilson, D.M. (2002) *Synthesis*, 1669–1674.

56 Wasserscheid, P. and Welton, T. (eds) (2007) *Ionic Liquids in Synthesis*, 2nd edn, Wiley-VCH, Weinheim.

57 Hoffmann, J., Nüchter, M., Ondruschka, B. and Wasserscheid, P. (2003) *Green Chemistry*, **5**, 296–299.

58 Kremsner, J.M. and Kappe, C.O. (2006) *The Journal of Organic Chemistry*, **71**, 4651–4658.

59 Leadbeater, N.E., Torenius, H.M. and Tye, H. (2004) *Combinatorial Chemistry & High Throughput Screening*, **7**, 511–528.

60 Habermann, J., Ponzi, S. and Ley, S.V. (2005) *Mini-Reviews in Organic Chemistry*, **2**, 125–137.

61 Leadbeater, N.E. and Torenius, H.M. (2006) in *Microwaves in Organic Synthesis*, 2nd edn (ed. A. Loupy), Wiley-VCH, Weinheim, pp. 327–361 (Chapter 7).

62 Varma, R.S. and Namboodiri, V.V. (2001) *Chemical Communications*, 643–644; Varma, R.S. and Namboodiri, V.V. (2002) *Tetrahedron Letters*, **43**, 5381–5383; Namboodiri, V.V. and Varma, R.S. (2002) *Chemical Communications*, 342–343.

63 Khadilkar, B.M. and Rebeiro, G.L. (2002) *Organic Process Research & Development*, **6**, 826–828.

64 Deetlefs, M. and Seddon, K.R. (2003) *Green Chemistry*, **5**, 181–186.

65 Vo Thanh, G., Pegot, B. and Loupy, A. (2004) *European Journal of Organic Chemistry*, 1112–1116.

66 Bica, K., Gmeiner, G., Reichel, C., Lendl, B. and Gaertner, P. (2007) *Synthesis*, 1333–1338.

67 Lee, S. (2006) *Chemical Communications*, 1049–1063; Davis, J.H. Jr, (2004) *Chemistry Letters*, **33**, 1072–1077.

68 de Kort, M., Tuin, A.W., Kuiper, S., Overkleeft, H.S., van der Marel, G.A. and Buijsman, R.C. (2004) *Tetrahedron Letters*, **45**, 2171–2175.

69 Fu, S.-K. and Liu, S.-T. (2006) *Synthetic Communications*, **36**, 2059–2067; Arfan, A. and Bazureau, J.P. (2005) *Organic Process Research & Development*, **9**, 743–748.

70 Legeay, J.-C., Vanden Eynde, J.J. and Bazureau, J.P. (2005) *Tetrahedron*, **61**, 12386–12397.

71 Vallin, K.S.A., Emilsson, P., Larhed, M. and Hallberg, A. (2002) *The Journal of Organic Chemistry*, **67**, 6243–6246.

72 Leadbeater, N.E., Torenius, H.M. and Tye, H. (2003) *Tetrahedron*, **59**, 2253–2258.

73 Alexander, M.V., Khandekar, A.C. and Samant, S.D. (2004) *Journal of Molecular Catalysis A-Chemical*, **223**, 75–83.

74 Estager, J., Lévêque, J.-M., Turgis, R. and Draye, M. (2006) *Journal of Molecular Catalysis A-Chemical*, **256**, 261–264.

75 Ley, S.V., Leach, A.G. and Storer, R.I. (2001) *Journal of the Chemical Society-Perkin Transactions 1*, 358–361.

76 Leadbeater, N.E. and Torenius, H.M. (2002) *The Journal of Organic Chemistry*, **67**, 3145–3148.

77 Lin, Y.L., Cheng, J.-Y. and Chu, Y.-H. (2007) *Tetrahedron*, **63**, 10949–10957.

78 Silva, A.M.G., Tomé, A.C., Neves, M.G.P.M.S., Cavaleiro, J.A.S. and Kappe, C.O. (2005) *Tetrahedron Letters*, **46**, 4723–4726.

79 Glenn, A.G. and Jones, P.B. (2004) *Tetrahedron Letters*, **45**, 6967–6969.

80 Moseley, J.D., Lenden, P., Thompson, A.D. and Gilday, J.P. (2007) *Tetrahedron Letters*, **48**, 6084–6087.

81 Davies, H.M. and Beckwith, R.E. (2004) *The Journal of Organic Chemistry*, **69**, 9241–9247.

82 Gonda, J., Martinková, M., Zadrošová, A., Šoteková, M., Raschmanová, J., Čonka, P., Gajdošíková, E. and Kappe, C.O. (2007) *Tetrahedron Letters*, **48**, 6912–6915.

83 Harris, G.L. (ed.) (1995) *Properties of Silicon Carbide*, Institute of Electrical Engineers, London. Choyke, W.J., Matsunami, H. and Pensl, G.(eds) (2004) *Silicon Carbide: Recent Major Advances*, Springer, Berlin; Saddow, S.E. and Agarwal, A. (eds) (2004) *Advances in Silicon Carbide Processing and Applications*, Artech House Inc., Norwood.

84 Thompson, L.A. and Ellman, J.A. (1996) *Chemical Reviews*, **96**, 555–600; Dolle, R.E., Le Bourdonnec, B., Morales, G.A., Moriarty, K.J. and Salvino, J.M. (2006) *Journal of Combinatorial Chemistry*, **8**, 597–635.

85 Kappe, C.O. and Dallinger, D. (2006) *Nature Reviews. Drug Discovery*, **5**, 51–64.

86 Stadler, A. and Kappe, C.O. (2001) *Journal of Combinatorial Chemistry*, **3**, 624–630; Stadler, A. and Kappe, C.O. (2003) *Methods in Enzymology*, **369**, 197–223; Dallinger, D. and Kappe, C.O. (2007) *Nature Protocols*, **2**, 1713–1721.

87 Desai, B., Dallinger, D. and Kappe, C.O. (2006) *Tetrahedron*, **62**, 4651–4664.

88 Selway, C.N. and Terret, N.K. (1996) *Bioorganic and Medicinal Chemistry*, **4**, 645–654.

89 Kappe, C.O., Kumar, D. and Varma, R.S. (1999) *Synthesis*, 1799–1803.

90 Varma, R.S. and Kumar, D. (1999) *Tetrahedron Letters*, **40**, 7665–7669; Usyatinsky, A.Ya. and Khmelnitsky, Y.L. (2000) *Tetrahedron Letters*, **41**, 5031–5034.

91 Olsson, R., Hansen, H.C. and Andersson, C.-M. (2000) *Tetrahedron Letters*, **41**, 7947–7950.

92 Quiroga, J., Cisneros, C., Insuasty, B., Abonía, R., Nogueras, M. and Sánchez, A. (2001) *Tetrahedron Letters*, **42**, 5625–5627.

93 Nüchter, M., Ondruschka, B., Tied, A., Lautenschläger, W. and Borowski, K.J. (2001) *American Genomic/Proteomic Technology*, **1**, 34–39.

94 Nüchter, M. and Ondruschka, B. (2003) *Molecular Diversity*, **7**, 253–264.

95 Stadler, A., Yousefi, B.H., Dallinger, D., Walla, P., Van der Eycken, E., Kaval, N. and Kappe, C.O. (2003) *Organic Process Research & Development*, **7**, 707–716.

96 Pisani, L., Prokopcová, H., Kremsner, J.M. and Kappe, C.O. (2007) *Journal of Combinatorial Chemistry*, **9**, 415–421; See also: Schön, U., Messinger, J., Eichner, S. and Kirschning, A. (2008) *Tetrahedron Letters*, **49**, 3204–3207.

97 Le Bas, M.-D.H. and O'Shea, D.F. (2005) *Journal of Combinatorial Chemistry*, **7**, 947–951.

98 Leadbeater, N.E. and Schmink, J.R. (2007) *Tetrahedron*, **63**, 6764–6773.

99 Amore, K.M., Leadbeater, N.E., Miller, T.A. and Schmink, J.R. (2006) *Tetrahedron Letters*, **47**, 8583–8586.

100 Matloobi, M. and Kappe, C.O. (2007) *Chimica Oggi-Chemistry Today*, **25**, 26–31; Matloobi, M. and Kappe, C.O. (2007) *Combinatorial Chemistry & High Throughput Screening*, **10**, 735–750.

101 Cotterill, I.C., Usyatinsky, A.Ya., Arnold, J.M., Clark, D.S., Dordick, J.S., Michels, P.C. and Khmelnitsky, Y.L. (1998) *Tetrahedron Letters*, **39**, 1117–1120.

102 Glass, B.M. and Combs, A.P. (2001) in *High-Throughput Synthesis Principles and Practices* (ed. I. Sucholeiki), Marcel Dekker, Inc., New York, pp. 123–128 (Chapter 4.6).

103 Glass, B.M. and Combs, A.P. (2001) Article E0027. *Fifth International Electronic Conference on Synthetic Organic Chemistry*, (eds C.O. Kappe, P. Merino, A. Marzinzik, H. Wennemers, T. Wirth, J.J. Vanden Eynde and S.-K. Lin), CD-ROM edition, ISBN 3-906980-06-5, MDPI, Basel, Switzerland.

104 Al-Obeidi, F.A.D. and Austin, R.E. (2002) U. S. Patent 20020187078 A1; *Chemical Abstracts*, **138**, (2002) 26204.

105 Sarko, C.R. (2005) in *Microwave Assisted Organic Synthesis* (eds P. Lidström and J.P. Tierney), Blackwell Publishing, Oxford, pp. 222–236 (Chapter 8).

106 Alcázar, J. (2005) *Journal of Combinatorial Chemistry*, **7**, 353–355.

107 Martinez-Teipel, B., Green, R.C. and Dolle, R.E. (2004) *The QSAR & Combinatorial Science*, **23**, 854–858.

108 Macleod, C., Martinez-Teipel, B., Barker, W.M. and Dolle, R.E. (2006) *Journal of Combinatorial Chemistry*, **8**, 132–140.

109 Murray, J.K. and Gellman, S.H. (2006) *Journal of Combinatorial Chemistry*, **8**, 58–65.

110 Kremsner, J.M., Stadler, A. and Kappe, C.O. (2007) *Journal of Combinatorial Chemistry*, **9**, 285–291.

111 High-throughput Synthesis Methods: Kappe, C.O. and Stadler, A. (2005) *Microwaves in Organic and Medicinal Chemistry*, Wiley-VCH, Weinheim, pp. 291–391 (Chapter 7); Stadler, A. and Kappe, C.O. (2006) in *Microwaves in Organic Synthesis*, 2nd edn, (ed. A. Loupy), Wiley-VCH, Weinheim, pp. 726–787 (Chapter 16); Erdélyi, M. (2006) *Topics in Heterocyclic Chemistry*, **1**, 79–128; Lange, T. and Lindell, S. (2005) *Combinatorial Chemistry & High Throughput Screening*, **8**, 595–606.

112 Solid-phase synthesis: Al-Obeidi, F., Austin, R.E., Okonya, J.F. and Bond, D.R.S. (2003) *Mini Reviews in Medicinal Chemistry*, **3**, 449–460.

113 Soluble polymer supports: Swamy, K.M.K., Yeh, W.-B., Lin, M.-J. and Sun,

C.-M. (2003) *Current Medicinal Chemistry*, **10**, 2403–2423.

114 Fluorous-phase synthesis: Zhang, W. (2006) *Topics in Current Chemistry*, **266**, 145–166; Zhang, W. (2004) *Chemical Reviews*, **104**, 2531–2556; Olofsson, K. and Larhed, M. (2004) in *Handbook of Fluorous Chemistry* (eds J.A. Gladysz, D.P. Curran and I.T. Horvath), Wiley-VCH, Weinheim, pp. 359–365 (Chapter 10.19).

115 Polymer-supported reagents, catalysts and scavengers: Crosignani, S. and Linclau, B., (2006) *Topics in Heterocyclic Chemistry*, **1**, 129–154; Baxendale, I.R., Lee, A.-L. and Ley, S.V. (2005) in *Microwave Assisted Organic Synthesis* (eds P. Lidström and J.P. Tierney), Blackwell Publishing Ltd, Oxford, UK, pp. 133–176 (Chapter 6); Bhattacharyya, S. (2005) *Molecular Diversity*, **9**, 253–257; Parlow, J.J. (2005) *Current Opinion in Drug Discovery & Development*, **8**, 757–775.

116 Erdélyi, M. and Gogoll, A. (2003) *The Journal of Organic Chemistry*, **68**, 6431–6434.

117 Collins, J.M. and Collins, M.J. (2006) *Microwaves in Organic Synthesis* 2nd edn, (ed. A. Loupy), Wiley-VCH, Weinheim, pp. 898–930 (Chapter 20); Collins, J.M. and Leadbeater, N.E. (2007) *Organic and Biomolecular Chemistry*, **5**, 1141–1150.

118 Yu, H.-M., Chen, S.-T. and Wang, K.-T. (1992) *The Journal of Organic Chemistry*, **57**, 4781–4784.

119 Erdélyi, M. and Gogoll, A. (2002) *Synthesis*, 1592–1596.

120 Bacsa, B., Desai, B., Dibó, G. and Kappe, C.O. (2006) *Journal of Peptide Science*, **12**, 633–638.

121 Feliu, L., Font, D., Soley, R., Tailhades, J., Martinez, J. and Amblard, M. (2007) *ARKIVOC*, **iv**, 65–72; Monroc, S., Feliu, L., Planas, M. and Bardají, E. (2006) *Synlett*, 1311–1314.

122 Bacsa, B. and Kappe, C.O. (2007) *Nature Protocols*, **2**, 2222–2227.

123 Rizzolo, F., Sabatino, G., Chelli, M., Rovero, P. and Papini, A.M. (2007)

International Journal of Peptide. Research and Therapeutics, **13**, 203–208.

124 Bowman, M.D., Jeske, R.C. and Blackwell, H.E. (2004) *Organic Letters*, **6**, 2019–2022; Bowman, M.D., Jacobson, M.M., Pujanauski, B.G. and Blackwell, H.E. (2006) *Tetrahedron*, **62**, 4715–4727.

125 Scharn, D., Wenschuh, H., Reineke, U., Schneider-Mergener, J. and Germeroth, L. (2000) *Journal of Combinatorial Chemistry*, **2**, 361–369.

126 Chen, J., Yang, G., Zhang, H. and Chen, Z. (2006) *Reactive & Functional Polymers*, **66**, 1434–1451; Lee, M.-J. and Sun, C.-M. (2003) *Chinese Pharmaceutical Journal*, **55**, 405–452; Sun, C.-M. (2002) Combinatorial Chemistry Methods and Protocols, in *Methods in Molecular Biology Series* (ed. L. Bellavance), The Humana Press Inc., New Jersey, pp. 345–371 (Chapter 10); Sun, C.-M. (1999) *Combinatorial Chemistry & High Throughput Screening*, **2**, 299–318.

127 Su, Y.-S., Lin, M.-J. and Sun, M.-C. (2005) *Tetrahedron Letters*, **46**, 177–180.

128 Wu, C.-H. and Sun, C.-M. (2006) *Tetrahedron Letters*, **47**, 2601–2604.

129 Yeh, W.-B., Lin, M.-J., Lee, M.-J. and Sun, C.-M. (2003) *Molecular Diversity*, **7**, 185–198; Lin, M.-J. and Sun, C.-M. (2003) *Tetrahedron Letters*, **44**, 8739–8742.

130 Chang, W.-J., Kulkarni, M.V. and Sun, C.-M. (2006) *Journal of Combinatorial Chemistry*, **8**, 141–144; Yeh, W.-P., Chang, W.-J., Sun, M.-L. and Sun, C.-M. (2007) *Tetrahedron*, **63**, 11809–11816.

131 Curran, D.P. (2004) in *Handbook of Fluorous Chemistry* (eds J.A. Gladysz, D.P. Curran and I.T. Horvath), Wiley-VCH, Weinheim, pp. 101–127 (Chapter 7).

132 Zhang, W., Chen, C.H.-T., Lu, Y. and Nagashima, T. (2004) *Organic Letters*, **6**, 1473–1476.

133 Zhang, W., Nagashima, T., Lu, Y. and Chen, C.H.-T. (2004) *Tetrahedron Letters*, **45**, 4611–4613.

134 Zhang, W., Lu, Y., Chen, C.H.-T., Curran, D.P. and Geib, S. (2006) *European Journal of Organic Chemistry*, 2055–2059; Zhang,

W. and Chen, C.H.-T. (2005) *Tetrahedron Letters*, **46**, 1807–1810; Lu, Y. and Zhang, W. (2004) *The QSAR & Combinatorial Science*, **23**, 827–835; Zhang, W. and Tempest, P. (2004) *Tetrahedron Letters*, **45**, 6757–6760; Zhang, W., Lu, Y. and Chen, C.H.-T. (2003) *Molecular Diversity*, **7**, 199–202.

135 Solinas, A. and Taddei, M. (2007) *Synthesis*, 2409–2453; Ley, S.V., Ladlow, M. and Vickerstaffe, E. (2006) in *Exploiting Chemical Diversity for Drug Discovery* (eds P.A. Bartlett and M. Entzeroth), RSC Publishing, Cambridge, pp. 3–32 (Chapter 1); Ley, S.V., Baxendale, I.R. and Myers, R.M. (2006) in *Combinatorial Synthesis of Natural Product-Based Libraries* (ed. A.M. Boldi), CRC Press, Boca Raton, pp. 131–163 (Chapter 6); Hinzen, B. and Hahn, M.G. (2006) in *Combinatorial Chemistry* (eds W. Bannwarth and B. Hinzen), Wiley-VCH, Weinheim, pp. 457–512 (Chapter 6); M.R. Buchmeiser (ed.), (2003) *Polymeric Materials in Organic Synthesis and Catalysis*, Wiley-VCH, Weinheim; Ley, S.V. and Baxendale, I.R. (2002) *Nature Reviews. Drug Discovery*, **1**, 573–586.

136 Donati, D., Morelli, C. and Taddei, M. (2005) *Tetrahedron Letters*, **46**, 2817–2819.

137 Kirschning, A., Solodenko, W. and Mennecke, K. (2006) *Chemistry – A European Journal*, **12**, 5972–5990; (2004) *Topics in Current Chemistry*, **244**, 1–317.

138 Wang, Y. and Sauer, D.R. (2004) *Organic Letters*, **6**, 2793–2796.

139 Dawood, K.M. and Kirschning, A. (2005) *Tetrahedron*, **61**, 12121–12130; See also: Michrowska, A., Mennecke, K., Kunz, U., Kirschning, A. and Grela, K. (2006) *Journal of the American Chemical Society*, **128**, 13261–13267.

140 Choudary, B.M., Madhi, S., Chowdari, N.S., Kantam, M.L. and Sreedhar, B. (2002) *Journal of the American Chemical Society*, **124**, 14127–14136.

141 Shore, G., Morin, S. and Organ, M.G. (2006) *Angewandte Chemie-International Edition*, **45**, 2761–2766.

142 Lipshutz, B.H., Frieman, B.A., Lee, C.-T., Lower, A., Nihan, D.M. and Taft, B.R. (2006) *Chemistry, an Asian Journal*, **1**, 417–429.

143 Dallinger, D., Gorobets, N.Yu. and Kappe, C.O. (2003) *Molecular Diversity*, **7**, 229–245.

144 Hinchcliff, A., Hughes, C., Pears, D.A. and Pitts, M.R. (2007) *Organic Process Research & Development*, **11**, 477–481.

145 Mukade, T., Dragoli, D.R. and Ellman, J.A. (2003) *Journal of Combinatorial Chemistry*, **5**, 590–596.

146 Baxendale, I.R., Ley, S.V. and Piutti, C. (2002) *Angewandte Chemie-International Edition*, **41**, 2194–2197; Baxendale, I.R., Ley, S.V., Nessi, M. and Piutti, C. (2002) *Tetrahedron*, **58**, 6285–6304.

147 Lehmann, H. (2007) in *Ernst Schering Foundation Symposium Proceedings* (eds P.H. Seeberger and T. Blume), Springer, Berlin Heidelberg, pp. 133–149.

148 Loones, K.T.J., Maes, B.U.W., Rombouts, G., Hostyn, S. and Diels, G. (2005) *Tetrahedron*, **61**, 10338–10348.

149 Kremsner, J.M., Stadler, A. and Kappe, C.O. (2006) *Topics in Current Chemistry*, **266**, 233–278; Ondruschka, B., Bonrath, W. and Stuerga, D. (2006) in *Microwaves in Organic Synthesis*, 2nd edn, (ed. A. Loupy), Wiley-VCH, Weinheim, pp. 62–107 (Chapter 2).

150 Iqbal, M., Vyse, N., Dauvergne, J. and Evans, P. (2002) *Tetrahedron Letters*, **43**, 7859–7862.

151 Lehmann, F., Pilotti, Å. and Luthman, K. (2003) *Molecular Diversity*, **7**, 145–152.

152 Shackelford, S.A., Anderson, M.B., Christie, L.C., Goetzen, T., Guzman, M.C., Hananel, M.A., Kornreich, W.D., Li, H., Pathak, V.P., Rabinovich, A.K., Rajapakse, R.J., Truesdale, L.K., Tsank, S.M. and Vazir, H.N. (2003) *The Journal of Organic Chemistry*, **68**, 267–275.

153 Leadbeater, N.E., Williams, V.A., Barnard, T.M. and Collins, M.J., Jr, (2006) *Organic Process Research & Development*, **10**, 833–837.

154 Leadbeater, N.E., Williams, V.A., Barnard, T.M. and Collins, M.J., Jr, (2006) *Synlett*, 2953–2958.

155 Barnard, T.M., Vanier, G.S. and Collins, M.J., Jr, (2006) *Organic Process Research & Development*, **10**, 1233–1237.

156 Carlsson, A.-C., Jam, F., Tullberg, M., Pilotti, Å., Ioannidis, P., Luthman, K. and Grøtli, M. (2006) *Tetrahedron Letters*, **47**, 5199–5201.

157 Ekström, J., Wettergren, J. and Adolfsson, H. (2007) *Advanced Synthesis and Catalysis*, **349**, 1609–1613.

158 Pawluczyk, J.M., McClain, R.T., Denicola, C., Mulhearn, J.J., Jr, Rudd, D.J. and Lindsley, C.W. (2007) *Tetrahedron Letters*, **48**, 1497–1501.

159 Glasnov, T.N. and Kappe, C.O. (2007) *Macromolecular Rapid Communications*, **28**, 395–410.

160 Baxendale, I.R. and Pitts, M.R. (2006) *Chimica Oggi-Chemistry Today*, **24**, 41–45; Baxendale, I.R., Hayward, J.J. and Ley, S.V. (2007) *Combinatorial Chemistry & High Throughput Screening*, **10**, 802–836.

161 Cablewski, T., Faux, A.F. and Strauss, C.R. (1994) *The Journal of Organic Chemistry*, **59**, 3408–3412.

162 Glasnov, T.N., Vugts, D.J., Koningstein, M.M., Desai, B., Fabian, W.M.F., Orru, R.V.A. and Kappe, C.O. (2006) *The QSAR & Combinatorial Science*, **25**, 509–518.

163 Bagley, M.C., Jenkins, R.L., Lubinu, M.C., Mason, C. and Wood, R. (2005) *The Journal of Organic Chemistry*, **70**, 7003–7006.

164 Smith, C.J., Iglesias-Sigüenza, F.J., Baxendale, I.R. and Ley, S.V. (2007) *Organic and Biomolecular Chemistry*, **5**, 2758–2761.

165 Shieh, W.-C., Dell, S. and Repič, O. (2001) *Organic Letters*, **3**, 4279–4281.

166 Shieh, W.-C., Dell, S. and Repič, O. (2002) *Tetrahedron Letters*, **43**, 5607–5609.

167 Shieh, W.-C., Lozanov, M. and Repič, O. (2003) *Tetrahedron Letters*, **44**, 6943–6945.

168 Moseley, J.D. and Lawton, S.J. (2007) *Chimica Oggi-Chemistry Today*, **25**, 16–19.

169 Moseley, J.D., Sankey, R.F., Tang, O.N. and Gilday, J.P. (2006) *Tetrahedron*, **62**,

4685–4689; Moseley, J.D. and Lenden, P. (2007) *Tetrahedron*, **63**, 4120–4125.

170 Paulus, R.M., Erdmenger, T., Becer, C.R., Hoogenboom, R. and Schubert, U.S. (2007) *Macromolecular Rapid Communications*, **28**, 484–491.

171 Barnard, T.M., Leadbeater, N.E., Boucher, M.B., Stencel, L.M. and Wilhite, B.A. (2007) *Energy Fuels*, **21**, 1777–1781.

172 Arvela, R.K., Leadbeater, N.E. and Collins, M.J. Jr, (2005) *Tetrahedron*, **61**, 9349–9355.

173 Pitts, M.R., McCormack, P. and Whittall, J. (2006) *Tetrahedron*, **62**, 4705–4708.

174 Moseley, J.D., Lenden, P., Lockwood, M., Ruda, K., Sherlock, J.-P., Thomson, A.D. and Gilday, J.P. (2008) *Organic Process Research & Development*, **12**, 30–40.

175 Bowman, M.D., Holcomb, J.L., Kormos, C.M., Leadbeater, N.E. and Williams, V.A. (2008) *Organic Process Research & Development*, **12**, 41–57.

176 Watts, P. (2004) *Current Opinion in Drug Discovery & Development*, **7**, 807–812; Watts, P. and Haswell, S.J. (2003) *Current Opinion in Chemical Biology*, **7**, 380–387.

177 He, P., Haswell, S.J. and Fletcher, P.D. (2004) *Lab on a Chip*, 4, 38–41; He, P., Haswell, S.J. and Fletcher, P.D. (2004) *Applied Catalysis A-General*, **274**, 111–114.

178 Comer, E. and Organ, M.G. (2005) *Journal of the American Chemical Society*, **127**, 8160–8167.

179 Comer, E. and Organ, M.G. (2005) *Chemistry – A European Journal*, **11**, 7223–7227.

180 Bremner, W.S. and Organ, M.G. (2007) *Journal of Combinatorial Chemistry*, **9**, 14–16.

181 Howarth, P. and Lockwood, M. (2004) *The Chemical Engineer*, **756**, 29–31.

182 Bierbaum, R., Nüchter, M. and Ondruschka, B. (2004) *Chemie Ingenieur Technik*, **76**, 961–965.

183 Razzaq, T. and Kappe, C.O. (2008) *Chemistry & Sustainability: Energy & Materials*, **1**, 123–132.

5
Starting With Microwave Chemistry

This chapter presents an overview of some of the more practical issues associated with performing organic synthesis under microwave conditions. Some general advice and useful recommendations will be discussed. In particular, for the microwave chemistry novice, valuable information will be presented on topics such as: how to design and set-up microwave experiments, how to optimize reaction conditions and which safety issues have to be considered. The most important features regarding MAOS that have been already covered in Chapters 2–4 are highlighted again in this chapter in order to give the reader a compact overview. At the end of this chapter, a list of Frequently Asked Questions (FAQs) can be found.

5.1
Why Use Microwave Reactors?

Since the first published experiments on MAOS in the mid-1980s, the use of microwave energy for heating chemical reactions has shown tremendous benefits in organic synthesis. Significant rate enhancements, improved yields, and cleaner reaction profiles have been reported for many different reaction types over the past two decades. Thus, when planning a reaction protocol or designing novel synthetic pathways, the use of microwave irradiation as heating source should be considered as first choice and not the last resort!

The major advantages in using dedicated microwave reactors in organic synthesis can be summarized as follows:

- *Rate enhancement:* Due to the higher temperatures used, reaction times are often drastically reduced from hours to minutes or even seconds.
- *Increased yield:* In many cases the short reaction times and carefully optimized reaction temperatures minimize the occurrence of unwanted side reactions.
- *Improved purity:* Cleaner reactions due to less by-product formation lead to simplified purification steps.
- *Greater reproducibility:* The homogeneous microwave field and exact temperature control in dedicated single-mode reactors promise comparable results for every experimental run.

Practical Microwave Synthesis for Organic Chemists: Strategies, Instruments, and Protocols
C. Oliver Kappe, Doris Dallinger, and S. Shaun Murphree
Copyright © 2009 WILEY-VCH Verlag GmbH & Co. KGaA, Weinheim
ISBN: 978-3-527-32097-4

- *Expanded reaction conditions ("reaction space"):* Access to transformations and reaction conditions which cannot be easily achieved under conventional reflux conditions.

Whilst in the early published reports in the area of MAOS the utilization of simple, domestic microwave ovens without any parameter control was described, it has become evident that the use of these instruments was not only dangerous but also sometimes led to irreproducible results (see Chapter 3). To meet the increasing demand of preparative chemists for precise reaction control, the development of dedicated microwave reactors for synthetic purposes was necessary in order to provide scientists with accurate measurements of the process-determining parameters. Modern microwave reactors enable the combination of rapid in-core heating with sealed-vessel (autoclave) conditions and on-line monitoring of reaction parameters. Performing reactions at elevated pressure has, therefore, become vastly easier as compared to conventional heating methods. Precise temperature and pressure sensors enable on-line reaction control at any stage of the process. Dedicated software packages allow stepwise method development and optimization. Additional tools for automatization and special processing techniques simplify the procedures used for new and improved compound syntheses (see Chapter 3).

Two philosophies of microwave reactor design exist: single-mode and multimode instruments (Chapter 3). The main and most important difference for the user consists in reaction scale. While single-mode instruments are suitable for small-scale reactions ranging from 60 μL up to 50 mL filling volume, larger volumes (up to about 3 L) can be processed with multimode microwave reactors, either in batch or parallel. For both systems larger volumes can be handled by using standard glassware (round-bottom flasks from 0.125–5 L) under open-vessel conditions or applying flow-through techniques (Chapter 4). With respect to the individual demands of the user, a choice can be made of either a single-mode instrument when the main focus is on discovery chemistry, method development, or reaction optimization, or a multimode instrument when larger scales are required, as in a process laboratory.

Throughout the following sections of this chapter a strong preference is given to the discussion of single-mode microwave technology using sealed-vessel processing. A survey of the currently published microwave chemistry literature (see Figure 1.1) indicates that 90–95% of all published microwave methods using dedicated instruments apply this experimental configuration. It will be stated explicitly when practical topics concerning other instruments or processing techniques will be presented. In any case, and this cannot be emphasized enough, the use of kitchen microwave ovens for laboratory experiments can under no circumstances be justified!

5.2
Getting Started – Method Development

Microwave heating is still often applied to already known conventional thermal reactions in order to accelerate the reaction and, therefore, to reduce the overall processing time. Given the current acceptance of microwave technology and the steadily increasing availability of appropriate instrumentation, microwave heating is nowadays often used directly as a starting point when beginning a new project. In many instances initial experiments in an oil-bath are no longer performed. To find a set of appropriate initial microwave reaction conditions when getting started on a new synthetic project, a search in commercially available microwave chemistry databases can be helpful. The Biotage PathFinder database represents a collection of detailed protocols for reactions performed with Biotage single-mode microwave instruments [1]. Currently, more than 4500 entries are available in this web-based tool, where keyword and/or structure searches can be carried out. A different database with similar search functions that covers MAOS protocols since 1996 was introduced in 2007 [2]. Since a large number of reliable published microwave protocols do exist – at the time of writing about 3000 publications since 2001 – suitable starting conditions can often be found in the generally accessible chemical literature.

After selecting appropriate initial reaction conditions with the help of one of the above-mentioned tools, information on the success or failure of the reaction can typically be obtained in a few minutes using standard reaction monitoring techniques such as HPLC, GC, LC-MS or TLC. Through a subsequent iterative process further optimization of the reaction conditions can be carried out. A guideline on how to perform microwave method development and subsequent optimization experiments is presented in this section and is shown in the Flowchart in Figure 5.1. For rapid method development and reaction optimization purposes, automated single-mode reactors are highly suitable.

5.2.1
Open or Closed Vessels?

Utilizing dedicated microwave reactors for organic synthesis allows a choice of open- or closed-vessel conditions. Both processing techniques have their merits but the main criterion when making a decision between the two is generally the reaction temperature that needs to be achieved.

Microwave heating in open vessels often enables reflux conditions to be reached in a shorter timeframe as compared to an oil-bath experiment if the reaction mixture is strongly microwave absorbing. If solvents are heated by microwave irradiation at atmospheric pressure in an open vessel, the boiling point of the solvent (as in an oil-bath experiment) typically limits the reaction temperature that can be achieved. The major benefit of performing microwave-assisted reactions at atmospheric pressure is the possibility of carrying out syntheses on a larger scale using standard laboratory glassware (round-bottom flasks, reflux condensers) in the microwave cavity. In contrast, pressurized reactions require special vessels and scale-up to more than 1 L of processing volume can generally only be achieved using parallel rotors or flow-through systems (see Sections 3.5 and 4.8). It should be noted that under atmospheric conditions the observed rate enhancements compared to an oil-bath reflux experiment at the same temperature cannot be expected to be very high. In those special cases where superheated solvents/reaction mixtures are generated under microwave

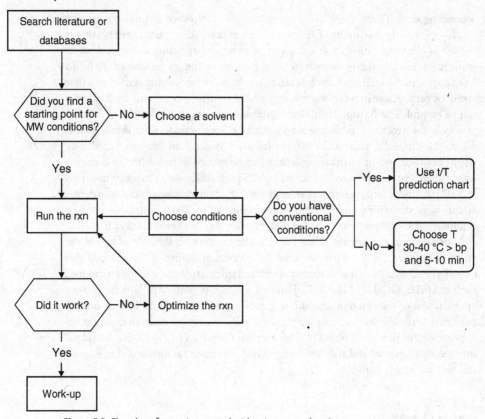

Figure 5.1 Flowchart for getting started with microwave chemistry.

reflux conditions (Section 2.5.3) rate enhancements by a factor of 10 can be observed. Another important aspect of open-vessel technology is safety, especially on a larger scale. It is inherently safer to perform reactions under atmospheric pressure conditions, rather than in a sealed vessel.

Nevertheless, most of the recently described microwave procedures operate under sealed-vessel conditions and, as a consequence, all available microwave reactors are mainly designed for these processing conditions. The higher reaction temperatures at elevated pressures have led to impressive results with respect to reduced reaction times and improved yields. When applying closed-vessel technology, solvents can be heated far above their boiling point under atmospheric pressure conditions. In some cases up to 100 °C higher temperatures can be reached (e.g., methanol with a boiling point of 65 °C can be heated up to 165 °C). In comparison to open-vessel microwave synthesis, pressurized reactions are generally smaller in scale due to the limited maximum filling volume of the utilized dedicated reaction vessels (up to 7 mL for the commonly used 10 mL microwave vials for single-mode microwave reactors from CEM and Biotage). For the vast number of chemical transformations, sealed-vessel conditions provide satisfactory results, but in some cases it is necessary to utilize open-vessel conditions – for example, when a volatile by-product has to be removed from the equilibrium in order to drive the reaction to completion (e.g., water removal in esterifications, see Section 4.3). However, in order to accomplish significant rate enhancements (>100-fold), the use of sealed vessels is preferred.

5.2.2
Choice of Solvent

The next step in method development is the proper choice of the solvent, which can be a crucial factor for the outcome of a chemical reaction (see Figure 5.1). In general, microwave-mediated transformations can often be performed using the same solvent as in a classical protocol. However, one needs to be aware that solvents interact differently with microwaves, according to their dielectric properties. The higher the loss tangent (tan δ), the better the conversion of microwave energy into heat, and the more effective the microwave heating will be (see Section 2.3 and Table 5.1). To put it in more general terms, the more polar a reaction mixture, the greater its ability to couple with microwave energy, leading to a more rapid rise in temperature. The main criterion for choosing a solvent in MAOS is not the boiling point as under thermal (reflux) conditions but rather the dielectric properties, in particular the tan δ value.

As already mentioned, in many instances there will be no need to make a change in the solvent selection compared to a conventionally heated experiment. Hence, for a first attempt, the same solvent that would normally be used for a specific transformation under conventional conditions can also be used in a microwave experiment (the few exceptions to this general rule include diethyl ether and carbon disulfide because of their low boiling points and associated safety issues). If the temperature in the microwave experiment will not rise to the desired target temperature in a reasonable time period (1–2 min), a change to a stronger microwave absorbing solvent should be considered. Common solvents for MAOS are acetonitrile, alcohols, N,N-dimethylformamide, or – with increasing popularity – water.

Table 5.1 Physical properties of common solvents (data from Ref. [3]).

Solvent	bp (°C)	ε'	ε''	tan δ	Microwave absorbance
Ethylene glycol	197	37.0	49.950	1.350	very good
Dimethylsulfoxide	189	45.0	37.125	0.825	good
Ethanol	78	24.3	22.866	0.941	good
Methanol	65	32.6	21.483	0.659	good
Water	100	80.4	9.889	0.123	medium
1-Methyl-2-pyrrolidone	204	32.2	8.855	0.275	medium
N,N-dimethylformamide	153	37.7	6.070	0.161	medium
1,2-Dichlorobenzene	180	9.9	2.772	0.280	medium
Acetonitrile	81	37.5	2.325	0.062	medium
Dichloromethane	40	9.1	0.382	0.042	low
Tetrahydrofuran	66	7.4	0.348	0.047	low
Toluene	110	2.4	0.096	0.040	very low

Low absorbing solvents – such as dichloromethane, tetrahydrofuran, dioxane or toluene (see Tables 5.1 and 2.3) – may also be processed with microwaves, but are poorly heated if they are used in pure form. However, as reaction mixtures are often composed of several different (polar) reagents, there should always be enough potential for efficient coupling with microwave energy. When employing nearly microwave-transparent reaction mixtures, the addition of strongly microwave-absorbing passive heating elements (PHEs) like silicon carbide, Carboflon (fluoropolymer doped with carbon black) or Weflon (Teflon doped with graphite) is recommended to achieve appropriate heat transfer within the mixture (see Section 4.6). These additives, available in different shapes such as cylinders or stir bars, heat the reaction mixture by thermal convection and also help to protect parts of the microwave hardware equipment from absorbing too much microwave energy. Alternatively, low absorbing solvents may be doped with small amounts of strongly absorbing media such as ionic liquids (see Section 4.5.2).

It also has to be kept in mind that solvents can behave differently at elevated temperatures – most of them become less polar with increasing temperature. This phenomenon results from the decrease in the dielectric loss (ε'') and loss tangent (tan δ) with increasing temperature (see Section 2.3). For example, the absorption of microwave radiation in water decreases at higher temperatures. While it is relatively easy to heat water from room temperature to 100 °C by microwave irradiation, it is significantly more difficult to further heat water to 200 °C and beyond in a sealed vessel (see Section 4.5.1).

It has to be emphasized that when utilizing microwave heating in sealed vessels, it is no longer necessary to use high-boiling solvents, as in a conventional reflux set-up, to achieve a high reaction temperature. With modern instrumentation, it is easily possible to carry out a reaction in methanol at 165 °C (see Figure 2.13), or a transformation in dichloromethane at 100 °C.

However, utilizing low-boiling solvents often results in the development of a significant pressure at elevated temperatures. Therefore, it is crucial not to exceed the operation limits of the equipment used. With some instruments, a solvent library will help the user to estimate the expected pressure in the reaction vessel. In the Biotage Pathfinder database, a vapor-pressure calculator for a variety of solvents and chemicals is incorporated, which allows prediction of the expected pressure at a certain temperature [1]. An overview of estimated vapor pressures at various temperatures for some solvents commonly used in microwave chemistry is presented in Figure 5.2. The limiting factor is, in general, the pressure stability of the vessel used. Common pressure vessels for single-mode reactors usually have a pressure limit of about 20 bar (CEM and Biotage), imposed by the sealing/capping technique. For multimode instruments, different materials and vessel designs are available, enabling reactions to be performed at up to 100 bar.

Due to the developing pressure build-up when solvents/reaction mixtures are heated to elevated temperatures, all microwave vessels have a defined maximum filling volume in order not to exceed the pressure limit of about 20 bar. The total filling volume of the standard vessels is 10 mL (max. recommended filling volume 5–7 mL), but a sufficient headspace to contain the resulting vapors has to be guaranteed in order to avoid an abort of the reaction or, in some cases, a vessel failure as a result of

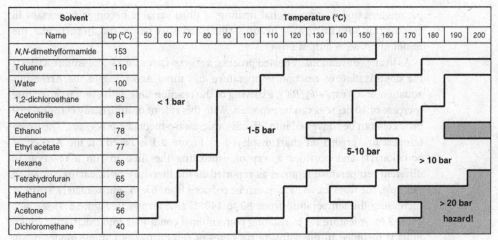

Solvent		Temperature (°C)															
Name	bp (°C)	50	60	70	80	90	100	110	120	130	140	150	160	170	180	190	200
N,N-dimethylformamide	153																
Toluene	110																
Water	100																
1,2-dichloroethane	83																
Acetonitrile	81																
Ethanol	78																
Ethyl acetate	77																
Hexane	69																
Tetrahydrofuran	65																
Methanol	65																
Acetone	56																
Dichloromethane	40																

< 1 bar
1–5 bar
5–10 bar
> 10 bar
> 20 bar
hazard!

Figure 5.2 Approximate vapor pressures attained by solvents at various temperatures.

exceeding the pressure limit. The headspace is a crucial factor that has to be kept in mind, in particular when gases are released during a reaction, as in decarboxylation protocols or hydrogen transfer reactions (see Section 4.4). When a higher temperature is desirable to achieve faster conversions, low-boiling solvents – where a high temperature can often not be obtained due to the pressure build-up – can be replaced by closely related solvents with a higher boiling point, for example, dichloromethane can be substituted by dichloroethane (40 versus 83 °C).

Despite the fact that solvents are selected on the basis of somewhat different criteria for MAOS as compared to conventional syntheses, the general rules governing solvent selection in synthetic organic chemistry still have to be followed.

Microwave-mediated reactions can, in principle, also be carried out without solvents (see Section 4.1). The requirements for these dry-media reactions are different to those for reactions in solution. Particularly in the early days of MAOS, the solvent-free approach was very popular since it allowed the safe use of domestic household microwave ovens and standard open-vessel technology. As no solvent is involved, the pressure build-up is rather low, and in most instances these transformations can be performed under open-vessel conditions in dedicated instruments. On the other hand, solvent-free and/or solid reaction mixtures can easily be locally overheated, even though the overall bulk temperature may be comparatively low (macroscopic hot spot formation). Stirring and accurate temperature measurement can prove rather difficult within such a matrix, impeding the investigation of certain reaction conditions (Section 4.1). Thus, degradation or decomposition of reagents in addition to reproducibility issues can be severe problems in solvent-free reactions.

5.2.3
Temperature, Time and Microwave Power

After having chosen a solvent, the reaction conditions have to be developed (see Figure 5.1). In addition to temperature and time, which are also important parameters under conventional heating, a third variable becomes important in conjunction with microwave heating – the microwave power and, in particular, the *initial* microwave output power.

As in any conventionally heated process, a crucial factor for a successful reaction is the optimization of reaction temperature and time. According to the Arrhenius equation, $k = A\exp(-E_a/RT)$, a halving of the reaction time with every temperature increase of 10 degrees can be expected. With this rule of thumb, many conventional protocols can be converted into effective microwave-mediated processes. The time–temperature prediction chart displayed in Figure 5.3 is based on the Arrhenius equation [1] and provides a way of estimating the time to run a reaction in different temperature regimes as reported in the literature or in databases. As an example, the time for a reaction can be reduced from 6 h to approximately 5 min by increasing the temperature from 80 to 140 °C (see arrows in Figure 5.3; see also Table 2.6). A feature for translating conventional conditions into microwave conditions is included in the software packages of several modern single-mode instruments. Depending on the installed software version with the CEM Discover, the option to convert from conventional conditions is present in the main window when programming a method in the temperature control mode. In addition, this feature is offered on a specific web site as an upgrade for the installed software program. When using the Biotage Initiator, this feature can be enabled by employing the Wizard for setting up a reaction.

T (°C)	Times – change in field color represents change in unit (h/min/s)									
20	1	2	4	6	8	12	24	48	96	172
30	30	1	2	3	4	6	12	24	48	86
40	15	30	1	1.5	2	3	6	12	24	43
50	8	15	30	45	1	1.5	3	6	12	22
60	4	8	15	23	30	45	1.5	3	6	11
70	2	4	8	11	15	23	45	1.5	3	5
80	56	2	4	6	8	11	23	45	1.5	3
90	28	56	2	3	4	6	11	23	45	1
100	14	28	56	1	2	3	6	11	23	40
110	7	14	28	42	56	1	3	6	11	20
120	4	7	14	21	28	42	1	3	6	10
130	2	4	7	11	14	21	42	1	3	5
140	53	2	4	5	7	11	21	42	1	3
150	26	53	2	3	4	5	11	21	42	1
160	13	26	53	1	2	3	5	11	21	38
170	7	13	26	40	53	1	3	5	11	19
180	3	7	13	20	26	40	1	3	5	9
190	2	3	7	10	13	20	40	1	3	5
200	1	2	3	5	7	10	20	40	1	2
210		1	2	2	3	5	10	20	40	1
220			1	1	2	2	5	10	20	35
230				1	1	2	5	10	18	
240						1	1	2	5	9
250								1	2	4

Figure 5.3 Time–temperature prediction chart. Adapted from Ref. [1].

When investigating completely new reactions for which no thermal protocols are available, a feasible starting point is approximately 30–40 °C above the boiling point of the solvent used (see Figure 5.1). Know-how on the way to proceed from here will improve with personal practical experience and varies with the type of chemistry to be accomplished. Performing the reaction initially for 5–10 min generally gives a good overview of its progress.

When selecting an appropriate starting temperature for a microwave experiment, the user also has to pay attention to the decomposition temperatures of the employed reagents and solvents. For example, it is well established that N,N-dimethylformamide decarbonylates at temperatures >180 °C to yield carbon monoxide and dimethylamine in the presence of base [4].

In general, the temperature limit of a single-mode microwave reactor is 250 °C (Biotage) or 300 °C (CEM). This allows the development of completely new reaction procedures at high temperatures and elevated pressures using sealed vessels. However, the operational limits of the respective instrument/vessel system as stated by the manufacturer should always be followed and not be exceeded (see Chapter 3).

As speeding up chemical transformations is one of the key motives for employing microwave heating, a microwave-assisted reaction can be expected to reach completion within a few minutes. In general, if a microwave reaction under sealed-vessel conditions is not complete within 60 min then further review of the reaction conditions (solvent, catalyst, molar ratios) is needed. The published record for the longest microwave-mediated reaction in a single-mode instrument is 22 h for a copper-catalyzed N-arylation (see Scheme 5.1a) [5]. The shortest ever reported microwave reaction requires a microwave pulse of 6 s to reach completion (ultrafast carbonylation chemistry; see Scheme 5.1b) [6].

With respect to the overall reaction time in a microwave-heated experiment, two important definitions must be considered: total irradiation time or hold time (Figure 5.4). When applying "hold time", the ramp time (gray line in Figure 5.4a)

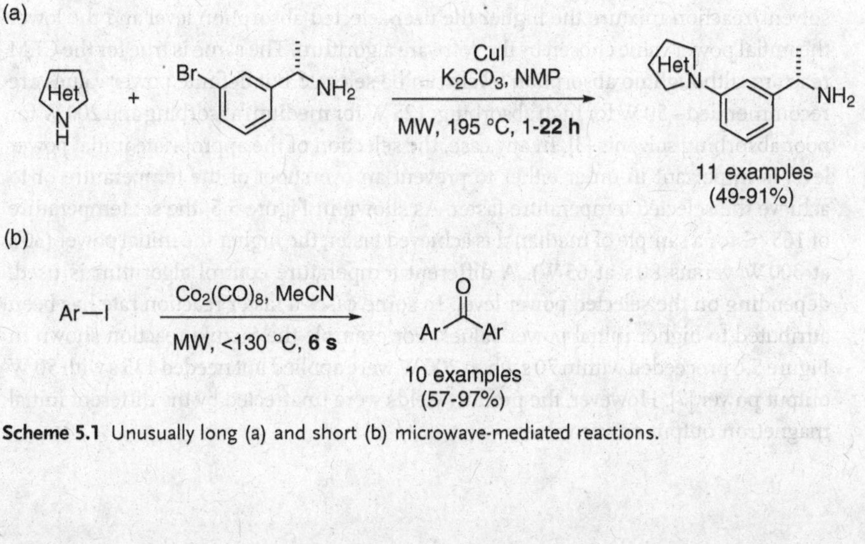

(a)

(b)

Scheme 5.1 Unusually long (a) and short (b) microwave-mediated reactions.

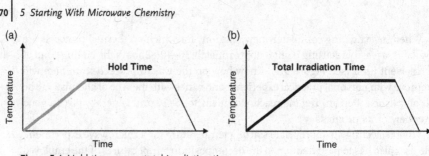

Figure 5.4 Hold time versus total irradiation time.

required to reach the desired set temperature is excluded, whereas the "total irradiation time" does include the ramp time. It can generally be recommended to carry out microwave-assisted transformations using the "hold time" feature because of reproducibility issues, in particular when going from small scale to a larger scale where the ramp time can change significantly – typically longer ramp times are necessary to reach the same target temperature as on a small scale. If small scale reactions in single-mode instruments are considered, the utilization of the hold time function is only an issue for comparatively short reaction times of less than 10 min. For transformations with reaction times >10 min, the "total irradiation time" mode can be applied, since in a high power-density single-mode instrument a relatively short ramp time (1–2 min) will typically be experienced. It should be stressed that any publication dealing with microwave chemistry experiments should specify whether the stated reaction time refers to "hold time" or "total irradiation time".

As a third parameter in addition to temperature and time – being unique to microwave chemistry – the microwave output power has to be considered when performing microwave-assisted reactions. Since the vast majority of MAOS today is carried out in temperature control mode (as opposed to power control, see Section 5.2.4), only the importance of the *initial* magnetron output power will be discussed. The initial power level is dependent on the absorptivity of the used solvent or reaction mixture, respectively. For the Biotage instruments three different absorption levels can be chosen: normal, high and very high. The better the absorptivity of the solvent/reaction mixture the higher the user-selected absorption level and the lower the initial power value chosen by the software algorithm. The same is true for the CEM reactors, although no absorption levels can be selected but definite power values are recommended – 50 W for high absorbing, 125 W for medium absorbing and 200 W for poor absorbing solvents [3]. In any case, the selection of the appropriate initial power level is important in order either to prevent an overshoot of the temperature or to achieve the selected temperature faster. As shown in Figure 5.5, the set temperature of 165 °C for a sample of methanol is achieved faster, the higher the initial power (30 s at 300 W versus 80 s at 65 W). A different temperature control algorithm is used, depending on the selected power level. In some cases a faster reaction rate has been attributed to higher initial power values. For example the Suzuki reaction shown in Figure 5.6 proceeded within 70 s when 200 W were applied but needed 135 s with 50 W output power [7]. However, the product yields were unaffected by the different initial magnetron output power values.

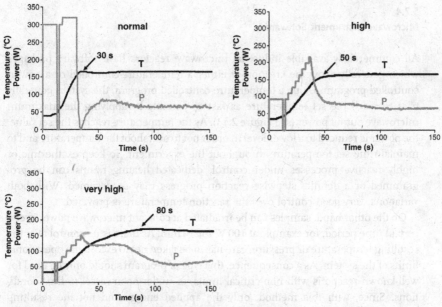

Figure 5.5 Temperature–power profiles of 3 mL MeOH at a set temperature of 165 °C in the Biotage Initiator 2.0 (400 W) at different initial power settings (IR temperature monitoring). Gray lines: microwave output power (*P*); black lines: temperature (*T*).

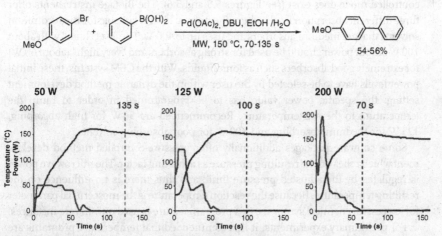

Figure 5.6 Correlation of initial output power and reaction time for a Suzuki reaction at 50, 125 and 200 W initial power and a set temperature of 150 °C in a CEM Discover (fiber-optic temperature measurement). The dotted lines represent the time required for achieving full conversion as monitored by on-line Raman spectroscopy. Adapted from Ref. [7].

5.2.4
Microwave Instrument Software

All commercially available industrial microwave reactors have software packages which generally allow the creation of either a temperature-controlled or a power-controlled program. Using a temperature-controlled program, the system generally tries to reach the set temperature as fast as possible by applying the maximum microwave output power (see Figure 2.13). As the temperature reaches the set value, the power is reduced to a lower level in order not to overshoot the temperature and to maintain the set temperature throughout the experiment. To keep exothermic or highly reactive processes under control, dedicated heating ramps can be programmed or a definite stepwise reaction progress may be designed. With both variations, very good control over the reaction temperature is provided.

On the other hand, samples can be irradiated at constant microwave power over a certain time period, for example at 100 W for 10 min. As there is no control over the resulting temperature or pressure, care has to be taken not to exceed the operational limits of the system. As a consequence, this type of program should only be used for well-known reactions with non-critical limits, or under open-vessel (reflux) conditions. Since, with this method, only the applied energy and not the resulting temperature is controlled, the quality of reaction control is often compromised compared to employing a temperature program. However, when applying the power-controlled mode, the attained temperature must also be specified in a publication in the interest of reproducibility.

In addition to the power-controlled option where the reaction is performed at a constant power level, the possibility of setting an initial power level in the temperature-controlled mode does exist (see Figures 5.5 and 5.6). The Biotage instruments offer three different absorption levels where the normal level is suitable for most common solvents that are medium absorbers and uses up to 400 W. The level "high" with about 100 W output power should be used for strong absorbers and "very high" (about 60 W) for extremely good absorbers such as ionic liquids. With the CEM systems, these initial power levels have to be selected by the user within the dynamic method development setting: the specific power value has to be programmed in order to ramp the temperature to the set temperature. Recommended are 50 W for high absorbing, 125 W for medium absorbing and 200 W for poor absorbing solvents.

Some software packages additionally offer pressure-controlled method development which relies on the resulting pressure as a limiting factor. The microwave power is regulated by the adjusted pressure limit, and thus there is no influence on the resulting temperature. Because the reaction temperature is the most crucial parameter for successful chemical synthesis, this program feature is used only in rare instances.

For preliminary experiments, it is recommended that temperature programs are used in order to investigate the operational limits of the reaction. Once the maximum suitable temperature and pressure are known, the variation of microwave output power can be a further means of optimizing the reaction outcome. The option of introducing more power into the reaction system by simultaneous external cooling of the reaction mixture ("heating-while-cooling") has been discussed in Section 2.5.4.

Consequently, the choice of which strategy is utilized for reaction optimization experiments is highly dependent on the type of instrument used. Whilst multimode reactors employ powerful magnetrons with up to 1600 W microwave output power, monomode reactors apply a maximum of only 400 W. This is due to the high density microwave field in a single-mode set-up and the smaller sample volumes that need to be heated. In principle, it is possible to translate optimized protocols from mono-mode to multimode instruments and to increase the scale by a factor of 100 without a loss of efficiency (see Section 4.8).

In the advanced edit mode (Biotage) or so-called dynamic mode (CEM) more sophisticated reaction set-ups can be programmed in addition to the normal temperature–time parameters. Several heating or cooling steps, change in the stirring speed, programming of several reactions with definite increments or cooling while heating can be selected. An additional feature that is available is to change the reaction parameters on-the-fly. Editing process values such as temperature, time, power or cooling during the run is generally rarely used but the option does exist to set, for example, a higher power or to reduce the temperature when the set temperature cannot be reached within an adequate ramp time.

For both types of the commercially available single-mode instruments (Biotage Initiator, CEM Discover) nearly identical software options do exist as described above, the only difference being the software management. Whereas Discover instruments in general require an external computer for reaction monitoring and data storage – a built-in keypad for programming and on-the-fly changes is existent – the Initiator instruments have a built-in touch screen.

5.2.5
Reaction Optimization

If the initially chosen set of reaction parameters in a microwave-heated experiment produced a satisfactory result, the reaction mixture can be worked-up. In most cases this will not be the case and further optimization of the reaction protocol will be necessary (see Figure 5.1). In general, optimizing a microwave synthesis is very similar to optimizing conventional reactions, the reaction time and temperature being the main parameters to be changed. When initial reaction optimizations with respect to temperature and time do not lead to the expected results, all remaining parameters that usually would be varied – solvent, reagent, catalyst, molar ratios, concentration – should be altered. It should be emphasized that to conduct a systematic optimization study, only one variable at a time should be modified, so that one can trace which parameter change led to a success.

After analysis of the reaction mixture by standard methods such as TLC, HPLC or GC further modifications to the procedure may be deemed necessary, depending on the obtained results. In Table 5.2, an overview of optimization problems along with suggested solutions for obtaining the optimum reaction conditions is presented.

In general, the first parameters that should be modified in an optimization study are reaction temperature and/or time. When the temperature has to be increased because of no reaction or incomplete conversion, the only limits are the pressure

Table 5.2 Problems occurring during optimization and suggested solutions.

Result	Options
No reaction or incomplete conversion	Increase reaction temperature
	Extend the reaction time
	Change solvent, catalyst, reagents, and so on
	Change molar ratios of reagents
	Increase concentration
	Increase the initial power
Decomposition of reagents/products	Decrease temperature
	Shorten reaction time
	Change to a more temperature stable reagent
	Decrease concentration
	Decrease initial power
Reaction complete	Reduce time until the conversion is still complete to maximize rate enhancement

build-up in the vial and the operating limit of 250 or 300 °C as long as all the involved reagents/catalysts and solvents can withstand the higher temperature. Sometimes a less reactive, but more temperature stable reagent is preferable. A simple way to test the temperature stability of the reagents used involves mixing the respective substrate with the solvent and running the experiment without other reagents under the developed conditions. When, after analysis of the sample, the same outcome is observed, the reagent is stable at the chosen temperature/time conditions. In some cases the desired product may be formed but then decomposes at the typically high temperatures used in a microwave experiment. It can therefore be advisable to shorten the reaction time in order to "trap" the product.

In order to get a satisfactory result, in certain instances the concentration of the reaction mixture needs to be varied. This can be accomplished easily by simply switching to the next smaller or larger vial within the single-mode instrument.

A typical optimization regime for a microwave-mediated Suzuki reaction performed in the CEM Discover is shown in Figure 5.7 [8]. All possible parameters such as reaction time and temperature, microwave power, molar ratios of reagents, catalyst loading and amount of solvent were investigated to find the optimum reaction conditions.

The short reaction times provided by microwave synthesis make it ideal for rapid reaction scouting and optimization of reaction conditions. Compared to conventional methods, more optimization experiments can therefore be performed in the same time period. Many reaction parameters, such as reaction temperature and time, variations in solvents, additives and catalysts, or the molar ratios of the substrates can be evaluated in a few hours in order to optimize the desired chemistry. Rapid reaction optimization is particularly convenient since microwave synthesis is highly suitable for automation – single-mode instruments can be equipped with vial racks with up to 96 positions and with a robotic arm that moves the vials in and out of the cavity. Many optimization experiments can thus be performed sequentially in an automated

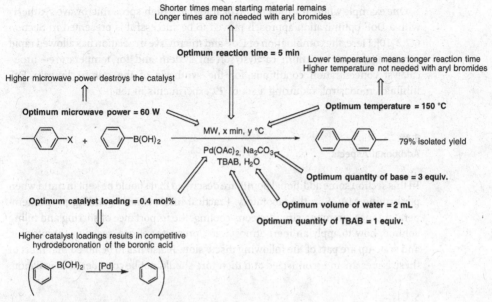

Figure 5.7 A typical optimization regime for a Suzuki reaction on a 1 mmol scale. Adapted from Ref. [8].

fashion without the necessity for user intervention (see Section 4.7.1). Due to the installed software options, reactions can be programmed with definite increment steps for either temperature or time. For example, a temperature optimization regime can be set up with only one programming step, for example, 100 °C starting temperature with an increment of 10 °C. The first reaction will be performed at a temperature of 100 °C. This temperature value will subsequently be constantly increased by 10 °C for each of the following experiments – 110 °C for the second, 120 °C for the third reaction, and so on. The above-mentioned advantages – faster reactions rates due to higher temperatures, more rapid heating, automation and software features – make optimization employing microwave heating in combination with modern analytical methods more time efficient and superior compared to conventional methods.

As an alternative to traditional reaction optimization, a statistics-based "Design of Experiments" (DoE) approach can be applied for investigating the optimum reaction conditions [9, 10]. DoE is a novel method for the optimization and screening of experimental parameters in organic synthesis. By integrating statistical DoE protocols into dedicated software packages, this method has become easily accessible and an efficient tool for reaction optimizations. A general issue when using DoE can be the potentially large number of experiments needed in order to obtain a plausible model. This concern can be solved by the use of appropriate robotics that is incorporated in modern single-mode microwave instruments. Due to the dramatically reduced reaction times, combining high-speed microwave synthesis with a DoE approach has already proven to be an ideal tool for obtaining optimized reaction conditions in a short period of time [9, 10].

One example where the integration of automated high-speed microwave synthesis with a DoE optimization approach proved to be successful is presented in Section 6.2.2 [10]. Here, the combination of DoE and microwave irradiation has allowed rapid screening for an optimum catalyst/solvent system and for temperature– time– catalyst concentration conditions for the synthesis of the mitotic kinesin Eg5 inhibitor monastrol, requiring a set of 29 experiments in total.

5.3
Additional Aspects

In this section some additional points are described that should be kept in mind when performing MAOS in the laboratory. Practical information regarding proper temperature measurement, simultaneous cooling, the importance of stirring and filling volume, how to apply an inert atmosphere, pressure release, reaction monitoring, and scale-up are part of the following discussion. It also has to be noted that most of these issues are interconnected and therefore should not be considered in isolation.

5.3.1
Temperature Measurement

Arguably one of the most important issues when discussing microwave-heated transformations is the measurement of the reaction temperature (see Section 2.5.1). Proper temperature monitoring is of the utmost importance in the context of reproducibility, in particular when moving from one microwave platform to another. Therefore, it has to be stressed again, that domestic microwave ovens (even if modified) should be banned from the chemical laboratory since, in addition to the associated safety risks, a correct temperature measurement cannot be provided. For example, the ACS organic chemistry journals will typically not consider work describing the use of kitchen microwave ovens or manuscripts without reaction temperature measurement, as specified in the relevant publication guidelines (see Box 5.1).

Box 5.1
ACS guidelines for publications describing microwave chemistry.

Reports of syntheses conducted in microwave reactors must clearly indicate whether sealed or open reaction vessels were used and *must document* the manufacturer and model of the reactor, the method of monitoring the reaction mixture temperature, and the temperature reached or maintained in each experiment. Reporting a wattage rating or power setting is not an acceptable alternative to providing temperature data.

Manuscripts describing work done with domestic (kitchen) microwave ovens will be accepted only when the reported studies have been conducted at atmospheric pressure.

In the most popular single-mode microwave reactors (Biotage Initiator, CEM Discover, see Chapter 3) the reaction temperature is generally determined by a calibrated external IR sensor, integrated into the cavity, that detects the surface temperature of the reaction vessel. It is assumed – if proper stirring is ensured and the reaction mixture is microwave absorbing – that the measured temperature on the outside of the reaction vessel will correspond more or less to the temperature of the reaction mixture contained inside. For routine synthetic applications the use of standard IR probes is, therefore, quite acceptable.

However, the user has to be aware of the inherent limitations of IR sensors and should recognize situations where the use of these external probes is not appropriate (see Sections 5.3.2 and 5.3.4 and Chapters 2 and 4). In such cases, internal temperature monitoring with a fiber-optic or related probe is more reliable (available for the CEM Discover systems).

It should be pointed out that when studying differences between microwave heating and conventional heating (in particular non-thermal microwave effects), it is particularly important to use highly accurate temperature monitoring devices. In order to accurately compare the results obtained by microwave heating with the outcome of a conventionally heated reaction, a reactor system should be used that allows one to perform both types of transformations *in the identical reaction vessel* and to monitor the internal reaction temperature in both experiments directly with the same fiber-optic probe device (see Section 2.5.1).

Another important issue is temperature monitoring in parallel microwave synthesis using multivessel rotor systems (see Section 4.7.1) where all reaction vessels are exposed to the same irradiation conditions. In general, when employing parallel rotors in multimode instruments, temperature monitoring is provided via a fiber-optic probe in one reference vessel and an external IR sensor that records the temperatures for all the rotor vessels. In particular, when low microwave absorbing solvents (or no solvent at all) are used in a library synthesis, the heating characteristics of the individual reaction mixtures are strongly dependent on the microwave absorptivity of the specific building blocks, thus leading to different temperatures in the reaction vessels, resulting in different conversions compared to a single-mode sequential method [11]. In such a case, the positioning of the fiber-optic sensor is crucial since the vessel with the fiber-optic probe is used as the control. For example, if the fiber-optic probe is placed in a vessel with high absorbing reagents, a lower microwave power will be used to reach and keep the temperature. As a consequence, the set temperature will not be attained for lower absorbing substrates, leading to lower yields. Therefore, when planning an experiment in parallel, careful consideration should be given as regards to in which rotor vessel the fiber-optic probe is immersed in order to obtain high quality results.

5.3.2
Simultaneous Cooling

The underlying principle of this technique is that the reaction vessel is cooled from the outside by compressed air or a microwave-transparent cooling fluid while being irradiated by microwaves. This allows a higher level of microwave power to be directly administered to the reaction mixture, but will prevent overheating by continuously removing latent heat.

The most important fact when considering employing simultaneous external cooling is to ensure an accurate temperature measurement. Importantly, external IR sensors should never be used in conjunction with simultaneous external cooling of the reaction vessel (see Section 2.5.1). Since the IR sensor will only provide the surface temperature of the reaction vessel (and not the "true" reaction temperature inside), cooling of the reaction vessel from the outside with compressed air will not afford a reliable temperature measurement and, therefore, affect the reproducibility of the results. It has been demonstrated by several research groups that by using the simultaneous cooling technique the internal reaction temperatures will be significantly higher than what is recorded by the IR sensor on the outside. When using simultaneous external cooling, an internal fiber-optic probe device must therefore be employed. Without knowing the actual reaction temperature, care must be taken not to misinterpret the results obtained with the simultaneous cooling approach (see Section 2.5.4).

5.3.3
Filling Volume

For all available microwave vessels, a certain minimum and maximum filling volume is specified by the microwave instrument manufacturers (see Chapter 3). The suggested minimum filling volumes are crucial to guarantee a precise temperature measurement and are dependent on the height at which the IR sensor is positioned or, on the length of the immersion temperature probe (fiber-optic or gas balloon). In the case of IR temperature measurement from the bottom of the vessel, a comparatively small amount of reaction mixture (200 µL for the 10 mL vial) may be sufficient to obtain a reliable temperature feedback for the CEM Discover series. On the other hand, an IR sensor mounted on the side wall of the cavity, as used in the Biotage instruments (see Figure 2.8), requires a specific minimum filling volume that is dependent on the type of reaction vessel (200 µL for the smallest reaction vial and 2 mL for the 10 mL vial; see Figure 3.8). Multimode instruments with an IR sensor mounted on the cavity side wall (see Section 3.5) need larger volumes for precise temperature monitoring. For immersion temperature probes, it must be ensured that the probe has sufficient contact with the reaction mixture, including when the mixture is stirred, in order to obtain reliable and reproducible results.

On the other hand, the maximum filling volumes (5–7 mL for standard 10 mL vials) are defined so as to leave enough headspace for pressure build-up. Too low a volume will give incorrect temperature monitoring, while too high a volume can lead to abortion of the experiment due to exceeding the pressure limit of the instrument (about 20 bar for single-mode instruments), leakage of volatile reagents/products/ solvents and, in the worst case, to a destructive vessel failure. For these reasons it is not advised to exceed or go below the specified volume range of the microwave vial.

5.3.4
The Importance of Stirring

With respect to correct and reliable temperature measurement, stirring is an important issue (see Sections 2.5.1 and 4.2). Inefficient agitation can lead to temperature gradients within the reaction mixture due to field inhomogeneities in the high-density single-mode microwave cavities. Therefore, external IR sensors will only represent the internal reaction temperature properly if efficient agitation – also of homogeneous reaction mixtures – is ensured. Extreme care must therefore be taken with heterogeneous reactions, such as solvent free, dry-media or highly viscous systems (see Sections 4.1 and 4.2).

Indicative for this aspect are for example unstirred biphasic mixtures of ionic liquids (ILs) and nonpolar organic solvents. The immiscibility creates a severe problem related to the accurate temperature measurement and reproducibility since differential heating will occur (see Section 2.5.3). In terms of the differential heating problem, an instrument-related issue comes into play: in the two most popular single-mode microwave reactors the temperature is measured at different positions (Biotage: at about 1.5 cm height from the side; CEM: from the bottom) of the otherwise more or less identical microwave vessels (Figure 2.8). Depending on the microwave system used, either the temperature of the very hot IL phase (IR from the bottom) or the temperature of the cooler organic layer (IR from the side) will be recorded (see also Figure 4.6). In order to avoid these problems it is recommend to use only ILs (e.g. as heating aids) that are soluble in the particular solvent in question.

Due to the field inhomogeneities in high power-density single-mode microwave reactors it should be emphasized that effective stirring should be ensured in any case, even for completely homogeneous reaction mixtures (see Section 2.5.1). In the case of a heterogeneous reaction mixture, a pre-stirring period for about 20 s before the heating process, as featured by the instrument software, should be considered. The stirring speed can also be varied: in the Biotage Initiator from 300 to 900 rpm, whereas three different modes (low, medium, high) can be chosen in the CEM Discover. If the user is not sure about the stirring efficiency, in particular when a large amount of starting material is not soluble in the solvent, an initial stirring test on a standard magnetic stirrer can be performed as a rough guide. The proper stirring speed can subsequently be set on the microwave instrument. If the stirring does not prove to be efficient enough larger or differently shaped stir bars can be employed, as long as the stir bar size is not 1/4 of the microwave wavelength (antenna effect). In most cases initially insoluble materials will dissolve in the chosen solvent as the temperature rises.

5.3.5
Pressure Release

As microwave heating in closed vessels is always accompanied with a certain autogenic pressure build-up due to the superheating of solvent to temperatures far above the boiling point under atmospheric conditions, a certain overpressure can

Figure 5.8 Different sealing mechanisms for CEM (left) and
Biotage (right) vials. Sealed crimp caps can be utilized also for
CEM vials in designated instruments.

remain in the vessel, even after the reaction mixture has been cooled down to about
40–50 °C using active gas jet cooling. In particular, an overpressure will persist when
gases are released during the reaction. This overpressure must be addressed only
when Biotage instruments are employed. Depending on the instrument type two
different vial-sealing mechanisms exist: "snap-on" IntelliVent caps for CEM and
permanently sealed crimp caps for Biotage vessels (Figure 5.8). Whereas no special
tool is necessary for attaching and removing the IntelliVent caps from CEM, a
crimper/decapper is needed for proper capping and decapping of the Biotage vials.

If the pressure in the vial exceeds 20 bar in the CEM instrument, the pressure
control system allows a controlled venting of the pressure and subsequently
automatically reseals to maintain optimum safety ("IntelliVent"). On the other
hand, in the Biotage reactors, as a safety feature, the cavity lid will not open when the
pressure in the vial exceeds 4 bar after the reaction mixture has been cooled. A
special venting feature is incorporated in order to release the remaining pressure
manually (Figure 5.9). A syringe needle can be inserted through the septum into the
vial, taking care not to insert the needle into the reaction mixture itself, since under
these circumstances the mixture may be ejected through the needle. If there is a risk
of emission of poisonous gases precautions have to be taken, for example, the gases
should be collected in a balloon. In addition, if the temperature is >60 °C the
overpressure should not be released due to safety issues.

In any event, after the reaction a slight overpressure of less than 4 bar can still
remain in the vial and can be detected by a bulge of the septum. If this is the case, the
pressure should also be released manually with a needle – for safety reasons, this
should be done in the fume hood.

5.3.6
Inert Atmosphere

In general, an inert atmosphere is often not necessarily required in a sealed-vessel
microwave experiment, even if the conventional reaction is carried out in this way.

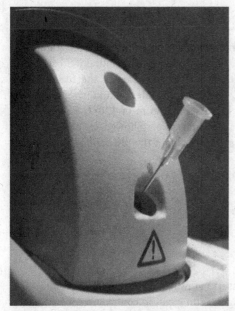

Figure 5.9 Overpressure release after the reaction (Biotage).

Many transition-metal catalyzed reactions that traditionally require an argon or nitrogen atmosphere due to air sensitive catalyst/ligand systems can be performed without inert atmosphere using microwave heating. An additional issue is the lifetime of catalysts – or heat sensitive species in general – that may be increased because of the minimization of wall effects under direct "in core" microwave irradiation preventing the decomposition of the catalysts on the hot vessel surface (see Section 2.5.3). The combination of these factors with the drastically reduced reaction times often avoids the necessity for an inert atmosphere.

Nevertheless, if an inert atmosphere is required, the microwave vial can be flushed either before capping, with an inert gas such as argon or nitrogen or, when the chemistry is more delicate, after capping, by simply piercing the septum with a needle that is connected, for example, to an argon line. Sometimes it is advantageous, as in conventional synthesis, to put the vial in the drying oven before use to eliminate traces of moisture.

Special equipment for gas loading is available from some manufacturers which, however, is typically used for charging the vessels with reactive gases (e.g., hydrogen, carbon monoxide, see Section 4.4).

5.3.7
Multiple Heating Steps

The application of a heating regime including several heating/ramping steps – without interim cooling – can sometimes be favorable for the reaction outcome, for example when the reaction pathway proceeds via intermediates from which different reaction products can be obtained dependening on the temperature. In addition, when multi-step reactions in one pot are performed, the programming of heating steps can be advantageous, for example when an initially chosen high temperature regime would be unsuitable for the first step of the reaction. An example is shown in Scheme 5.2, where a cascade Suzuki coupling/aldol condensation was performed [12]. A lower temperature of 120 °C for 5 min for the Suzuki coupling was chosen for the first step, followed by an additional 10 min at 150 °C for the aldol condensation to obtain the phenanthrene products in good yields.

The programming of several heating steps is feasible with all dedicated microwave instruments as this feature is typically included within the respective software.

5.3.8
Reaction Monitoring

One apparent disadvantage of sealed-vessel microwave synthesis is that monitoring the reaction progress is not possible during the run itself, in contrast to conventionally heated reactions under atmospheric/reflux conditions where reaction sampling is typically very easy. Sample withdrawal is only possible after interrupting the reaction by cooling and opening the reaction vessel. The only exceptions being the Advancer (Biotage, see Section 3.5.2, Figure 3.23) and the BatchSYNTH (Milestone, see Section 3.5.4.4, Figure 3.39) multimode instruments where special inlet and outlet ports enable this feature. Subsequent analysis of the reaction mixture can thus be conducted by standard methods such as TLC, HPLC or GC. It has been demonstrated that by using *in situ* Raman spectroscopy the progress of a variety of chemical transformations performed under sealed-vessel microwave conditions can be directly monitored online [7]. However, this kind of instrumentation is not commercially available to date and only custom-built systems exist. There are also several limitations as the reaction mixtures need to be completely homogeneous and quite concentrated to obtain high quality spectra.

Similarly, in a closed vessel the operator is not able, as in a conventional experiment, to visually follow the progress of the reaction while it proceeds. Color or viscosity changes, the formation of precipitation or the efficiency of the magnetic stirring, can therefore not easily be monitored. The latest development toward the elimination of some of these problems is a built-in camera for *in situ* reaction monitoring, available for the CEM Discover S-Class instruments [13].

5.3.9
Reaction Scale versus Instrument

Depending on the desired reaction scale the user has to decide on the appropriate instrumentation, the choice being either a single-mode or a multimode microwave

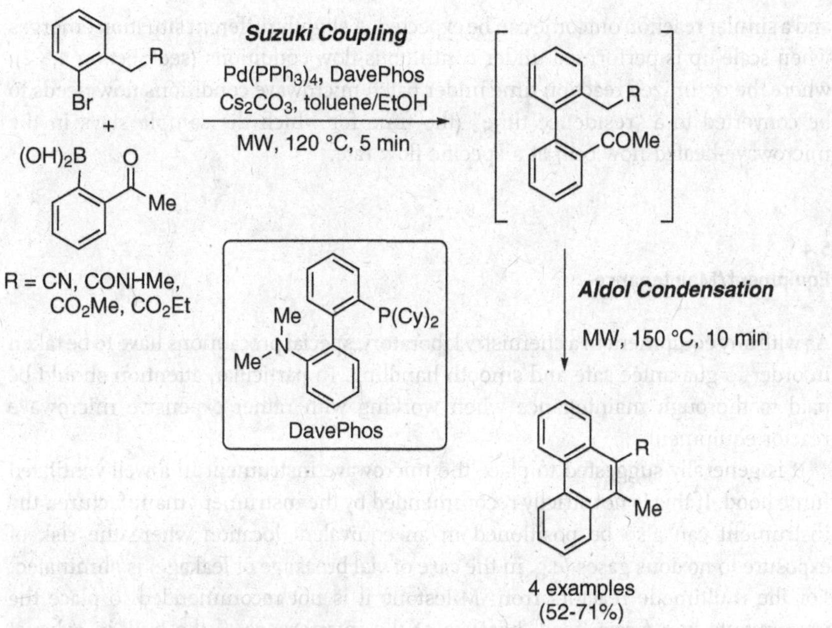

Scheme 5.2 Suzuki coupling/aldol condensation cascade reaction including a two-step heating regime.

reactor. Most examples of microwave-assisted chemistry published to date have been performed on a less than 1 g scale with a typical reaction volume of 1–5 mL in single-mode instruments, whereas multimode reactors are mainly considered for scale-up purposes. As already mentioned in Chapter 3, the available microwave power in multimode instruments is higher than in single-mode systems (1000–1600 versus 300–400 W), but the power density of the field is comparatively low, making heating of small individual samples (<3 mL) difficult. A constant high microwave output power would need to be employed when attempting to heat a too small sample in a nearly empty microwave cavity, in particular if the sample is low microwave absorbing. This could lead to damage and a shorter lifetime of the magnetron by overheating. Therefore, multimode instruments should generally be used for larger scale synthesis.

In contrast, single-mode instruments generate a single, comparatively homogeneous energy field of high power intensity, and therefore couple efficiently with small samples.

When considering the scale-up of MAOS, reactions should ideally be directly scalable from small to large scale. The previously optimized reaction conditions on a small scale using a single-mode instrument can be transferred to a larger scale synthesis for multimode reactors without the need for changing any parameters. The only issue that typically has to be considered is the longer ramp time the reaction will need in order to attain the desired temperature. When the small scale reaction was optimized using the hold time mode (Figure 5.4a), reproducibility should be assured and a similar reaction outcome can be expected. A slightly different situation emerges when scale-up is performed under continuous-flow conditions (see Section 4.8.2), where the optimized reaction time under batch microwave conditions now needs to be converted to a "residence time" (the time for which the sample stays in the microwave-heated flow cell) at a specific flow rate.

5.4
Equipment/Maintenance

As with any equipment in a chemistry laboratory, special precautions have to be taken in order to guarantee safe and smooth handling. In particular, attention should be paid to thorough maintenance when working with rather expensive microwave reactor equipment.

It is generally suggested to place the microwave instrument in a well-ventilated fume hood. If this is not strictly recommended by the instrument manufacturer, the instrument can also be positioned in an equivalent location where the risk of exposure to noxious gases (e.g. in the case of vial breakage or leakage) is eliminated. For the multimode reactors from Milestone it is not recommended to place the instruments in a fume hood, because of the interference of the built-in exhaust system with the ventilation of the fume hood. Only the built-in exhaust system should be connected to the hood. Sometimes a high level of dust in the air can cause problems to the built-in ventilation systems due to the accumulation of dust and therefore instrument performance could be slowed down.

The main issue related to maintenance is cleaning the microwave cavity after a vial breakage or leakage. All glass particles and spill have to be removed from the cavity. In addition, the waste tray insert (Biotage Initiator) or spill cup (CEM Discover) needs to be emptied or replaced. The cavity is best cleaned by employing a tissue soaked with an appropriate solvent that is able to dissolve the chemicals used in the experiment. When suitable, vacuuming could be helpful to completely remove the debris.

Special attention must be paid to the IR sensor, which should be cleaned with a soft tissue dampened with distilled water or alcohol. Care has to be taken not to scratch the surface, since the temperature measurement could thus be distorted. For instruments where the IR sensor is located at the bottom (CEM Discover), it must be ensured that the path between the sensor and the reaction vessel is unobstructed, as debris and solvent can easily collect at or near the IR sensor.

After a careful cleaning of the microwave cavity and IR sensor, the proper performance of the system has to be checked, that is, the accurate temperature and pressure measurement. Either the IR sensor has to be re-calibrated (CEM) after every vial breakage or a reference run with a water sample (Biotage) has to be performed where the temperature and pressure readings have to be within defined limits.

For a more detailed description on what specific steps have to be taken for maintenance, the user is referred to the respective instrument manuals and to the instructions given by the manufacturer. In any event, if the microwave instrument is seriously damaged after a vessel failure or other malfunction, or the temperature and pressure measurement is not correct, a trained service engineer should be contacted.

5.5
General Suggestions

In this section, a selection of general suggestions and tricks that should be considered in order to exploit the full potential of MAOS is discussed (see Box 5.2). Since microwave-mediated organic synthesis nowadays is a well accepted and mature technology and has lost its stigma as a laboratory curiosity, this enabling technology should be used for all possible reaction steps, not only for difficult ones – it should truly be considered the "first choice" rather than "last resort" method of heating reactions in the laboratory. In addition, the short reaction times associated with MAOS offer a tremendous timesaving when applying microwave heating right from the start, not only making use of this technology when the reaction is not working well under conventional heating conditions. In the past, microwave conditions were often used only when all other options to perform a particular reaction had failed, or when exceedingly long reaction times or high temperatures were required to complete a reaction. At present, microwave heating is often still used as the "last resort" to obtain a meaningful conversion in a reasonable timeframe for one single difficult transformation in a long sequence of steps. The fact that, for example, natural product synthesis can also be carried out in a different way, was recently reported for the 17-step asymmetric total synthesis of the fungal metabolite (−)-stephacidin A where, out of the 17 steps, six reaction steps were performed under microwave heating [14]. The authors reported that the use of microwave technology helped in reducing reaction times from hours to minutes for most transformations and also increased the yields of several key transformations.

Before getting started with a microwave chemistry project, the literature or the available web-based databases should be consulted in order to find proper reaction conditions. More than 3000 publications covering syntheses applying microwave heating in dedicated instruments can be found in the literature covering the period from 2001 to 2008. In the Biotage PathFinder database a collection of detailed protocols (4500 entries) for reactions performed with Biotage single-mode microwave instruments is available [1] whereas other databases cover published MAOS protocols since 1996 [2].

Microwave heating is the ideal technology to try out "crazy" ideas, while still bearing safety considerations in mind (see Section 5.6). Due to the possibility of rapid reaction scouting, failure only costs a couple of minutes, whereas success may save hours or days.

However, it should also be obvious that microwave chemistry cannot be the solution for every synthetic problem and will simply not work in all cases. For instance, any user should be aware of thermodynamically and kinetically controlled reaction pathways for a specific transformation. Since microwave heating – or heating in general – will generally favor the synthesis of the thermodynamically

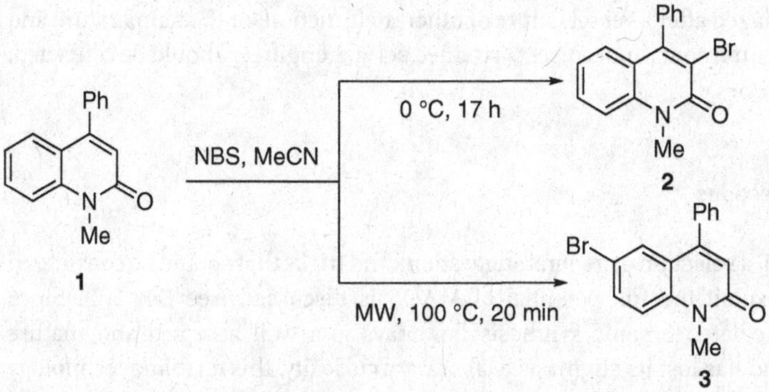

Scheme 5.3 Thermodynamic and kinetic control in the bromination of quinolones.

more stable product, high temperature microwave heating will not always lead to the desired product. An example is shown in Scheme 5.3. In the bromination of quinolinone 1, either the 3-bromo or 6-bromo isomer, 2 or 3, can be obtained, depending on the reaction temperature [15]. Performing the bromination at 100 °C (microwaves, 20 min) allowed the selective preparation of the 6-bromoquinolinone isomer, whereas conducting the reaction at a temperature of 0 °C (17 h) led to a 92 : 8 selectivity in favor of the 3-bromo derivative. Apparently, the higher reaction temperatures used under microwave conditions favored the formation of the thermodynamically more stable 6-bromoquinolin-2(1*H*)-one isomer. Here, microwave chemistry was therefore able to speed up an otherwise slow reaction, but ultimately led to an undesired product!

Another important concern is the use of new microwave vessels, when critical chemistry is performed. In particular, when transition metal-catalyzed reactions with low catalyst loadings are conducted, the use of new vials is recommended in order to ensure that trace impurities impregnated on the inside of the glass vessel are not interfering with the chemistry. The reason for this lies in the fact that, even after proper cleaning of a used microwave vial, transition metal-catalysts can still adhere to the glass surface, thus leading to incorrect presumptions on catalyst loading and

Box 5.2
General tips.

- Using MAOS for all possible reaction steps, not only for difficult ones: should be your "first choice" rather than "last resort"
- Check literature or a database before getting started
- Try out "crazy ideas". Failure only costs a couple of minutes, success may save hours or days!
- Think about your chemistry!
- Use new microwave vessels for critical chemistry
- Report proper experimental procedures (follow ACS guidelines).

reaction outcome. In this context it should be mentioned that both CEM and Biotage recommend using their microwave reaction vessels only once. Vials used in reactions with a tendency to build up persistent residues should also not be allowed to remain long in circulation. Similarly, stir bars with noticeable discolorations or defects should be discarded.

In the interest of reproducibility it is strongly recommended that microwave procedures should be properly described following ACS guidelines (see Box 5.1). Most importantly, the reaction temperature has to be documented – even when a reaction was performed under power control. In addition, it should be stated whether the synthesis was conducted under open- or closed-vessel conditions, if total irradiation or hold time was used and the model and manufacturer of the microwave instrument should be declared.

5.6
Safety Aspects

Although commercially available dedicated microwave reactors are designed in such a way that a safe performance of organic reactions is virtually guaranteed, some precautions have to be taken when planning and preparing microwave experiments. The safety aspects that have to be considered, in particular for beginners in microwave chemistry, are presented in this section (see Box 5.3).

Operating with chemicals and pressurized vessels always carries a certain risk, but the safety features and the precise reaction control of commercially available

---

Box 5.3
Safety considerations for performing MAOS.

- Use dedicated instrumentation only!
- Do not operate damaged equipment
- Be aware of microwave leakage
- Check reaction vessels for damage before use
- Always start on a small scale
- Do not fall below minimum or exceed maximum filling volume
- Understand your chemistry
- Be aware of gas-producing transformations (headspace)
- Be aware of the expected pressures (solvent library)
- Be aware of solvent and reagent stability
- Be careful when employing compounds with risky functional groups (N_3, NO_2)
- Avoid corrosive reagents/reactants
- Be aware of reaction kinetics (exothermic reactions/thermal runaways)
- Be careful with strongly absorbing materials (heterogeneous catalysts)
- Always ensure adequate stirring (localized superheating) but . . . . . . do not use 3 cm stir bars (1/4 wavelengths = antenna).

microwave reactors protect the users from accidents, perhaps more so than with any classical heating source. The use of domestic microwave ovens in conjunction with flammable organic solvents is hazardous and must be strictly avoided as these instruments are not designed to withstand the resulting conditions when performing chemical transformations.

No damaged equipment should be operated. Also be aware that by misuse of the equipment, damage can occur. An example would be the use of very low microwave-absorbing mixtures: if the reaction mixture (including the used reagents or catalysts), is not sufficiently polar to couple efficiently with microwaves, there is a risk of damaging the instrument or parts of the equipment. This is mainly a problem associated with multimode instruments, if low-absorbing materials or too small samples are irradiated with high microwave output power for extended time periods. The introduced microwave energy has to be absorbed somehow, and if it is not absorbed by the reagents then vessel sensors or other accessories may inadvertently be heated. To avoid such destructive processes, it can be advantageous to add highly microwave-absorbing components as heating aids to the reaction mixtures (see Sections 4.5.2 and 4.6). Modern microwave instruments are equipped with comprehensive safety features to protect the magnetrons from damage. Specially designed reflector systems prevent microwaves from being reflected to the magnetrons or a double-check temperature measurement avoids overheating by promptly switching off the magnetrons if the output power is not efficiently converted.

On the other hand, extremely polar and thus strongly microwave absorbing mixtures may heat too rapidly and, therefore, the safety limits of the instrument or of the vessels are exceeded (rate of pressure and temperature increase), causing damage to the equipment. This may occur if the vessels are overloaded with too high a volume or too concentrated mixtures. The use of more dilute mixtures, the programming of heating ramps or the setting of an appropriate initial power level (selection of a lower power value) can avoid these problems.

The user should be aware that when operating damaged instruments, microwave leakage can occur. When the instrument is in proper working condition no microwave leakage generally arises since all commercially available instruments are designed for safe use. It should be emphasized that under open-vessel conditions, where it could be conceived that microwave irradiation could "escape" via the reflux condenser, it is safe for the operator to perform MAOS because of the attached attenuator (see Figure 5.10) that prevents microwave leakage. For most systems the attenuator is a cylindrical device, attached at the top of the instrument, that has to have defined dimensions – the height of the cylinder must be 2.5 times the diameter and the diameter should be smaller than 7.18 cm to provide a sufficient barrier for the microwaves. The maximum allowed MW leakage at 5 cm distance is $5 \, \text{mW cm}^{-2}$. Relatively inexpensive digital microwave leak detectors are readily available and it is advisable to have such a device on hand to periodically check for stray microwave radiation.

Microwave vials should always be carefully checked prior to their use to see if they are damaged by scratches or cracks, in particular when the vials have been reused very often. Generally, a single usage of the vials is recommended by the instrument

Figure 5.10 Microwave attenuator for open-vessel microwave conditions (CEM Discover).

manufacturers but for economic reasons – especially in academic research – the vials are often reused several times. If damage is noticed, the vial has to be discarded, since the pressure stability of the glass is reduced by even very small scratches.

To reduce the risk of vessel failure, the microwave pressure vials come with several safety features. These can include an effective self-venting system where unforeseen overpressure is released by a quick open-resealing step, or the use of safety disks which rupture when their pressure limit is reached. The vials (0.2–20 mL) of the Biotage single-mode reactors are protected by the pressure limit (20 bar) of the caps used, which is significantly lower than the operating limit of the vials themselves (40–50 bar).

If an unknown reaction is intended to be performed under microwave irradiation it is recommended to start on a small scale – less than 5 mL filling volume – since it can not be predicted precisely how the reaction will behave under microwave conditions. Similarly, a scale-up should not be conducted at more than a volume factor of 5–10 at a time.

Another piece of advice is to stay within the limits of the manufacturer-specified vessel filling volume (see Section 5.3.3). A correct temperature measurement cannot be guaranteed when the minimum filling volume is not reached. Since some IR sensors are positioned at a defined height, a specific filling level is required. On the other hand, when the temperature is monitored internally with a fiber-optic probe, proper contact with the reaction mixture has to be assured. More important with respect to safety is the maximum filling volume since, when exceeding this limit, vessel breakage can occur due to the lack of head space for volatiles.

When planning an experiment under sealed-vessel microwave conditions, the chemistry needs to be properly considered, since most of the occurring vial explosions can be ascribed to operators being unfamiliar with the chemistry. Care has to be taken when reactions are performed that lead to the evolution of gaseous components, for example catalytic transfer hydrogenations involving ammonium formate. Here, significant amounts of ammonia, carbon dioxide and hydrogen are released. The volume of gas that will be formed has to be kept in mind, in order not to exceed the operational limits through the build-up of pressure (consider the headspace in the vial). Otherwise, migrating to an open vessel is another alternative for gas-producing reactions, but this at the same time significantly reduces the available temperature range, since solvent superheating, as under closed-vessel conditions, is no longer possible.

When performing sealed-vessel microwave synthesis, the expected pressure of the solvent system (see Section 5.2.2) and the pressure limit of the vial or the employed instrument, respectively, have to be matched. With some instruments, a solvent library will help the user to estimate the expected pressure in the reaction vessel. Included in the web-based Biotage Pathfinder is also a vapor-pressure calculator for a variety of solvents and chemicals where the expected pressure at a certain temperature can be estimated [1].

In general, the pressure build-up should not be a real safety issue for well-designed systems, but can certainly impede the experimental progress. If overpressurization occurs either the venting mechanisms release the overpressure, or the microwave power is shut off when the operational limit is reached, thus protecting the equipment from damage. Even so, if the pressure build-up occurs too fast, this can result in vessel failure. In case of any venting activity, it is essential that the exhaust system immediately starts to withdraw the solvent vapors. Especially in large-scale syntheses, the formation of ignitable mixtures can be critical. Organic solvents are often highly flammable and the microwave radiation might act as an ignition source for the system and cause severe explosions. Thus, proper protection of the microwave instrumentation used is absolutely necessary to minimize the potential hazard to the operator. All multimode instruments can be fitted with electronic solvent sensors, which measure the concentration of organic solvents in the atmosphere within the cavity. If a critical limit is reached, the magnetrons are turned off and the solvent vapors are exhausted.

Another crucial point is the thermal stability of the solvents and reagents used. In general, thermally labile compounds benefit from the short reaction times when employing microwave heating, but on the other hand the applied temperatures are often higher than with classical heating. Thus, when exposed to high temperatures, several solvents and reagents may decompose, forming hazardous compounds. Potential high risk compounds include those containing nitro or azide groups, as well as low molecular weight ethers, as these are known to cause explosions when heated. For instance, diethyl ether, which forms explosive peroxides and has a very low boiling point, should not be used in microwave-assisted reactions. A recommended approach to find the optimum reaction conditions for chemistry involving labile starting materials such as azide- or nitro-compounds is to start at low temperatures and low substrate concentrations.

Corrosive reagents – or reactions where corrosive by-products are formed – such as for example phosphorous oxychloride, highly concentrated acids (e.g., hydrofluoric acid or sulfuric acid) or the use of strong bases like sodium hydroxide should be avoided in microwave chemistry. The safety risk when employing these reagents in high concentrations lies in the corrosion of metallic cavity parts in the case of vial breakage or leakage. Another issue is the degradation of the inner surface of the glass vial. If this is observed, the vial has to be discarded. A somewhat safer use of highly corrosive reagents can be ensured by dilution with appropriate solvents.

Precautions should be taken, especially in a scale-up approach, when dealing with exothermic reactions under microwave irradiation. Due to the rapid energy transfer of microwaves, any uncontrolled exothermic reaction is potentially hazardous (thermal runaway). The temperature increase and pressure rise may occur too rapidly for the instrument's safety mechanisms and vessel rupture may occur.

The use of strongly microwave-absorbing materials that are insoluble in the solvent system – in particular heterogeneous metal catalysts – forms a distinct hazard and often leads to damage of the reaction vessels. As metals are typically very strong microwave absorbers, the formation of extreme hot spots may occur, which may weaken the stability of the vessels due to the onset of melting processes and may cause explosive destruction of the reaction vials. Therefore, care has to be taken when heterogeneous metal catalysts such as palladium-on-charcoal are employed. It has to be ensured that all palladium is covered with solvent, so that no dry particles stick to the inner wall of the vial. The choice of solvent is also crucial when heterogeneous catalysts are utilized – the catalyst should be well suspended in the solvent, otherwise the catalyst will adhere to the vessel wall and an explosion can occur. The probability that these problems may arise is increased when large amounts of the heterogeneous catalyst are used. If homogeneous metal catalysts which can produce elemental metal precipitates (e.g., palladium or copper) are applied, stirring is highly recommended to avoid the deposition of thin metal layers on the inner surface of the reaction vessel. It is generally recommended to use metal/ligand complexes that have a high temperature stability when performing microwave-assisted transition metal-catalyzed reactions. The use of labile metal catalysts such as palladium acetate at high temperatures in a solvent such as toluene will almost certainly ensure the formation of a palladium metal film on the inner wall of the reaction vessel. If the film gets too thick when using high catalyst concentration, absorbance of microwave irradiation coupled with arcing phenomena will occur, leading to destruction of the reaction vessel. Such arcing phenomena have been recorded by utilizing built-in digital cameras [13] and can also be observed when irradiating, for example, magnesium turnings in a microwave reactor [16].

A common phenomenon that is encountered when using microwave irradiation to heat solvents under atmospheric conditions is so-called superheating (see Section 2.5.3), whereby the solvents are heated well above their boiling points under open-vessel conditions. This effect is exploited for some synthetic strategies, but it may also be hazardous as superheated mixtures can start to boil spontaneously. Localized superheating, that is the selective heating of strongly microwave absorbing heterogeneous catalysts or reagents in a less polar reaction medium, can also be an issue. Therefore, stirring is always recommended in MAOS, but the right size of the stir bar is crucial. If the metallic core is 1/4 of the wavelength of the radiation (3 cm; $\lambda = 12.25$ cm) the stir bar will operate as an antenna, causing spark discharges and destruction of the stir bar or even explosion of the vessel.

5.7
Advantages and Limitations

As with any technology, there are benefits but also limitations – the main advantages and limitations encountered with microwave heating are listed in Box 5.4.

In general, most reactions that can be carried out under thermal heating can also be performed, and most likely accelerated, by microwave irradiation. It has to be stressed that most published results suggesting rate enhancements and improved yields on going from conventional heating to microwave heating can be explained in terms of purely thermal/kinetic effects (see Section 2.5).

Microwave-assisted reactions are generally performed at substantially higher reaction temperatures than conventional reflux experiments. The temperature profiles may in some cases be difficult – if not impossible – to reproduce under standard thermal heating. These high temperatures – up to 100 °C above the boiling point – can be reached very quickly due to the direct "in core" microwave heating as a result of a rapid energy transfer. The higher temperatures can lead to several different effects apart from the expected increase in reaction speed – in many instances cleaner transformations, leading to less by-products compared to conventionally heated processes will be experienced. These beneficial effects can be quite substantial considering the rapid heating and cooling (see Figure 2.13). The minimization of wall effects due to the direct "in core" heating, and cases where yields have increased dramatically are not uncommon.

Due to the advantages mentioned above, microwave synthesis is a valuable tool for rapid reaction scouting and optimization of reaction conditions. Compared to

---

Box 5.4
Advantages and limitations of microwave heating.

Advantages

- Energy efficient direct "in core heating", rapid energy transfer
- Rapid superheating of solvents in sealed vessels
- Reduced reaction times (from hours to minutes)
- Higher yields/cleaner reactions
- Rapid reaction scouting and optimization of conditions
- Can do things that you cannot do conventionally
- Excellent control over reaction parameters
- Ideally suited for automation and high-throughput synthesis.

Limitations

- No direct reaction monitoring or visual inspection
- No reagent addition during the reaction (closed vessel)
- Not applicable for production scale
- Equipment cost.

conventional heating, more optimization reactions can be conducted in the same timeframe and a "Yes" or "No" answer is obtained considerably faster.

Selectivities in chemical transformations are quite often influenced by the reaction temperature and since microwave-heated reactions are often performed in a comparatively high temperature regime, altered product distributions and selectivities as compared to a conventionally heated experiment will sometimes be experienced. Thus, unexpected reaction products are occasionally observed under microwave irradiation, favoring reaction pathways not seen under conventional processing at lower temperatures.

By employing commercially available dedicated microwave instruments excellent control over the reaction parameters is assured – an exact temperature and pressure measurement is provided by the corresponding monitoring devices. Note that, in contrast to traditional heating, a more precise reaction control is often experienced when applying microwave heating in dedicated systems.

In addition, MAOS is highly suitable for automation due to the instruments' incorporated robotics. As speed is a critical factor in the field of drug discovery and medicinal chemistry, the combination of microwave heating with high-throughput techniques for compound library generation is nowadays a popular and rather convenient application.

However, one has to keep in mind that the dramatic rate enhancements compared to a classical, thermal process cannot be achieved in all cases. On the other hand, considering all the presented benefits associated with microwave heating, the simple convenience of using microwave technology makes this non-classical heating method almost a "must have" tool in modern synthetic chemistry.

One of the main limitations associated with MAOS is analytics – more precisely, reaction monitoring (see Section 5.3.8). In general, the withdrawal of samples for analytical purposes is not possible while the pressurized reaction vial is in the microwave cavity. Since no sample withdrawal is possible, it is obvious that this is also true for reagent addition. The standard operating procedure "dropwise addition of a reagent" that is heavily used in traditional organic synthesis cannot, therefore, be applied in a sealed-vessel microwave experiment. To visually follow the progress of a reaction while being processed under sealed-vessel microwave conditions, a digital camera can be integrated into a microwave reactor (CEM Discover S-Class). However, this clearly adds additional cost to the microwave system.

Another drawback of microwave heating technology is the scale-up from the g to the kg scale, where the limiting issue is the penetration depth (see Sections 2.3 and 4.8). At present, performing MAOS on a multi-kilogram scale is still difficult to accomplish and publications in this area are very rare. To overcome this severe limitation, microwave instrument manufacturers are currently involved in designing and developing new equipment to manage some of the hurdles associated with scale-up beyond the traditional laboratory scale.

Another issue of this relatively new technology is equipment cost. Although prices for dedicated microwave reactors for organic synthesis have come down since their first introduction in early 2000, the current price range for microwave reactors is still many times higher than that of conventional heating equipment. This fact has severely limited the penetration of microwave synthesis in academic laboratories around the world.

5.8
Frequently Asked Questions (FAQs)

1. *Is it okay to use kitchen microwave ovens for chemistry?*
   No. Domestic microwave ovens are not designed for performing chemical transformations. In particular, the lack of safety controls makes these devices inappropriate tools for conducting MAOS. In contrast to dedicated instruments, the cavities are not produced to withstand explosions from a vessel breakage. In addition, there is no possibility of stirring the reaction mixture and of monitoring temperature and pressure. Problems of reproducibility will occur because of the nonexistence of these monitoring features and due to the inhomogeneities of the microwave field (formation of hot- and cold spots).

2. *I want to get started with microwave chemistry. Should I buy a single- or multimode instrument?*
   This depends mainly on the type of research that is being performed and on the required reaction scale. For small scale discovery chemistry applications in academia and industry, method development or reaction optimization where reaction volumes of up to 50 mL are sufficient, a single-mode instrument is preferable. In combination with automation features such as robotic vial transfer, single-mode reactors are also very useful for library production purposes. With the currently available single-mode instruments, both closed- and open-vessel synthesis as well as flow synthesis can be performed.
   On the other hand, if reactions on a >100 mL scale need to be performed on a routine basis, or more advanced processing techniques (high temperature/pressure, pre-pressurized, parallel synthesis) are required, the use of a multimode instrument is suggested. For all commercially available multimode instruments a variety of different reactor configurations is available.

3. *Do I need to place my microwave instrument in the fume hood?*
   It is generally suggested to place the instrument in a well-ventilated fume hood. If this is not explicitly required by the manufacturer, the instrument can also be placed in an equivalent location where the risk of exposure to noxious gases (e.g., in the case of vial failure or leakage) is minimized.

4. *Can microwaves leak from the instrument?*
   All the instruments are designed in such a way that no microwave leakage can occur when in proper working conditions. Even under open-vessel reflux microwave conditions an attached "attenuator" prevents microwave leakage from the cavity (see Figure 5.10).

5. *How do I translate conditions from published kitchen microwave experiments to dedicated systems?*
This is not possible. Since a reliable temperature measurement is typically not feasible when using a kitchen microwave oven it is necessary to completely reoptimize the chemistry under temperature control in a dedicated system. The given power ratings (e.g., 5 min at 400 W, or 3 min at 60% power level) often found in publications describing kitchen microwave experiments are more or less meaningless.

6. *How do I translate oil-bath chemistry into microwave chemistry?*
If the reaction temperature and time are known, it is comparatively simple to convert from conventional to microwave conditions (Section 5.2). By using the rule of thumb that a 10 °C increase in reaction temperature results in a halving of the reaction time, the proper conditions can quickly be estimated. Another way is to employ a time/temperature prediction chart that is based on the Arrhenius equation and is available on the Biotage web site. In addition, the "translating oil-bath into microwave conditions feature" included with the CEM and Biotage instruments can be used. The conversion is calculated automatically by the software (standard method programming with the Synergy software at CEM and method creation with the Wizard at Biotage).

7. *What is a good temperature/time starting point for an unknown reaction?*
We would recommend starting at a temperature that is about 30–40 °C higher than the boiling point of the solvent and a reaction time of 5–10 min. Further optimizations can subsequently be performed, for example, by increasing the temperature, keeping in mind the microwave absorptivity of the solvent and the decomposition temperature of reagents and solvents.

8. *Should I run my reactions in temperature- or power-controlled mode?*
It is most common and advisable to perform reactions in the temperature-controlled mode where the software system attempts to reach the set temperature as fast as possible by applying an appropriate level of microwave magnetron output power. In the power-controlled mode the temperature will rise until the maximum temperature for the set power value is reached. This feature should only be used by more experienced microwave users. As there is no control over the resulting temperature or pressure, care has to be taken not to exceed the operating limits of the vessel/reactor system.

9. *How do I select the initial power level/absorbance level?*
The initial power level is dependent on the absorptivity of the solvent or reaction mixture used. For the Biotage instruments three different levels can be chosen: normal, high and very high. The normal level is suitable for most common solvents that are medium absorbers and uses up to 400 W of magnetron power. The level "high" with about 100 W output power should be used for strong absorbers and very high (about 60 W) for extremely good absorbers such as ionic liquids. For the CEM instruments 50 W are recommended for high, 150 W for medium and 250 W for poor absorbers. The selection of the appropriate initial

power level is important in order to prevent an overshoot of the temperature or to achieve the selected temperature faster.

10. *Is it necessary to stir the reaction mixture?*
Yes. Since the field homogeneity in single-mode instruments is not as good as often portrayed it is necessary to stir even completely homogeneous reaction mixtures (see Section 2.5.1). If the reaction mixture is not stirred or efficient stirring cannot be ensured – for example, for solvent-free or dry-media reactions and for very viscous or biphasic reaction systems – temperature gradients may develop, and thus the temperature measurement will not be reliable and representative of the whole reaction volume.

11. *Do I have to use a solvent?*
Not necessarily. Solvent-free or dry-media reactions are reported to be "green" processes, the pressure build-up is rather low, and in most instances these reactions can be performed under open-vessel conditions (see Section 4.1). However, some severe disadvantages exist since stirring and accurate temperature measurement can prove rather difficult. Moreover, degradation or decomposition of reagents can be severe problems for these kind of reactions. Since MAOS under sealed-vessel conditions is nowadays a safe and reliable technology, there is essentially no reason not to use a solvent for small scale organic synthesis.

12. *Can I use an inert atmosphere?*
Yes, but often an inert atmosphere is not necessarily needed, even if the reaction is carried out in this way conventionally (see Section 5.3.6). Many transition metal-catalyzed reactions that traditionally require an inert atmosphere due to sensitive catalyst/ligand systems can be performed without an argon or nitrogen atmosphere by applying rapid sealed-vessel microwave processing.
To apply an inert atmosphere, the microwave vial – if necessary oven-dried – can be flushed with an inert gas such as argon or nitrogen, either before capping or, when the chemistry is more delicate, after capping by using a syringe needle through the septum.

13. *What is the difference between hold time and total irradiation time?*
Applying "hold time", the ramp time to reach the desired set temperature is excluded from the processing time, whereas the "total irradiation time" includes the ramp time as well. It is recommended to set a hold time due to reproducibility issues, in particular when going from small scale to large scale where the ramp time often takes significantly longer. In any report or publication it should be stated if the reaction times given under microwave conditions refer to "hold time" or "total irradiation time".

14. *My reaction does not reach the selected temperature, what should I do?*
Different solutions are possible, all following the same principle of doping the solvent/reaction mixture with polar and thus more strongly absorbing materials

(Sections 4.5.2 and 4.6). Water, for example, can be doped with small amounts of inorganic salts such as sodium chloride to reach a higher temperature under microwave conditions. Adding small quantities of extremely strongly absorbing ionic liquids to organic solvents has the same effect. In contrast to these invasive additives that change the polarity of the reaction mixture, more convenient passive (non-invasive) heating elements such as silicon carbide cylinders can also be introduced to the reaction mixture.

15. *Can I use resins in the microwave?*
    Yes. Successful peptide synthesis on solid-phase has been performed at temperatures around 60 to 70 °C (Section 4.7.2.1). The commercially available microwave peptide synthesizers guarantee a proper handling of the resin beads, for example, Teflon vessels and agitation via gas bubbling. Solid-phase organic synthesis or polymer-assisted solution phase synthesis that is normally performed at higher temperatures has also been carried out (Section 4.7.2). The higher temperatures and the applied stirring generally do not affect the resin beads, probably due to the associated short reaction times when applying microwave heating. Stability studies revealed that degradation of polystyrene resin starts at 220 °C but reactions up to 200 °C in some cases proceeded smoothly.

16. *Can I use molecular sieves as water scavengers in conjunction with microwave heating?*
    Some publications report on the use of molecular sieves (typically 4 Å) for the scavenging of water in microwave-assisted reactions. However, no detailed investigations on the impact of strongly microwave-absorbing molecular sieves in these processes have been published.

17. *Can I use gaseous reagents under microwave conditions?*
    Yes. Special accessory tools make it possible to pre-pressurize a microwave vial with gaseous reagents (e.g., gaseous hydrogen for hydrogenation reactions or carbon monoxide for carbonylations, see Section 4.4). Instead of employing gaseous reagents, solid reagents that liberate the reactive gas are a popular alternative. For example $Mo(CO)_6$ liberates carbon monoxide at higher temperatures for carbonylation reactions. In any event, sufficient headspace to guarantee a safe reaction performance needs to be left – consider the autogenic pressure of the solvent in addition to the pre-pressurization!

18. *I often get an explosion running my palladium-catalyzed reactions. What should I do?*
    Ensure that all palladium is covered with solvent, so that no dry particles stick on the glass wall of the vial (if necessary use a spatula to move the palladium into the reaction mixture). Palladium is an extremely good absorber of microwave energy, will melt the vial and cause explosions if traces are attached to the glass surface. The same is true for other transition metal catalysts and related highly microwave-absorbing materials. Also be careful with the employed solvent, not every solvent is compatible with transition metal catalysts, in particular when heterogeneous catalysts are used. For example, do not use toluene in combination with palladium-on-charcoal since toluene does not suspend this catalyst well, the

palladium catalyst can stick to the glass wall and thus explosions can occur. Be careful with the amount of heterogeneous catalyst as well. The more catalyst, the higher the possibility for an explosion.

19. *Why can water not be heated above 200°C?*
    The dielectric loss ε'' and loss tangent $\tan\delta$ of pure water decrease with increasing temperature. This means that the absorption of microwave irradiation in water decreases at higher temperatures. While it is relatively easy to heat water from room temperature to 100 °C, it is significantly more difficult to further heat water to 200 °C and beyond in a sealed vessel. In fact, supercritical water ($T > 374$ °C) is transparent to microwave irradiation (Sections 2.3 and 4.5.1). Most organic materials and solvents behave similarly to water, in the sense that the dielectric loss ε'' will decrease with increasing temperature.

20. *Is there any way to monitor reactions while they are in the closed microwave cavity?*
    In the Discover S-Class reactor (CEM) the reaction can be monitored via a built-in camera for *in situ* reaction monitoring. Otherwise it is rather difficult to monitor the reaction under closed-vessel conditions while the reaction is running. Typically, the reaction needs to be stopped, cooled, the vessel removed from the cavity and subsequently opened before an analysis can be performed.

21. *Can I withdraw samples while the reaction is going on?*
    When performing reactions under closed-vessel conditions in single-mode instruments, no. Sample withdrawal is only possible after stopping the reaction by cooling and opening the reaction vessel, and in some of the multimode reactor systems.

22. *I have optimized my reaction on a small scale but do not get the same result on a larger scale. What is the reason?*
    One possible reason could be that the required reaction temperature was not reached in the large scale run (Section 4.8). Due to the larger volume that needs to be heated, the microwave power of a single-mode instrument may not be sufficient to heat the reaction mixture to the identical temperature when going to larger scales, for example from 5 to 50 mL. To overcome this problem, a switch to a better absorbing solvent (e.g., from toluene to benzotrifluoride) may be recommended (see Figure 4.20).
    Another point to consider is stirring – an appropriately sized stir bar should be used that is capable of stirring the larger volume efficiently (see FAQ 10). When the scale-up experiment is conducted in a multimode instrument, be aware of the longer ramp time that is necessary to reach the set temperature and keep in mind the difference between hold time and total irradiation time.

23. *Does the pressure in a sealed-vessel microwave experiment have a direct influence on the reaction?*
    No. The comparatively moderate pressure of 3–20 bar inside a typical microwave vial is far too low to have a direct influence on the kinetics of a specific reaction

pathway as in true high pressure chemistry where pressures >10 kbar are generally applied.

24. *Will a fiber-optic sensor give me a more accurate temperature measurement?*
Generally, yes. Especially when low absorbing solvents (hexane, toluene) are employed, the internal temperature measurement with a fiber-optic probe gives a more reliable result than the external IR temperature sensor (Section 2.5.1). Since the standard Pyrex microwave vials for use in single-mode instruments do absorb some microwave irradiation, the external temperature that is measured on the glass surface will be higher than the actual temperature inside the vessel. In contrast, for medium to good microwave absorbing solvents/reaction mixtures the difference between internal (fiber-optic) and a properly calibrated external (IR) temperature measurement will be small. When reactions are performed applying simultaneous cooling, however, the internal temperature must be measured with a fiber-optic sensor.

25. *Should I use the simultaneous cooling option?*
If you have the possibility on your instrument and you do have an internal fiber-optic temperature probe you may want to give it a try. Results differing from conventional microwave heating have been reported for those transformations where either the starting material or the product was temperature sensitive (Section 2.5.4).

26. *Does the IR temperature sensor need to be calibrated?*
This depends on the microwave instrument. The IR sensor for the CEM instruments needs to be calibrated from time to time (particularly after vessel failure or other malfunction in the cavity). For the Biotage reactors, a reference run with a water sample has to be performed to ensure that the IR sensor is working properly, since the sensor cannot be calibrated by the user. If the temperature and pressure values are not within the specified limits (vapor pressure in correlation to the temperature), the instrument sensors need to be calibrated by a service engineer. After a destructive vessel failure or vial leakage, the IR sensor has to be cleaned properly and checked if it is in proper working condition.

27. *My laboratory does not have a compressed air supply. Can I still use the cooling feature?*
Instead of the compressed air supply, a standard compressor can be connected to the instruments, which is sufficient for efficient active gas jet cooling after the reaction.

28. *Can I use the same glass vial for both the CEM and Biotage single-mode instruments?*
No. The vials look very similar but are different in dimension and therefore not interchangeable. The CEM vial is somewhat smaller in diameter but larger in length and the user must ensure that the proper vial for the respective microwave instrument is employed.

29. *Can I reuse the glass vials for my CEM/Biotage single-mode reactors?*
Single vial usage is recommended by both manufacturers. If the vials are not damaged by any scratches or cracks, the vials may be reused at the users own

risk and will have to be checked very carefully before starting a new reaction. It is recommended to use new vials when sensitive chemistry is performed since trace amounts of, for example, metal catalysts can be impregnated into the vessel surface and can influence the outcome of a chemical transformation.

30. *Is microwave heating more energy efficient than conventional heating?*
Microwave processing under sealed-vessel conditions (taking advantage of increased reaction rates at higher temperatures) will be significantly more energy efficient than conventional heating in open vessels at the solvent reflux temperature (Section 4.8). It is important to note that the energy savings in these cases are mainly the result of the significantly shortened reaction times and are not directly connected to the heating mode. On the other hand, when reactions are compared using both microwave and oil-bath open-vessel conditions for the same time period, the thermal runs are typically more energy efficient than the microwave experiments due to the limited energy efficiency of magnetrons in converting electric energy into microwave irradiation.

31. *Can I use other frequencies apart from 2.45 GHz?*
In principle, yes. Although five permitted microwave frequencies for industrial, scientific and medical use do exist, all commercially available microwave instruments operate at a frequency of 2.45 GHz in order not to interfere with telecommunication, wireless networks and cellular phone frequencies. MAOS at other frequencies than 2.45 GHz (e.g., 5.8 GHz) has rarely been reported in the literature.

32. *Why is not everybody using this technology if it is so valuable?*
One of the major drawbacks of this technology is equipment cost. As with any new technology, the current situation is bound to change over the next several years and less expensive equipment should become available. At this point microwave instruments will probably be standard equipment in every chemical laboratory.

References

1 www.biotagepathfinder.com.

2 www.mwchemdb.com.

3 Hayes, B.L. (2002) *Microwave Synthesis: Chemistry at the Speed of Light*, CEM Publishing, Matthews, NC.

4 Wan, Y.Q., Alterman, M., Larhed, M. and Hallberg, A. (2002) *The Journal of Organic Chemistry*, 67, 6232–6235.

5 Wu, Y.-J., He, H. and L'Heureux, A. (2003) *Tetrahedron Letters*, 44, 4217–4218.

6 Enquist, P.-A., Nilsson, P. and Larhed, M. (2003) *Organic Letters*, 5, 4875–4878.

7 Leadbeater, N.E. and Smith, R.J. (2007) *Organic and Biomolecular Chemistry*, 5, 2770–2774.

8 Leadbeater, N.E. (2005) *Chemical Communications*, 2881–2902.

9 Tye, H. (2004) *Drug Discovery Today*, 9, 485–491; Evans, M.D., Ring, J., Schoen, A., Bell, A., Edwards, P., Berthelot, D., Nicewonger, R. and Baldino, C.M. (2003) *Tetrahedron Letters*, 44, 9337–9341; Tye, H. and Whittaker, M. (2004) *Organic and Biomolecular*

Chemistry, **2**, 813–815; McLean, N.J., Tye, H. and Whittaker, M. (2004) *Tetrahedron Letters*, **45**, 993–995; Gopalsamy, A., Shi, M. and Nilakantan, R. (2007) *Organic Process Research & Development*, **11**, 450–454; Sarotti, A.M., Spanevello, R.A. and Suárez, A.G. (2007) *Green Chemistry*, **9**, 1137–1140.

10 Glasnov, T.N., Tye, H. and Kappe, C.O. (2008) *Tetrahedron*, **64**, 2035–2041.

11 Amore, K.M., Leadbeater, N.E., Miller, T.A. and Schmink, J.R. (2006) *Tetrahedron Letters*, **47**, 8583–8586.

12 Kim, Y.H., Lee, H., Kim, Y.J., Kim, B.T. and Heo, J.-N. (2008) *The Journal of Organic Chemistry*, **73**, 495–501.

13 Bowman, M.D., Leadbeater, N.E. and Barnard, T.M. (2008) *Tetrahedron Letters*, **49**, 195–198; Leadbeater, N.E. and Shoemaker, K.M. (2008) *Organometallics*, **27**, 1254–1258.

14 Artman, G.D., III, Grubbs, A.W. and Williams, R.M. (2007) *Journal of the American Chemical Society*, **129**, 6336–6342.

15 Glasnov, T.N., Stadlbauer, W. and Kappe, C.O. (2005) *The Journal of Organic Chemistry*, **70**, 3864–3870.

16 Dressen, M.H., Kruijs, B.H., Meuldijk, J., Vekemans, J.A. and Hulshof, L.A. (2007) *Organic Process Research & Development*, **11**, 865–869.

6
Experimental Protocols

The final chapter in this book provides a collection of 27 experimental protocols encompassing a variety of different microwave processing techniques. Although most of the published microwave procedures today are performed in single-mode reactors using standard sealed-vessel conditions, there are cases where it is appropriate to use, for example, multimode instruments or open-vessel conditions. This chapter, therefore, to some extent reflects the contents of Chapter 4 where the most popular microwave processing techniques were discussed.

Each of the 27 experimental protocols describes the exact reaction conditions and provides information on the settings used on the particular microwave instrument. For each of the employed reagents, appropriate safety data have been included, which makes it easy to adapt these experiments for a practical course on microwave-assisted organic synthesis. It should be noted that in the reagent and hazard information tables at the beginning of each experiment the amounts of the chemicals needed for the complete protocol are listed – including also the work-up procedure. Reaction times range from 1 min to 2 h, and yields are consistently high. In general, starting materials are inexpensive and readily available, and straightforward work-up and purification (via extraction or precipitation/filtration) provides products of acceptable purity without the need for chromatography. Most of the microwave protocols in this chapter are adapted from published procedures originating from the author's own laboratory or other reliable sources. Each of the protocols was further evaluated by a team of advanced undergraduate students and checked for reproducibility and practicability.

Section 6.1 contains a collection of 13 protocols that utilize single-mode sealed-vessel conditions on a small reaction scale. Although all examples in this section have been performed in a Biotage Initiator Eight EXP instrument in 10 mL reaction vessels, the experiments can alternatively be carried out in other single-mode reactors that allow operation on the same scale (CEM Discover, Milestone MultiSYNTH). The first eight protocols (Section 6.1.1) involve classical organic transformations that are performed in standard organic solvents and include, for example, Diels–Alder cycloadditions, the formation and transformation of Grignard reagents, a palladium-catalyzed Heck reaction using a heterogeneous catalyst, an oxidation reaction with manganese dioxide and general carbonyl group chemistry. These examples represent the large majority of microwave chemistry experiments published today.

Practical Microwave Synthesis for Organic Chemists: Strategies, Instruments, and Protocols
C. Oliver Kappe, Doris Dallinger, and S. Shaun Murphree
Copyright © 2009 WILEY-VCH Verlag GmbH & Co. KGaA, Weinheim
ISBN: 978-3-527-32097-4

In Section 6.1.2 the *N*-alkylation of pyrazole in toluene is described that, due to the low microwave-absorbing properties of the solvent, necessitates the use of a heating aid such as an ionic liquid (see Section 4.5.2) or a so-called passive heating element (see Section 4.6). Given the importance of sustainable and "green" processing techniques today, several microwave chemistry examples are presented that involve water as solvent (Section 6.1.3, see Section 4.5.1), or are performed without solvent (Section 6.1.4, see Sections 4.1 and 4.2).

The high degree of control afforded by dedicated microwave reactors – and the rapid reaction times possible in most cases – make this technology ideal for optimizing processes with respect to cycle time, yield, and/or purity. For all single-mode microwave reactors automated sample handling capabilities are available, as well as convenient software functions to set up an experimental array. Whether through trial-and-error, full-parameter, or statistical "Design of Experiment" (DoE) approaches, microwave-assisted synthesis offers significant advantages over conventional technology. In Section 6.2 two different strategies for optimization are exemplified: one variable at a time (OVAT) and DoE. The former is more useful with a limited number of parameters or for systems that yield to intuitive navigation. The latter is preferred for more complex multivariate synthetic applications (see Section 5.2.5).

For the same reason that automated robotic microwave reactors are well-suited to the task of process optimization, they are also invaluable tools for the generation of compound libraries in the discovery phase. There are essentially two ways to carry out this task: through automated sequential synthesis or simultaneous (parallel) preparation (see Section 4.7.1). The examples highlighted in Section 6.3 showcase library generations of each type: a six-member Hantzsch library prepared in sequential mode using a Biotage Initiator Eight, a 12- member Biginelli library using a fully automated sequential set-up (Biotage Liberator), and an eight-member Biginelli library synthesized in parallel using the Milestone MicroSYNTH multimode system equipped with a 16-vessel rotor.

Although the present book and the covered microwave chemistry examples in this chapter have been selected keeping the microwave novice in mind, the scale-up of conventional small scale microwave chemistry is also addressed in this chapter. In general, the scale-up of chemical processes is a non-trivial endeavor, the success of which depends upon the proper consideration of agitation efficiency, thermal transfer, and mass transport phenomena. Microwave-assisted synthetic techniques are certainly not immune to these challenges, and pilot-plant scale microwave reactors are not yet a commercial reality. However, dielectric heating can be an indispensable tool for scaling up across the four orders of magnitude between the 0.1 mmol and 1 mol scales. The practical examples covered in Section 6.4 illustrate three common methods used for scale-up with microwave reactors (Section 4.8): (i) the batch-wise scale-up of an aspirin synthesis from a 10 mmol scale in a single-mode instrument to a 100 mmol scale in a multimode cavity, and the Suzuki reaction from a 1.5 mmol sealed-vessel experiment to a 150 mmol open-vessel preparation; (ii) parallel scale-up of a Biginelli reaction using an eight-vessel rotor in a multimode reactor; and (iii) the Fischer indole synthesis adapted to continuous flow conditions.

The final microwave chemistry examples presented in this chapter are designed for the advanced user and touch upon some unusual or specialized reaction techniques with microwave reactors: (i) removing volatile by-products during a reaction (Section 4.3); (ii) experiments using gaseous reagents (Section 4.4); (iii) working at extremely high temperatures and/or pressures (Section 4.5.1); and (iv) solid-phase peptide synthesis (Section 4.7.2.1). Most of these protocols require access to a particular microwave instrument or set of hardware.

For all experiments described in this chapter the reader is strongly advised to consult the original reference and the appropriate sections in Chapter 4 (Microwave Processing), Chapter 3 (Equipment) and Chapter 5 (Practical Tips) before starting the experiment.

6.1
General Small-Scale Sealed-Vessel Microwave Processing

6.1.1
Organic Medium

6.1.1.1 Base-Catalyzed Ester Formation

Table 6.1 Reagents and hazard information for the base-catalyzed ester formation.

Reagent [CAS No.]	mmol	Quantity
2,4,6-trimethylbenzoic acid [480-63-7] caution: avoid contact and inhalation	1.0	164 mg
potassium carbonate [584-08-7] harmful; harmful if swallowed; irritating to eyes, respiratory system and skin	1.81	250 mg
bromoethane [74-96-4] flammable; harmful; harmful by inhalation and if swallowed; irritating to eyes, respiratory system and skin; limited evidence of a carcinogenic effect; possible carcinogen; target organs: nerves, heart; may develop pressure; California Prop. 65 carcinogen	2.0	150 µL
acetone, anhydrous [67-64-1] flammable liquid; delayed target organ effects; moderate skin irritant; moderate eye irritant		3.0 mL
dichloromethane [75-09-2] toxic; harmful if swallowed; irritating to eyes, respiratory system and skin; may cause cancer; OSHA carcinogen; California Prop. 65 carcinogen; readily absorbed through skin; target organ: heart because methylene chloride is converted to carbon monoxide in the body; target organ: central nervous system because of possible dizziness, headache, loss of consciousness and death at high concentrations		30 mL
water		30 mL
sodium sulfate [7757-82-6] may be harmful if swallowed, inhaled, or absorbed through skin; may cause eye, skin, or respiratory tract irritation		0.5 g

Equipment	Biotage Initiator Eight 2.0 EXP
Vial preparation	Into a 10 mL process vial equipped with a stirring bar are placed 2,4,6-trimethylbenzoic acid (164 mg, 1.0 mmol), finely powdered potassium carbonate (250 mg, 1.81 mmol), bromoethane (150 µL, 2.0 mmol), and anhydrous acetone (3.0 mL). The vial is sealed by capping with a Teflon septum fitted in an aluminum crimp top.

Microwave processing		
	Time:	10 min
	Temperature:	150 °C
	Pre-stirring:	10 s
	Absorption level:	normal
	Fixed hold time:	on

Work-up and isolation	After cooling, the reaction mixture is transferred to a round-bottom flask and the solvent is evaporated *in vacuo*. The residue is partitioned between dichloromethane (10 mL) and water (10 mL). The organic phase is extracted 2× with 10 mL of water and the combined aqueous phases are re-extracted 2× with 10 mL of dichloromethane. The combined organic phases are dried over sodium sulfate and concentrated *in vacuo*.
Yield and physical data	Ethyl 2,4,6-trimethylbenzoate is obtained as a clear oil (192 mg, 100%). ^{1}H NMR (DMSO-d_6, 360 MHz) $\delta = 6.89$ (s, 2H), 4.30 (q, $J = 7.1$ Hz, 2H), 2.24 (s, 3H), 2.20 (s, 6H), 1.29 (t, $J = 7.1$ Hz, 3H).
Literature	[1]

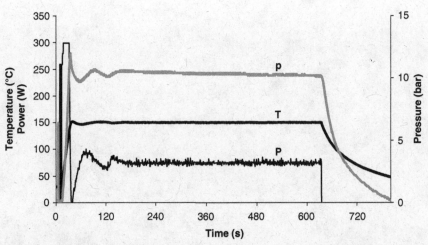

Figure 6.1 Power/pressure/temperature profile for the base-catalyzed ester formation.

Background Mesitoic acid (2,4,6-trimethylbenzoic acid) is notoriously recalcitrant to esterification, and previously reported microwave-mediated preparations involve neat mixtures of mesitoic acid and an alcohol in the presence of a strong acid catalyst, such as *p*-toluenesulfonic acid [2] or sulfuric acid [3]. Furthermore, it is usually difficult to alkylate carboxylic acids with non-activated alkyl halides, especially when the reaction suffers from steric hindrance.

The current method, therefore, represents a workaround for this synthetic challenge using mildly basic conditions and readily available alkyl bromides [1].

6.1.1.2 Knoevenagel Condensation

89%

Caution: Since one equivalent of gaseous carbon dioxide is formed, care must be taken to provide sufficient headspace volume in the reaction vessel (about 1.8 mL headspace per mmol product, see also Chapter 5).

Table 6.2 Reagents and hazard information for the Knoevenagel condensation.

Reagent [CAS No.]	mmol	Quantity
benzaldehyde [100-52-7] harmful; harmful in contact with skin and if swallowed; irritating to skin; may cause sensitization by inhalation and skin contact; combustible; readily adsorbed through skin; target organs: central nervous system; liver	4.0	410 µL
malonic acid [141-82-2] toxic; harmful if swallowed; toxic by inhalation; risk of serious damage to eyes	6.0	624 mg
piperidine [110-89-4] flammable; toxic; harmful if swallowed; toxic by inhalation and in contact with skin; causes burns	6.0	590 µL
ethanol [64-17-5] flammable; irritant; irritating to eyes, respiratory system and skin; target organs: nerves, liver		1.0 mL
water		15 mL
hydrochloric acid, 1 M [7647-01-0] may cause skin irritation; may be harmful if absorbed through the skin; may cause eye irritation; material may be irritating to mucous membranes and upper respiratory tract; may be harmful if inhaled; may be harmful if swallowed		6.0 mL

Equipment	Biotage Initiator Eight 2.0 EXP
Vial preparation	Into a 10 mL process vial equipped with a stirring bar are placed benzaldehyde (410 µL, 4.0 mmol), malonic acid (624 mg, 6.0 mmol), piperidine (590 µL, 6.0 mmol), and ethanol (1.0 mL). The vial is sealed by capping with a Teflon septum fitted in an aluminum crimp top.

Microwave processing	Time:	10 min
	Temperature:	140 °C
	Pre-stirring:	10 s
	Absorption level:	normal
	Fixed hold time:	on

Caution: Residual pressure must be relieved before removing the vessel from the microwave cavity by carefully inserting a needle into the pressure relief port. Please consult the manufacturer's instructions for proper procedure.

Work-up and isolation	The reaction mixture is poured into water (10 mL) and acidified under agitation by the addition of 1 M hydrochloric acid (6.0 mL). The resulting slurry is cooled in an ice-bath for 30 min, after which the precipitate is isolated by filtration and washed with cold water (5.0 mL). The product is dried overnight at 50 °C.
Yield and physical data	Cinnamic acid is obtained as a white solid (526 mg, 89%). mp 133 °C; ^{1}H NMR (DMSO-d_6, 360 MHz) $\delta = 12.41$ (br s, 1H), 7.69-7.67 (m, 2H), 7.59 (d, $J = 16$ Hz, 1H), 7.42–7.40 (m, 3H), 6.53 (d, $J = 16$ Hz, 1H).
Literature	[1]

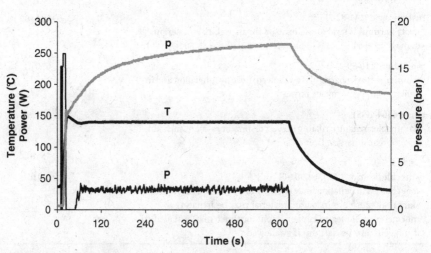

Figure 6.2 Power/pressure/temperature profile for the Knoevenagel condensation.

Background Although there have been a number of reports on microwave-assisted open-vessel Knoevenagel reaction protocols (mostly using domestic microwave ovens), no sealed-vessel microwave procedure has so far been reported. The current method thus represents a novel entry into sealed-vessel Knoevenagel chemistry in very good yield and excellent purity with a simple reaction and work-up sequence [1].

6.1.1.3 Williamson Ether Synthesis

Table 6.3 Reagents and hazard information for the Williamson ether synthesis.

Reagent [CAS No.]	mmol	Quantity
4-nitrophenol [100-02-7] harmful by inhalation, in contact with skin and if swallowed; readily absorbed through skin; target organs: blood, nerves	5.0	695 mg
potassium carbonate [584-08-7] harmful; harmful if swallowed; irritating to eyes, respiratory system and skin	5.0	690 mg
bromoethane [74-96-4] flammable; harmful; harmful by inhalation and if swallowed; irritating to eyes, respiratory system and skin; limited evidence of a carcinogenic effect; possible carcinogen; target organs: nerves, heart; may develop pressure; California Prop. 65 carcinogen	6.0	450 µL
methanol [67-56-1] flammable liquid; toxic by inhalation, ingestion, and skin absorption; irritant; target organs: eyes, kidney, liver, heart, central nervous system		1.5 mL
sodium hydroxide, 2% solution [1310-73-2] corrosive; causes burns		15 mL
diethyl ether [60-29-7] toxic; extremely flammable; may form explosive peroxides; harmful if swallowed; irritating to eyes, respiratory system, and skin; repeated exposure may cause skin dryness or cracking; vapors may cause drowsiness and dizziness; target organs: central nervous system, kidneys		65 mL
sodium sulfate [7757-82-6] may be harmful if swallowed, inhaled, or absorbed through skin; may cause eye, skin, or respiratory tract irritation		0.5 g

Equipment	Biotage Initiator Eight 2.0 EXP
Vial preparation	Into a 10 mL process vial equipped with a stirring bar are placed 4-nitrophenol (695 mg, 5.0 mmol), potassium carbonate (690 mg, 5.0 mmol), bromoethane (450 µL, 6.0 mmol), and methanol (1.5 mL). The vial is sealed by capping with a Teflon septum fitted in an aluminum crimp top.

Microwave processing	Time:	10 min
	Temperature:	120 °C
	Pre-stirring:	10 s
	Absorption level:	normal
	Fixed hold time:	on

Caution: Residual pressure must be relieved before removing the vessel from the microwave cavity by carefully inserting a needle into the pressure relief port. Please consult the manufacturer's instructions for proper procedure.

Work-up and isolation — The contents of the vial are transferred to a separatory funnel and partitioned between diethyl ether (15 mL) and 2% aqueous sodium hydroxide (15 mL). The aqueous layer is again extracted 2× with diethyl ether (25 mL); the organic layers are then combined, dried over sodium sulfate, and concentrated *in vacuo*.

Yield and physical data — 4-Nitrophenetol is obtained as a white solid (739 mg, 89%). mp 60 °C; ^{1}H NMR (DMSO-d$_6$, 360 MHz) $\delta = 8.19$ (d, $J = 9.2$ Hz, 2H), 7.12 (d, $J = 9.2$ Hz, 2H), 4.17 (q, $J = 6.9$ Hz, 2H), 1.36 (t, $J = 6.9$ Hz, 3H).

Literature — [5]

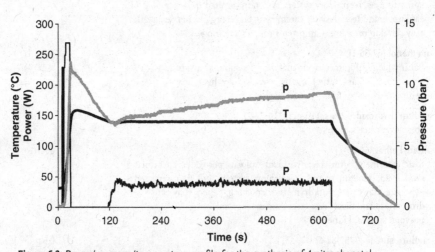

Figure 6.3 Power/pressure/temperature profile for the synthesis of 4-nitrophenetol.

Background The acceleration of this reaction by microwave irradiation was the subject of Gedye's pioneering work in the field over 20 years ago [4]. More recently, aryl alkyl ether synthesis has been adapted to the single-mode sealed-vessel microwave reactor. A biphasic system of potassium carbonate in methanol was found to be effective with phenols bearing an electron-withdrawing substituent in the para-position and non-activated alkyl bromides and chlorides [5]. The current procedure was adapted from this methodology.

6.1.1.4 Grignard Reaction

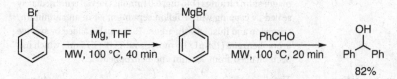

Table 6.4 Reagents and hazard information for the Grignard reaction.

Reagent [CAS No.]	mmol	Quantity
magnesium [7439-95-4] (Alfa Aesar 010232) flammable; irritant; contact with water liberates extremely flammable gases; target organs: kidneys; central nervous system	6.0	146 mg
bromobenzene [108-86-1] combustible; irritant; dangerous for the environment; irritating to skin; toxic to aquatic organisms; may cause long-term adverse effects in the aquatic environment; target organs: liver, kidneys	1.5	158 μL
tetrahydrofuran [109-99-9] flammable; irritant; may form explosive peroxides; irritating to eyes and respiratory system; possible carcinogen; target organs: nerves; liver		3.75 mL
benzaldehyde [100-52-7] harmful; harmful in contact with skin and if swallowed; irritating to skin; may cause sensitization by inhalation and skin contact; combustible; readily adsorbed through skin; target organs: central nervous system; liver	1.5	152 μL
hydrochloric acid, 0.1 M [7647-01-0] may cause skin irritation; may be harmful if absorbed through the skin; may cause eye irritation; material may be irritating to mucous membranes and upper respiratory tract; may be harmful if inhaled; may be harmful if swallowed		45 mL
dichloromethane [75-09-2] toxic; harmful if swallowed; irritating to eyes, respiratory system and skin; may cause cancer; OSHA carcinogen; California Prop. 65 carcinogen; readily adsorbed through skin; target organ: heart because methylene chloride is converted to carbon monoxide in the body; target organ: central nervous system because of possible dizziness, headache, loss of consciousness and death at high concentrations		90 mL
sodium sulfate [7757-82-6] may be harmful if swallowed, inhaled, or absorbed through skin; may cause eye, skin, or respiratory tract irritation		0.5 g
hexane [110-54-3] flammable liquid; irritant; reproductive hazard; target organs: peripheral nervous system, kidney, testes		3.0 mL

Equipment	Biotage Initiator Eight 2.0 EXP
Vial preparation, part I	Into a 10 mL process vial equipped with a stirring bar are placed magnesium turnings (146 mg, 6.0 mmol). The vial is immediately sealed by capping with a Teflon septum fitted in an aluminum crimp top and flushed with argon. To the vial is added by syringe bromobenzene (158 µL, 1.5 mmol) and THF (3.75 mL) which has been distilled from sodium and benzophenone.

Microwave processing

Time:	40 min
Temperature:	100 °C
Pre-stirring:	10 s
Absorption level:	normal
Fixed hold time:	on

Vial preparation, part II	To the reaction mixture is added by syringe benzaldehyde (152 µL, 1.5 mmol).

Microwave processing

Time:	20 min
Temperature:	100 °C
Pre-stirring:	10 s
Absorption level:	normal
Fixed hold time:	on

Work-up and isolation	The reaction mixture is poured onto 0.1 M hydrochloric acid (45 mL) and the resulting mixture is stirred until the residual magnesium has decomposed, then the resulting mixture is extracted with dichloromethane (3×30 mL). The combined organic extracts are dried over sodium sulfate and concentrated *in vacuo*. The residue is recrystallized in hexane (about 3 mL), and the product is isolated by filtration.
Yield and physical data	Diphenylmethanol is obtained as an off-white solid (226 mg, 82%). mp 66 °C; ^{1}H NMR (DMSO-d_6, 360 MHz) $\delta = 7.36$ (d, $J = 7$ Hz, 4H), 7.29 (t, $J = 7$ Hz, 4H), 7.19 (t, $J = 7$ Hz, 2H), 5.87 (d, $J = 4$ Hz, 1H), 5.68 (d, $J = 4$ Hz, 1H).
Literature	[1, 8, 9]

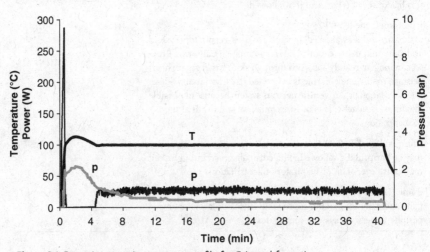

Figure 6.4 Power/pressure/temperature profile for Grignard formation.

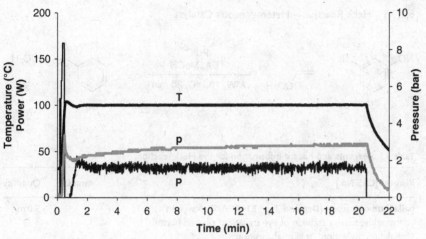

Figure 6.5 Power/pressure/temperature profile for the Grignard reaction.

Background The Grignard reaction is an important and well-characterized classical technique for the generation of alcohols with the concomitant formation of carbon–carbon bonds [6]. One of the most notorious limitations of the method, however, is the sometimes capricious behavior of the organometallic reagent preparation, especially with non-activated alkyl- and aryl halides. Various techniques have been developed to make these reactions more rapid and robust, usually by modifying the physical form of the magnesium metal [7]. However, microwave irradiation can be used to realize even greater kinetic gains, and in some cases the time-consuming magnesium pre-treatments can be bypassed [8–10]. Additionally, arcing phenomena of magnesium turnings in the Grignard reagent formation were reported when employing microwave irradiation [10]. This phenomenon seems to show a beneficial effect on the initiation times due to the removal of the MgO layer from the surface of the magnesium turnings.

The current procedure was adapted from the reports by Suna and Nilsson, compressing the reaction times so that the synthesis and isolation can be conducted within a three-hour period [8, 9].

6.1.1.5 Heck Reaction – Heterogeneous Catalyst

Table 6.5 Reagents and hazard information for the Heck reaction.

Reagent [CAS No.]	mmol	Quantity
palladium-on-carbon (Degussa Type E105CA/W, 5 weight%) flammable; may cause skin or eye irritation; may be harmful by inhalation, ingestion, or skin absorption		5.0 mg
triethylamine [121-44-8] flammable liquid; corrosive; highly toxic by skin absorption; delayed target organ effects; target organs: heart, kidney, liver, central nervous system	2.0	278 µL
acetonitrile [75-05-8] flammable liquid; delayed target organ effects (lungs, blood, kidney, liver, central nervous system); mild skin irritant; severe eye irritant		3.0 mL
3-bromobenzonitrile [6952-59-6] harmful by inhalation, in contact with skin, and if swallowed; irritating to eyes, respiratory system and skin	2.0	364 mg
acrylic acid [79-10-7] combustible liquid; toxic by inhalation, ingestion, and skin absorption; corrosive; mutagen; target organ effect; target organs: liver, kidney	2.0	137 µL
ethyl acetate [141-78-6] flammable liquid; irritant; target organ effect; target organs: blood, kidney, liver, central nervous system		40 mL
water		10 mL
sodium sulfate [7757-82-6] may be harmful if swallowed, inhaled, or absorbed through skin; may cause eye, skin, or respiratory tract irritation		0.5 g

Equipment	Biotage Initiator Eight 2.0 EXP
Vial preparation	Into a 10 mL process vial equipped with a stirring bar are placed palladium-on-carbon (5.0 mg), triethylamine (278 µL, 2.0 mmol), and acetonitrile (3.0 mL). The mixture is stirred for 10 min and then treated with 3-bromobenzonitrile (364 mg, 2.0 mmol) and acrylic acid (137 µL, 2.0 mmol). The vial is sealed by capping with a Teflon septum fitted in an aluminum crimp top.

Microwave processing	Time:	20 min
	Temperature:	180 °C
	Pre-stirring:	10 s
	Absorption level:	normal
	Fixed hold time:	on

Work-up and isolation	The contents of the vial are transferred to a round-bottom flask and the solvent is removed by rotary evaporation. To the residue is added water (10 mL) and the mixture is extracted 2 × with ethyl acetate (20 mL). The organic extracts are combined, filtered through a plug of Celite, dried over sodium sulfate, and concentrated *in vacuo* to give the product.

Yield and physical data	3-Cyanocinnamic acid is obtained as a buff solid (234 mg, 68%). mp 247 °C; ^{1}H NMR (DMSO-d_6, 360 MHz) $\delta = 12.6$ (br s, 1H), 8.23 (s, 1H), 8.04 (d, $J = 7.8$ Hz, 1H), 7.86 (d, $J = 7.8$ Hz, 1H), 7.58–7.64 (m, 2H), 6.70 (d, $J = 16.3$ Hz, 1H).

Literature	[13]

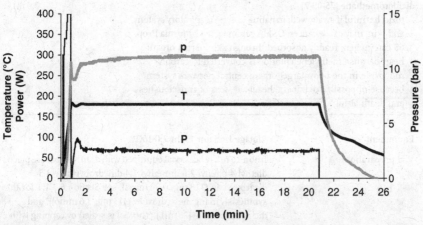

Figure 6.6 Power/pressure/temperature profile for the Heck reaction.

Background The palladium-catalyzed Heck coupling under microwave conditions has been studied extensively and reaction times could typically be reduced from hours to minutes [11–13]. By using water as solvent, the Leadbeater group was able to perform Heck reactions successfully employing ultra-low palladium catalyst concentrations (0.5–5 ppm) under homogeneous conditions [12]. As an alternative, the use of heterogeneous palladium-on-carbon (Pd/C) as a simple and inexpensive catalyst system for the Heck reaction has been reported [13]. The current procedure is an adaptation of the latter conditions. It has to be emphasized, that the quality of the Pd/C plays a crucial role in the outcome of the reaction.

6.1.1.6 Oxidative Aromatization

Table 6.6 Reagents and hazard information for the oxidative aromatization.

Reagent [CAS No.]	mmol	Quantity
diethyl 4-phenyl-2,6-dimethyl-1,4-dihydropyridine-3,5-dicarboxylate [1165-06-6] avoid contact with skin and eyes	0.36	118 mg
manganese dioxide [1313-13-9] delayed target organ effects; target organs: nerves, lungs	3.6	313 mg
dichloromethane [75-09-2] toxic; harmful if swallowed; irritating to eyes, respiratory system and skin; may cause cancer; OSHA carcinogen; California Prop. 65 carcinogen; readily absorbed through skin; target organ: heart because methylene chloride is converted to carbon monoxide in the body; target organ: central nervous system because of possible dizziness, headache, loss of consciousness and death at high concentrations		7.0 mL

Equipment	Biotage Initiator Eight 2.0 EXP
Vial preparation	Into a 10 mL process vial equipped with a stirring bar are placed diethyl 4-phenyl-2,6-dimethyl-1,4-dihydropyridine-3,5-dicarboxylate (118 mg, 0.36 mmol, see Section 6.3.1.1 for the synthesis), manganese dioxide (313 mg, 3.6 mmol), and dichloromethane (2.0 mL). The vial is sealed by capping with a Teflon septum fitted in an aluminum crimp top.

Microwave processing		
	Time:	1 min
	Temperature:	100 °C
	Pre-stirring:	10 s
	Absorption level:	normal
	Fixed hold time:	on

Work-up and isolation	The contents are filtered through a short Celite plug and rinsed forward with dichloromethane (5.0 mL). The filtrate is concentrated *in vacuo* to give the product.
Yield and physical data	Diethyl 4-phenyl-2,6-dimethylpyridine-3,5-dicarboxylate is obtained as a white solid (112 mg, 95%). mp 64 °C; ^{1}H NMR (DMSO-d_6, 360 MHz) $\delta = 7.41$–7.46 (m, 3H), 7.15–7.19 (m, 2H), 3.97 (q, $J = 7.1$ Hz, 4H), 3.33 (s, 1H), 2.50 (s, 6H), 0.83 (t, $J = 7.1$ Hz, 6H).
Literature	[15]

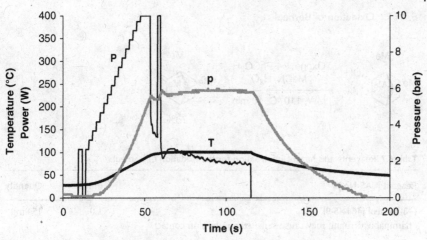

Figure 6.7 Power/pressure/temperature profile for oxidative aromatization.

Background Aside from the general interest in oxidative aromatization, the dehydrogenation of Hantzsch dihydropyridines has commanded particular attention due to its pharmaceutical potential [14]. The present procedure uses the very straightforward conditions worked out by Bagley and Lubinu [15].

6.1.1.7 Oxidation of Borneol

78%

Table 6.7 Reagents and hazard information for the oxidation of borneol.

Reagent [CAS No.]	mmol	Quantity
(−)-borneol [464-45-9] flammable; irritant; may cause sensitization by skin contact	1.0	154 mg
2-iodobenzoic acid [88-67-5] harmful; harmful if swallowed; irritating to respiratory system and skin; risk of serious damage to eyes	0.3	74 mg
Oxone [70693-62-8] oxidizing; corrosive; contact with combustible material may cause fire; harmful if swallowed; causes burns; may cause sensitization by inhalation and skin contact	1.0	615 mg
acetonitrile [75-05-8] flammable liquid; delayed target organ effects (lungs, blood, kidney, liver, central nervous system); mild skin irritant; severe eye irritant		2.7 mL
water		3.33 mL
dichloromethane [75-09-2] toxic; harmful if swallowed; irritating to eyes, respiratory system and skin; may cause cancer; OSHA carcinogen; California Prop. 65 carcinogen; readily absorbed through skin; target organ: heart because methylene chloride is converted to carbon monoxide in the body; target organ: central nervous system because of possible dizziness, headache, loss of consciousness and death at high concentrations		32 mL
sodium bicarbonate, 15% solution [144-55-8] may cause skin irritation; may be harmful if absorbed through the skin; may cause eye irritation; may be harmful if inhaled; material may be irritating to mucous membranes and upper respiratory tract; may be harmful if swallowed		30 mL
sodium sulfate [7757-82-6] may be harmful if swallowed, inhaled, or absorbed through skin; may cause eye, skin, or respiratory tract irritation		0.5 g

Equipment	Biotage Initiator Eight 2.0 EXP
Vial preparation	Into a 10 mL process vial equipped with a stirring bar are placed (−)-borneol (154 mg, 1.0 mmol), 2-iodobenzoic acid (74 mg, 0.3 mmol), Oxone (615 mg, 1.0 mmol), and a 2 : 1 mixture of acetonitrile in water (4.0 mL). The vial is sealed by capping with a Teflon septum fitted in an aluminum crimp top.

Microwave processing		
	Time:	20 min
	Temperature:	110 °C
	Pre-stirring:	10 s
	Absorption level:	high
	Fixed hold time:	on

Work-up and isolation	To the vial is added water (2.0 mL) and dichloromethane (2.0 mL), and the combined contents are briefly agitated. The heterogeneous mixture is filtered and the filtercake is washed with dichloromethane (30 mL). The filtrate is transferred to a separatory funnel and extracted 3× with 15% sodium bicarbonate solution (10 mL). The organic phase is collected, dried over sodium sulfate, and carefully concentrated *in vacuo* to provide the product. If necessary, the product can be recrystallized from diethyl ether or ethanol.
Yield and physical data	Camphor is obtained as an off-white solid (118 mg, 78%). mp 176 °C; ^{1}H NMR (DMSO-d$_6$, 360 MHz) $\delta = 2.29$ (app dt, $J = 18$ Hz; $J = 4$ Hz, 1H), 2.05 (app t, $J = 4$ Hz, 1H), 1.91–1.84 (m, 1H), 1.80 (d, $J = 18$ Hz, 1H), 1.68–1.61 (m, 1H), 1.35–1.22 (m, 2H), 0.91 (s, 3H), 0.79 (s, 3H), 0.76 (s, 3H).
Literature	[1]

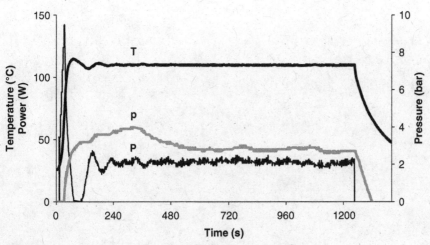

Figure 6.8 Power/pressure/temperature profile for the oxidation of borneol.

Background   The oxidation of alcohols to ketones comprises a vast body of work in the synthetic literature [16], and recent efforts have focused on catalytic methods that improve atom economy and environmental compatibility of the processes. Several reports have emerged using hypervalent iodine catalysts in the presence of Oxone® as a terminal oxidant [17]. The methodology enjoys a broad substrate range, and yields are good to excellent. This combination is also well suited for microwave heating in a sealed vessel, since no gases are liberated and the reagents are relatively thermally stable, which is not always the case in an oxidative environment. The present protocol thus represents the adaptation of this catalytic system for the microwave reactor, which enables a 20-fold reduction in reaction time [1].

6.1.1.8 Diels–Alder Reaction

Table 6.8 Reagents and hazard information for the Diels–Alder reaction.

Reagent [CAS No.]	mmol	Quantity
anthracene [120-12-7] irritant; dangerous for the environment; irritating to eyes, respiratory system, and skin; very toxic to aquatic organisms; may cause long-term adverse effects in the aquatic environment; photosensitizer; lachrymator; target organs: lungs, kidneys	3.7	660 mg
maleic anhydride [108-31-6] corrosive; harmful if swallowed; causes burns; may cause sensitization by inhalation and skin contact; sternutator	3.7	363 mg
toluene [108-88-3] flammable liquid; irritant; teratogen		4.0 mL

Equipment	Biotage Initiator Eight 2.0 EXP
Vial preparation	Into a 10 mL process vial equipped with a stirring bar are placed anthracene (660 mg, 3.7 mmol), maleic anhydride (365 mg, 3.7 mmol), and toluene (3.0 mL). The vial is sealed by capping with a Teflon septum fitted in an aluminum crimp top.
Microwave processing	Time: 5 min Temperature: 180 °C Pre-stirring: 10 s Absorption level: normal Fixed hold time: on
Work-up and isolation	The vial is kept at 4 °C for 3 h, then the contents of the reaction vial are filtered, washed with toluene (1.0 mL) and dried overnight at 50 °C.
Yield and physical data	9,10,11,15-Tetrahydro-9,10[3′,4′]-furanoanthracene-12,14-dione is obtained as a light tan solid (909 mg, 89%). mp 260 °C; ^{1}H NMR (DMSO-d_6, 360 MHz) $\delta = 7.45$–7.51 (m, 2H), 7.31–7.37 (m, 2H), 7.15–7.24 (m, 4H), 4.88 (s, 2H), 3.66 (s, 2H).
Literature	[19]

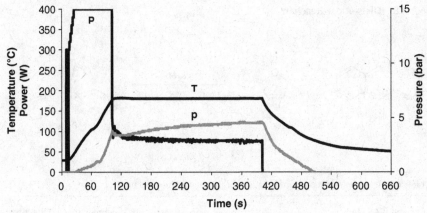

Figure 6.9 Power/pressure/temperature profile for the Diels–Alder reaction.

Background The pentacyclic adduct of maleic anhydride and anthracene has been achieved under conventional reflux conditions [18]. The current procedure adapts the reported protocol to the microwave reactor, reducing the reaction time from 2 h to 5 min [19].

6.1.2
Low-Absorbing Organic Medium – Passive Heating Elements

For more information on passive heating elements as heating aids for low absorbing solvents see Section 4.6.

6.1.2.1 *N*-Alkylation of Pyrazole

Table 6.9 Reagents and hazard information for the *N*-alkylation of pyrazole.

Reagent [CAS No.]	mmol	Quantity

pyrazole [288-13-1]
  harmful; harmful in contact with skin; irritating to eyes,
  respiratory system, and skin; harmful to aquatic organisms
	2.0	136 mg

(2-bromoethyl)benzene [103-63-9] harmful; harmful if swallowed; irritating to the eyes	1.0	137 µL
sodium bicarbonate [144-55-8]	1.0	84 mg
toluene [108-88-3] flammable liquid; irritant; teratogen		2.0 mL

ethyl acetate [141-78-6]
  flammable liquid; irritant; target organ effect; target organs:
  blood, kidney, liver, central nervous system
		250 mL

petroleum ether [101316-46-5]
  flammable liquid; may be harmful if inhaled; may cause respi-
  ratory tract irritation; vapors may cause drowsiness and dizziness;
  may cause skin irritation; may be harmful if absorbed through
  skin; target organs: central nervous system; lungs; liver; ears
		250 mL

Equipment	Biotage Emrys Liberator
Vial preparation	Into a 10 mL process vial equipped with a stirring bar are placed pyrazole (136 mg, 2.0 mmol), (2-bromoethyl)benzene (137 µL, 1.0 mmol), NaHCO₃ (84 mg, 1.0 mmol), toluene (2.0 mL), and a silicon carbide (SiC, 10 × 18 mm cylinder) passive heating element. The vial is sealed by capping with a Teflon septum fitted in an aluminum crimp top.
Microwave processing	Time: 30 min Temperature: 250 °C Fixed hold time: on Absorption level: normal

Work-up and isolation   The SiC cylinder is removed, the reaction mixture is transferred to a round-bottom flask and the toluene is removed by rotary evaporation. The crude product is purified by dry flash column chromatography with a 1 : 1 mixture of ethyl acetate and petroleum ether as eluent.

Yield and physical data   1-Phenethyl-1*H*-pyrazole is obtained as a pale yellow oil (152 mg, 88%). ^{1}H NMR (DMSO-d$_6$, 360 MHz) δ = 7.57 (d, J = 1.7 Hz, 1H), 7.43 (s, 1H), 7.13–7.27 (m, 5H), 6.16 (t, J = 1.7 Hz, 1H), 4.33 (t, J = 7.4 Hz, 2H), 3.08 (t, J = 7.4 Hz, 2H).

Literature   [22]

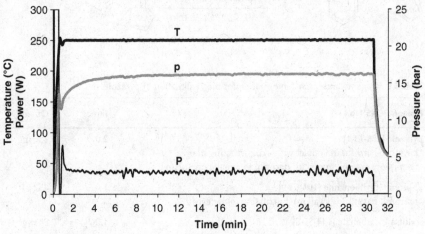

Figure 6.10 Power/pressure/temperature profile for the *N*-alkylation of pyrazole with SiC.

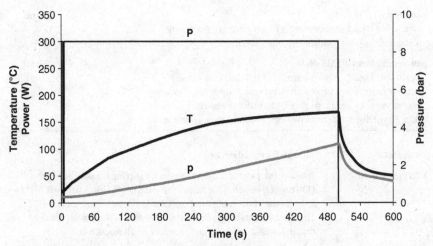

Figure 6.11 Power/pressure/temperature profile for the *N*-alkylation of pyrazole without SiC. The experiment was aborted by the instrument after 500 s because the set 250 °C could not be reached in an adequate time period (only 167 °C could be attained).

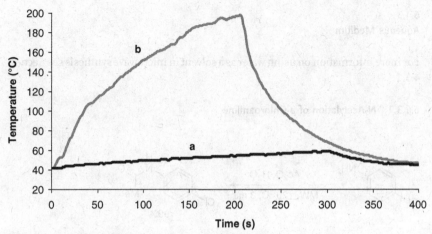

Figure 6.12 Heating profiles for 4 mL of dioxane doped with (bmimBF$_4$) at 150 W constant power in a quartz vial (IR temperature measurement): (a) neat dioxane; (b) dioxane doped with 30 mg of bmimBF$_4$.

Background The N-alkylation of pyrazole is one example where a low-absorbing solvent (toluene) is employed but the reaction mixture is not polar enough to couple efficiently with microwave irradiation to reach the desired temperature. The heating profile for this reaction – in the absence of any heating aids – is shown in Figure 6.11 where the set reaction temperature of 250 °C could not be reached. One approach to overcome this problem is to add small amounts of ionic liquids as heating aids which absorb the microwave radiation very strongly (see Section 4.5.2) [20, 21]. As Figure 6.12 illustrates, by adding 30 mg 1-butyl-3-methylimidazolium tetrafluoroborate (bmimBF$_4$) to 4 mL of dioxane much higher temperatures can be reached compared to heating without an ionic liquid (200 °C versus 50 °C). However, for the N-alkylation of pyrazole the addition of an ionic liquid is not possible due to destruction of the ionic liquid at elevated temperatures [20]. In contrast, when a noninvasive SiC heating element is added (see Section 4.6), the reaction proceeds smoothly and gives the alkylated pyrazole in 88% isolated yield [22].

6.1.3
Aqueous Medium

For more information on using water as a solvent in microwave synthesis see Section 4.5.1.

6.1.3.1 *N*-Acetylation of *p*-Chloroaniline

Table 6.10 Reagents and hazard information for the *N*-acetylation of *p*-chloroaniline.

Reagent [CAS No.]	mmol	Quantity
p-chloroaniline [106-47-8] toxic by inhalation, in contact with the skin and if swallowed; may cause sensitization by skin contact; very toxic to aquatic organisms; may cause long-term adverse effects in the aquatic environment; probable carcinogen; target organs: blood, central nervous system; Calif. Prop. 65 carcinogen.	5.0	638 mg
acetic anhydride [108-24-7] toxic if inhaled; material is extremely destructive to the tissue of the mucous membranes and upper respiratory tract; may cause respiratory tract irritation; may be harmful if absorbed through skin; causes skin burns; may cause eye irritation; causes eye burns; harmful if swallowed; causes burns	10.0	950 μL
water		8.0 mL

The *N*-acetylation of *p*-chloroaniline was performed in the Biotage Initiator and in the Milestone MultiSYNTH microwave instruments.

Equipment	Biotage Initiator Eight 2.0 EXP
Vial preparation	Into a 10 mL process vial equipped with a stirring bar are placed *p*-chloroaniline (638 mg, 5.0 mmol), acetic anhydride (0.95 mL, 10.0 mmol), and water (3.5 mL). The vial is sealed by capping with a Teflon septum fitted in an aluminum crimp top.

Microwave processing	Time:	3 min
	Temperature:	120 °C
	Pre-stirring:	10 s
	Absorption level:	high
	Fixed hold time:	on

Work-up and isolation: The reaction mixture is placed in an ice-bath for 10 min. The resulting precipitate is collected by filtration, washed with water, and dried overnight at 50 °C.

Yield and physical data: *p*-Chloroacetanilide is obtained as a white solid (837 mg, 99%). mp 177 °C; ^{1}H NMR (DMSO-d_6, 360 MHz) $\delta = 10.1$ (br s, 1H), 7.60 (d, $J = 8.6$ Hz, 2H), 7.33 (d, $J = 8.6$ Hz, 2 Hz), 2.03 (s, 3H).

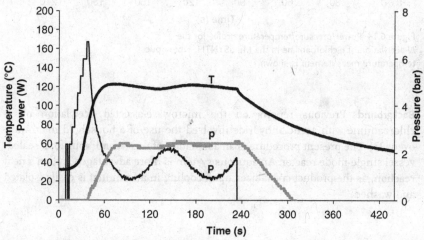

Figure 6.13 Power/pressure/temperature profile for the *N*-acetylation of *p*-chloroaniline in the Initiator.

Equipment	Milestone MultiSYNTH
Vial preparation	Into a 10 mL process vial are placed *p*-chloroaniline (638 mg, 5.0 mmol), acetic anhydride (0.95 mL, 10.0 mmol), and water (3.5 mL). The vial is sealed with the appropriate closure and placed into the position for the single-mode set-up.

Microwave processing	Mode:	Monomode irradiation
	Time:	3 min
	Temperature:	120 °C
	Preset power:	150 W
	Vibration:	50%

Work-up and isolation: This is identical to the procedure described above.

Yield and physical data: *p*-Chloroacetanilide is obtained as a white solid (805 mg, 95%).

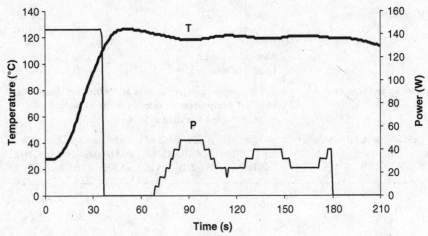

Figure 6.14 Power/pressure/temperature profile for the
N-acetylation of p-chloroaniline in the MultiSYNTH. Fiber-optic
temperature measurement is shown.

Background Previous reports on the microwave-assisted acetylation of p-chloroaniline with acetic anhydride involved the use of a household microwave oven [23]. The present procedure is an adaptation of these conditions for a sealed-vessel single-mode reactor. An aqueous system is more advantageous than a neat reaction, as the product crystallizes upon cooling in a form that is easily isolated and washed.

6.1.3.2 Dihydroindazolone Synthesis

Table 6.11 Reagents and hazard information for the dihydroindazolone synthesis.

Reagent [CAS No.]	mmol	Quantity
1,3-cyclohexanedione [504-02-9] may cause skin irritation; may be harmful if absorbed through the skin; may cause eye irritation; material may be irritating to mucous membranes and upper respiratory tract; may be harmful if inhaled; may be harmful if swallowed	1.25	140 mg
N,N-dimethylformamide dimethyl acetal [4637-24-5] flammable; harmful if swallowed; irritating to eyes, respiratory system and skin	1.5	200 µL
phenylhydrazine [100-63-0] toxic; dangerous for the environment; may cause cancer; also toxic by inhalation, in contact with skin and if swallowed; irritating to eyes and skin; may cause sensitization by skin contact; also toxic: danger of serious damage to health by prolonged exposure through inhalation, in contact with skin and if swallowed; possible risk of irreversible effects; very toxic to aquatic organisms; Calif. Prop. 65 carcinogen; combustible; target organs: blood, liver	1.25	125 µL
water		18 mL
acetic acid, glacial [64-19-7] combustible; corrosive; harmful in contact with skin; causes severe burns; lachrymator; target organs: kidneys, teeth	3.3	190 µL
dichloromethane [75-09-2] toxic; harmful if swallowed; irritating to eyes, respiratory system and skin; may cause cancer; OSHA carcinogen; California Prop. 65 carcinogen; readily absorbed through skin; target organ: heart because methylene chloride is converted to carbon monoxide in the body; target organ: central nervous system because of possible dizziness, headache, loss of consciousness and death at high concentrations		15 mL
sodium sulfate [7757-82-6] may be harmful if swallowed, inhaled, or absorbed through skin; may cause eye, skin, or respiratory tract irritation		0.5 g

(Continued)

Table 6.11 (*Continued*)

Reagent [CAS No.]	mmol	Quantity
diethyl ether [60-29-7] toxic; extremely flammable; may form explosive peroxides; harmful if swallowed; irritating to eyes, respiratory system, and skin; repeated exposure may cause skin dryness or cracking; vapors may cause drowsiness and dizziness; target organs: central nervous system, kidneys		2.0 mL
petroleum ether [101316-46-5] flammable liquid; may be harmful if inhaled; may cause respiratory tract irritation; vapors may cause drowsiness and dizziness; may cause skin irritation; may be harmful if absorbed through skin; target organs: central nervous system; lungs; liver; ears		2.0 mL

Equipment	Biotage Initiator Eight 2.0 EXP
Vial preparation	Into a 10 mL process vial equipped with a stirring bar are placed 1,3-cyclohexanedione (140 mg, 1.25 mmol) and *N,N*-dimethylformamide dimethyl acetal (200 µL, 1.5 mmol). This mixture is allowed to stir for 2 min and then treated with phenylhydrazine (125 µL, 1.25 mmol), water (3.0 mL), and glacial acetic acid (190 µL, 3.3 mmol). The vial is sealed by capping with a Teflon septum fitted in an aluminum crimp top.

Microwave processing	Time:	2 min
	Temperature:	200 °C
	Pre-stirring:	10 s
	Absorption level:	normal
	Fixed hold time:	on

Caution: Residual pressure must be relieved before removing the vessel from the microwave cavity by carefully inserting a needle into the pressure relief port. Please consult the manufacturer's instructions for proper procedure.

Work-up and isolation	The contents of the vial are transferred to a separatory funnel and partitioned between water (15 mL) and dichloromethane (15 mL). The organic layer is collected and dried over sodium sulfate and concentrated *in vacuo* to give a dark oil which solidified upon standing. This residue is triturated with a hot 1 : 1 mixture of diethyl ether and petroleum ether (2.0 mL). The mixture is allowed to cool to room temperature and the supernatant is removed with a syringe. This process is repeated, and the remaining solid is dried *in vacuo*.
Yield and physical data	1-Phenyl-6,7-dihydro-1*H*-indazol-4(5*H*)-one is obtained as a buff solid (247 mg, 93%). mp 137 °C; ^{1}H NMR (CDCl$_3$, 360 MHz) $\delta = 8.06$ (s, 1H), 7.48–7.51 (m, 4H), 7.38-7.45 (m, 1H), 2.96 (t, $J = 6.3$ Hz, 2H), 2.54 (t, $J = 6.3$ Hz, 2H), 2.16 (quint, $J = 6.3$ Hz, 2H).
Literature	[24]

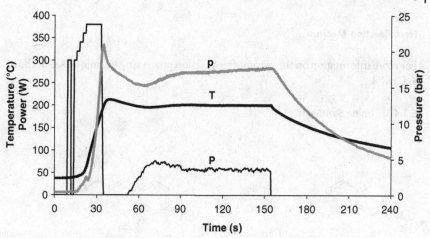

Figure 6.15 Power/pressure/temperature profile for the indazolone synthesis.

Background   As a novel entrée into the fused bicyclic pyrazole system, Molteni and coworkers [24] have developed a one-pot synthesis whereby a cyclic diketone undergoes condensation with dimethylformamide dimethylacetal (DMFDMA) to form an intermediate enaminoketone, which is then treated with a hydrazine derivative and heated under microwave irradiation to afford the corresponding dihydroindazolone in good yields. The current procedure is essentially taken from their model.

6.1.4
Neat Reaction Medium

For more information on the solvent-free reaction processing technique see Sections
4.1 and 4.2.

6.1.4.1 Imine Synthesis

Table 6.12 Reagents and hazard information for the imine synthesis.

Reagent [CAS No.]	mmol	Quantity
4-nitrobenzaldehyde [555-16-8] harmful; irritating to eyes; may cause sensitization by skin contact; harmful to aquatic organisms; may cause long-term adverse effects in the aquatic environment; possible mutagen	15.0	2.27 g
isopropylamine [75-31-0] flammable; harmful by inhalation; toxic in contact with skin and if swallowed; causes burns; readily absorbed through skin	15.0	1.29 mL
diethyl ether [60-29-7] toxic; extremely flammable; may form explosive peroxides; harmful if swallowed; irritating to eyes, respiratory system, and skin; repeated exposure may cause skin dryness or cracking; vapors may cause drowsiness and dizziness; target organs: central nervous system, kidneys		3.0 mL

Equipment	Biotage Initiator Eight 2.0 EXP
Vial preparation	Into a 10 mL process vial equipped with a stirring bar are placed 4-nitrobenzaldehyde (2.27 g, 15.0 mmol) and isopropylamine (1.29 mL, 15.0 mmol). The vial is sealed by capping with a Teflon septum fitted in an aluminum crimp top.
Microwave processing	Time: 5 min Temperature: 100 °C Pre-stirring: 10 s Absorption level: normal Fixed hold time: on

Work-up and isolation	Upon cooling to room temperature, the reaction mixture solidifies to form a brown solid, which is dissolved in ether (3.0 mL). The vial is stored in the freezer (−20 °C) overnight, and the filtrate is carefully removed with a pasteur pipette from the resulting precipitate which is subsequently dried *in vacuo*.
Yield and physical data	N-(4-Nitrobenzylidene)propan-2-amine is obtained as a tan solid (1.38 g, 72%). mp 62 °C; ^{1}H NMR (DMSO-d$_6$, 360 MHz) δ = 8.50 (s, 1H), 8.28 (d, J = 8.6 Hz, 2H), 7.97 (d, J = 8.6 Hz, 2H), 3.62 (sept, J 6.3 Hz, 1H), 1.20 (d, J = 6.3 Hz, 6H).
Literature	[25]

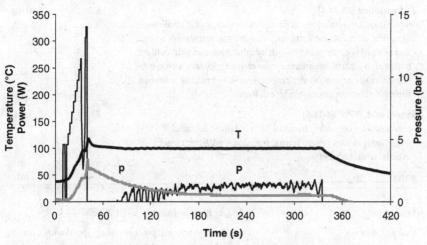

Figure 6.16 Power/pressure/temperature profile for the imine synthesis.

Background An imine synthesis protocol has been reported by Paquin and co-workers [25], whereby equimolar amounts of amines and aldehydes are irradiated for short periods in the absence of catalyst, solid support, or solvent. The current procedure is taken essentially from the literature, with slight modifications to the work-up and isolation steps.

6.1.4.2 N-Formylation of 2-Nitroaniline

82%

Table 6.13 Reagents and hazard information for the N-formylation of 2-nitroaniline.

Reagent [CAS No.]	mmol	Quantity
2-nitroaniline [88-74-4] toxic; toxic by inhalation, in contact with skin and if swallowed; danger of cumulative effects; irritating to eyes, respiratory system and skin; may cause sensitization by inhalation and skin contact; harmful to aquatic organisms; may cause long-term adverse effects in the aquatic environment; sensitizer; readily absorbed through skin; target organs: blood, liver	5.0	690 mg
formic acid, 95% [64-18-6] combustible; corrosive; harmful by inhalation; harmful if swallowed; causes severe burns; possible sensitizer; target organs: liver, kidneys	73.1	2.9 mL
water		55 mL

Equipment	Biotage Initiator Eight 2.0 EXP
Vial preparation	Into a 10 mL process vial equipped with a stirring bar are placed 2-nitroaniline (690 mg, 5.0 mmol) and 95% formic acid (2.9 mL, 73.1 mmol). The vial is sealed by capping with a Teflon septum fitted in an aluminum crimp top.
Microwave processing	Time: 3 min Temperature: 160 °C Pre-stirring: 10 s Absorption level: normal Fixed hold time: on
Work-up and isolation	The reaction mixture is poured into water (25 mL), placed in an ice-bath for 30 min and the resulting precipitate is isolated by filtration, washed with water (30 mL), and dried overnight at 50 °C.
Yield and physical data	N-(2-Nitrophenyl)formamide is obtained as a light yellow solid (681 mg, 82%). mp 62 °C; ^{1}H NMR (DMSO-d$_6$, 360 MHz) $\delta =$ 10.55 (br s, 1H), 8.39 (br s, 1H), 8.05 (d, $J = 8$ Hz, 2H), 7.73 (t, $J = 8$ Hz, 1H), 7.36 (t, $J = 8$ Hz, 1H).
Literature	[1]

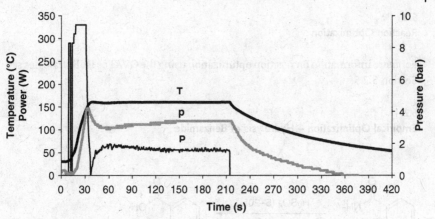

Figure 6.17 Power/pressure/temperature profile for the *N*-formylation of 2-nitroaniline.

Background Bose and coworkers [26] have reported a protocol for the *N*-formylation of otherwise recalcitrant aromatic and aryl amines by microwave irradiation in concentrated formic acid under open-vessel conditions at sub-reflux temperatures. The current example adapts this protocol for sealed-vessels with comparable yield [1].

6.2
Reaction Optimization

For more information on reaction optimization using the OVAT or DoE strategies see Section 5.2.5.

6.2.1
Empirical Optimization – Hydrolysis of Benzamide

Table 6.14 Reagents and hazard information for the hydrolysis of benzamide.

Reagent [CAS No.]	mmol	Quantity
benzamide [55-21-0] harmful; harmful if swallowed	12×2.0	2.91 g
sulfuric acid, 5% solution [7664-93-9] corrosive; causes severe burns; target organs: teeth, cardiovascular system		10.8 mL
sulfuric acid, 10% solution		10.8 mL
sulfuric acid, 20% solution		10.8 mL
water		12.0 mL

Equipment	Biotage Initiator Eight 2.0 EXP
Vial preparation	Into a 10 mL process vial equipped with a magnetic stirrer are placed benzamide (242 mg, 2.0 mmol) and aqueous sulfuric acid (2.7 mL, see Table 6.14 for concentrations). The vial is sealed by capping with a Teflon septum fitted in an aluminum crimp top.
Microwave processing	Time: see Table 6.15 Temperature: 160 °C Pre-stirring: 10 s Absorption level: normal Fixed hold time: on
Work-up and isolation	The vial is placed in an ice-bath for 1 h. The resulting precipitate is isolated by filtration, washed with ice-cold water (1.0 mL), and dried overnight at 50 °C.

Yield and physical data	Benzoic acid is obtained as a white solid (see Table 6.15 for conversion data). mp 123 °C; ^{1}H NMR (DMSO-d_6, 360 MHz) $\delta = 12.9$ (br s, 1H), 7.95 (d, $J = 8$ Hz, 2H), 7.63 (t, $J = 8$ Hz, 1H), 7.49 (t, $J = 8$ Hz, 2H).
Conversion determination	5 mg of the dried filtercake is dissolved in 1.0 mL acetonitrile and diluted to 10 mL with water. 5 µL of this solution is injected onto the HPLC.
Literature	[27]

Table 6.15 Conversion of benzamide at 160 °C as a function of time and acid concentration.^{a}

Time (min)	5% H_2SO_4	10% H_2SO_4	20% H_2SO_4
1	26	46	83
2	30	53	95
5	53	86	100
10	57	94	100

^{a}% HPLC conversion of isolated product.

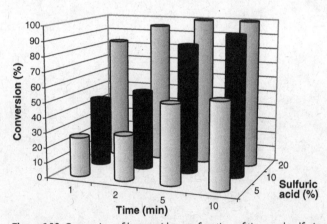

Figure 6.18 Conversion of benzamide as a function of time and sulfuric acid concentration.

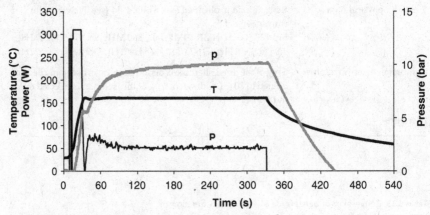

Figure 6.19 Representative power/pressure/temperature profile for the benzamide hydrolysis (5 min at 10% H_2SO_4).

Background The current example is adapted from a literature study [27] and designed to serve as an instructional laboratory optimization experiment, in which the two parameters of acid concentration and reaction time are varied (reaction temperature could also be included as a variable, if desired).

6.2.2
Optimization using Statistical Methods (DoE) – Synthesis of Monastrol

Table 6.16 Reagents and hazard information for the synthesis of monastrol.[a]

Reagent [CAS No.]	mmol	Quantity
3-hydroxybenzaldehyde [100-83-4] irritating to eyes, respiratory system and skin; may be harmful if swallowed, inhaled, or absorbed through the skin	0.82	100 mg
ethyl acetoacetate [141-97-9] irritant; irritating to eyes; combustible liquid	1.23	156 μL
thiourea [62-56-6] toxic by ingestion; skin sensitizer, irritant; carcinogen; target organ effect; target organs: liver, thyroid, bone marrow	0.82	62 mg
lanthanum(III) chloride [10099-58-8] causes severe irritation; harmful if swallowed, inhaled, or absorbed through skin; high concentrations are extremely destructive to tissues of the mucous membranes and upper respiratory tract, eyes, and skin	0.098	24 mg
ethanol, anhydrous [64-17-5] flammable; irritant; irritating to eyes, respiratory system and skin; target organs: nerves, liver		1.0 mL
chloroform [67-66-3] harmful by ingestion; irritant; carcinogen; target organ effect; target organs: central nervous system, blood, liver, cardiovascular system, kidney		150 mL
acetone [67-64-1] flammable liquid; delayed target organ effects; moderate skin irritant; moderate eye irritant		30 mL

[a] Only the quantities for the optimized conditions are given.

Only the experimental procedure for the optimized reaction conditions is described.

Equipment	CEM Discover LabMate
Vial preparation	Into a 10 mL process vial equipped with a stirring bar are placed 3-hydroxybenzaldehyde (100 mg, 0.82 mmol), ethyl acetoacetate (156 µL, 1.23 mmol), thiourea (62 mg, 0.82 mmol), lanthanum(III) chloride (24 mg, 0.098 mmol), and anhydrous ethanol (1.0 mL). The contents are stirred for 2 min, after which the vial is fitted with a snap-on cap.
Microwave processing	Temperature: 140 °C Time: 30 min Max. Power: 100 W
Work-up and isolation	Ethanol is removed *in vacuo* and the residue is subjected to silica gel chromatography with chloroform/acetone (5 : 1) as the eluent system. The appropriate fractions are then concentrated under reduced pressure.
Yield and physical data	Ethyl 4-(3-hydroxyphenyl)-6-methyl-2-thioxo-1,2,3,4-tetrahydropyrimidine-5-carboxylate (monastrol) is obtained as a light yellow solid (197 mg, 82%). mp 183–185 °C; ^{1}H NMR (DMSO-d_6, 360 MHz) $\delta = 10.29$ (s, 1H), 9.60 (s, 1H), 9.44 (s, 1H), 7.06–7.17 (m, 1H), 6.61–6.70 (m, 3H), 5.51 (d, $J = 1.44$ Hz, 1H), 4.02 (q, $J = 6.6$ Hz, 2H), 2.27 (s, 3H), 1.12 (t, $J = 6.9$ Hz, 3H).
Literature	[29]

Table 6.17 Reaction parameters and conversion data for the DoE optimization.

Entry	T (°C)	t (min)	Cat. (mol%)	Conv. (%)	Purity (%)
Matrix 1 (temperature–time)					
1.1	100	10	10	67	51
1.2	140	10	10	86	68
1.3	100	30	10	84	64
1.4	140	30	10	94	74
1.5	100	20	10	77	60
1.6	140	20	10	95	79
1.7	120	30	10	78	60
1.8	120	30	10	96	71
1.9	120	20	10	89	69
1.10	120	20	10	85	66
Matrix 2 (temperature–catalyst)					
2.1	100	30	5	88	77
2.2	140	30	5	91	86
2.3	100	30	15	93	77
2.4	140	30	15	92	90
2.5	100	30	10	91	77
2.6	140	30	10	88	86

Table 6.17 (Continued)

Entry	T (°C)	t (min)	Cat. (mol%)	Conv. (%)	Purity (%)
2.7	120	30	5	92	70
2.8	120	30	15	93	86
2.9	120	30	10	93	82
2.10	120	30	10	93	81
2.11	120	30	10	93	85
2.12	100	30	5	87	73
2.13	140	30	5	93	91
2.14	100	30	15	93	78
2.15	140	30	15	95	91
2.16	120	30	10	92	83
Optimized conditions					
	140	30	12	94	>99

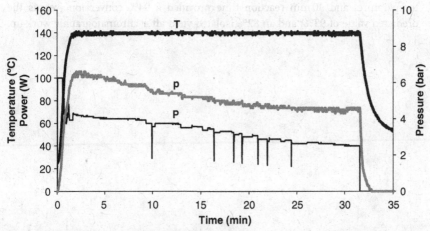

Figure 6.20 Power/pressure/temperature profile for the monastrol synthesis (optimized conditions).

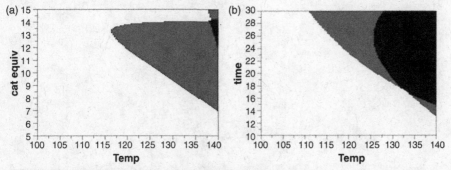

Figure 6.21 Sweet-spot prediction plots for: (a) temperature (°C) vs. catalyst loading, and (b) temperature (°C) vs. time (gray = conversion or purity criterion is met; black = both criteria are met). Reproduced with permission from Ref. [29].

Background The utility of the DoE technique can be illustrated using the dihydropyrimidine derivative monastrol, a specific mitotic kinesin Eg5 inhibitor, which has been previously synthesized via a one-pot, three-component Biginelli condensation in a sealed vessel under microwave irradiation [28]. However, the optimization of such a complex system could not be addressed reasonably using the OVAT approach. Thus, to fully optimize the formation of monastrol, a number of experimental variables were examined using the MODDE8 statistical DoE software package from Umetrics [29].

In the current example, taken entirely from the literature [29], the experimental design begins with the pre-optimized parameters of catalyst and solvent identity (lanthanum chloride and ethanol, respectively) and constructs one matrix of experiments to explore time–temperature space and another to probe catalyst loading effects (Table 6.17). Globally optimized conditions are then established using sweet-spot analysis (Figure 6.21), which shows the overlapping maxima of yield and purity. A verification experiment using the optimal values of 12 mol% $LaCl_3$, 140 °C reaction temperature, and 30 min reaction time provided a 94% conversion (versus the predicted value of 93%) and an 82% isolated yield after chromatographic work-up.

6.3
Library Generation

For more information on library generation in an automated sequential or parallel fashion see Section 4.7.1.

6.3.1
Automated Sequential Library Generation

6.3.1.1 Hantzsch Synthesis

6 examples
(47–99%)

Table 6.18 Reagents and hazard information for the Hantzsch library synthesis.

Reagent [CAS No.]	mmol	Quantity
ethyl acetoacetate [141-97-9] irritant; irritating to eyes; combustible liquid	3×5.0	1.9 mL
5,5-dimethyl-1,3-cyclohexanedione [126-81-8] may cause skin irritation; may be harmful if absorbed through the skin; may cause eye irritation; may be harmful if inhaled; material may be irritating to mucous membranes and upper respiratory tract; may be harmful if swallowed	3×5.0	2.1 g
1,3-cyclohexanecarboxaldehyde [2043-61-0] combustible; irritating to eyes, respiratory system and skin; lachrymator; unpleasant odor	1.0	121 µL
4-methoxybenzaldehyde [123-11-5] may be harmful if inhaled; may cause respiratory tract irritation	2×1.0	244 µL
isobutyraldehyde [78-84-2] highly flammable	2×1.0	182 µL
2-thiophenecarboxaldehyde [98-03-3] harmful if swallowed; irritating to eyes, respiratory system and skin; combustible	1.0	93 µL
ammonium hydroxide [1336-21-6] corrosive; dangerous for the environment; harmful if swallowed; causes burns; very toxic to aquatic organisms; lachrymator	6×4.0	1.8 mL

(Continued)

Table 6.18 (Continued)

Reagent [CAS No.]	mmol	Quantity
ethyl acetate [141-78-6] flammable liquid; irritant; target organ effect; target organs: blood, kidney, liver, central nervous system		100 mL
petroleum ether [101316-46-5] flammable liquid; may be harmful if inhaled; may cause respiratory tract irritation; vapors may cause drowsiness and dizziness; may cause skin irritation; may be harmful if absorbed through skin; target organs: central nervous system; lungs; liver; ears		100 mL
ethanol [64-17-5] flammable; irritant; irritating to eyes, respiratory system and skin; target organs: nerves, liver		13.5 mL
water		16.5 mL

Equipment	Biotage Initiator Eight 2.0 EXP
Vial preparation	Into a 5 mL process vial equipped with a stirring bar are combined the aldehyde component (1.0 mmol, see Table 6.19), the active methylene component (5.0 mmol, see Table 6.19), and 25% ammonium hydroxide (300 µL, 4.0 mmol). The vial is sealed by capping with a Teflon septum fitted in an aluminum crimp top.
Microwave processing	Time: 10 min Temperature: 140 °C Pre-stirring: 10 s Absorption level: normal Fixed hold time: on
Work-up and isolation	For examples (a)–(c), 4 mL water are added to the reaction mixture followed by extraction with dichloromethane (3×10 mL). The solvent of the combined organic phases is evaporated and the resulting yellow oil is purified by chromatography using a Biotage SP-1 chromatography system (25+ cartridge and samplet). The TLC to gradient method with ethyl acetate/petroleum ether 1 : 5 as eluent is employed. The appropriate fractions are then concentrated *in vacuo*. For examples (d)–(f), the reaction mixture is treated with ethanol (3.0 mL), heated to reflux, and allowed to cool slowly. The resulting precipitate is isolated by filtration, washed with an ice-cold 1 : 1 mixture of water and ethanol (3.0 mL), and dried overnight at 50 °C.
Yield and physical data	See Table 6.20 for yield data. (a) light yellow solid; mp 127–129 °C; ^{1}H NMR (DMSO-d$_6$, 360 MHz) $\delta = 8.64$ (s, 1H), 3.98–4.15 (m, 4H), 3.75 (d, $J = 6.0$ Hz, 1H), 2.21 (s, 6H), 1.40–1.65 (m, 5H), 1.19 (t, $J = 7.2$ Hz, 6H), 0.75–1.10 (m, 6H).

(b) yellow solid; mp 158–160 °C; ^{1}H NMR (DMSO-d$_6$, 360 MHz) δ = 8.74 (s, 1H), 7.04 (d, J = 8.6 Hz, 2H), 6.75 (d, J = 8.6 Hz, 2H), 4.79 (s, 1H), 3.93–4.06 (m, 4H), 3.67 (s, 3H), 2.24 (s, 6H), 1.13 (t, J = 7.2 Hz, 6H).

(c) yellow solid; mp 97–99 °C; ^{1}H NMR (DMSO-d$_6$, 360 MHz) δ = 8.65 (s, 1H), 3.98–4.16 (m, 4H), 3.74 (d, J = 5.3 Hz, 1H), 2.21 (s, 6H), 1.36–1.48 (m, 1H), 1.19 (t, J = 7.0 Hz, 6 H), 0.65 (d, J = 7.0 Hz, 6H).

(d) buff solid; mp 243–244 °C; ^{1}H NMR (DMSO-d$_6$, 360 MHz) δ = 9.23 (s, 1H), 7.04 (d, J = 8.6 Hz, 2H), 6.71 (d, J = 8.6 Hz, 2H), 4.74 (s, 1H), 3.65 (s, 3H), 2.44 (d, J 17.1 Hz, 2H), 2.31 (d, J = 17.1 Hz, 2H), 2.16 (d, J = 15.9 Hz, 2H), 1.97 (d, J = 15.9 Hz, 2H), 1.00 (s, 6H), 0.87 (s, 6H).

(e) white solid; mp 238–240 °C; ^{1}H NMR (DMSO-d$_6$, 360 MHz) δ = 9.10 (s, 1H), 3.79 (d, J = 4.0 Hz, 1H), 2.38 (d, J = 17.3 Hz, 2H), 2.28 (d, J = 17.3 Hz, 2H), 2.07–2.20 (m, 4H), 1.48–1.57 (m, 1H), 1.04 (s, 6H), 1.02 (s, 6H), 0.64 (d, J = 7.0 Hz, 6H).

(f) off-white solid; mp 278–279 °C; ^{1}H NMR (DMSO-d$_6$, 360 MHz) δ = 9.43 (s, 1H), 7.13 (dd, J = 5.0, 1.0 Hz), 6.79 (dd, J = 5.0, 3.5 Hz), 6.65 (d, J = 3.5 Hz), 5.14 (s, 1H), 2.44 (d, J = 17.4 Hz, 2H), 2.32 (d, J = 17.4 Hz, 2H), 2.22 (d, J = 15.9 Hz, 2H), 2.07 (d, J = 15.9 Hz, 2H), 1.02 (s, 6H), 0.94 (s, 6H).

Literature   [15, 30]

Table 6.19 Reagent charge amounts.

Entry	Active methylene	Amount	Aldehyde	Amount
a	ethyl acetoacetate	632 µL	cyclohexanecarboxaldehyde	121 µL
b	ethyl acetoacetate	632 µL	4-methoxybenzaldehyde	122 µL
c	ethyl acetoacetate	632 µL	isobutyraldehyde	91 µL
d	5,5-dimethyl-1,3-cyclohexanedione	700 mg	4-methoxybenzaldehyde	122 µL
e	5,5-dimethyl-1,3-cyclohexanedione	700 mg	isobutyraldehyde	91 µL
f	5,5-dimethyl-1,3-cyclohexanedione	700 mg	2-thiophenecarboxaldehyde	93 µL

Table 6.20 Isolated yields.

Entry	R^1	R^2	R^3	Yield (%)
a	Me	OEt	cyclohexyl	62
b	Me	OEt	4-MeO-Ph	60
c	Me	OEt	i-Pr	47
d	—CH$_2$C(Me)$_2$CH$_2$—		4-MeO-Ph	99
e	—CH$_2$C(Me)$_2$CH$_2$—		i-Pr	69
f	—CH$_2$C(Me)$_2$CH$_2$—		2-thienyl	95

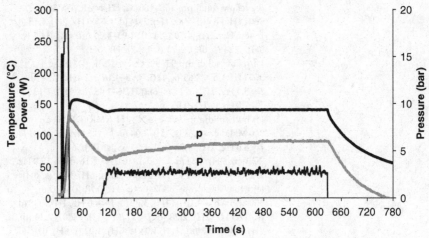

Figure 6.22 Representative power/pressure/temperature profile for the Hantzsch library synthesis.

Background The protocol involves the one-pot condensation of ammonia, an aldehyde, and two molecules of an active methylene compound. The Hantzsch multi-component reaction (MCR) is attractive for the generation of chemical libraries, since a wide array of substitution patterns can be achieved by variation of the aldehyde and active methylene components.

The groups of Bagley and Westman [15, 30] reported an optimized procedure for the Hantzsch synthesis using a single-mode microwave reactor, and the current example is taken essentially from the literature. The products derived from dimedone as the active methylene component were obtained in crystalline form and in generally high yield. The ethyl acetoacetate derivatives, on the other hand, provided oily crude mixtures that required silica gel chromatography for purification.

6.3.1.2 Biginelli Synthesis

Table 6.21 Reagents and hazard information for the Biginelli synthesis.[a]

Reagent [CAS No.]	mmol	Quantity
urea [57-13-6]	9×4.0	2.16 g
may be harmful if inhaled; may cause respiratory tract irritation; may be harmful if absorbed through skin; may cause skin irritation		
thiourea [62-56-6]	3×4.0	912 mg
harmful; dangerous for the environment; harmful if swallowed; limited evidence of a carcinogenic effect; toxic to aquatic organisms, may cause long-term adverse effects in the aquatic environment; possible risk of harm to the unborn child		
ytterbium triflate [252976-51-5]	12×0.4	2.98 g
irritating to eyes, respiratory system and skin; may be harmful if inhaled; may be harmful if swallowed		
ethyl acetoacetate [141-97-9]	5×6.0	~4 mL
irritant; irritating to eyes; combustible liquid		
methyl acetoacetate [105-45-3]	2×6.0	~1.5 mL
irritant; risk of serious damage to eyes; combustible		
isopropyl acetoacetate [542-08-5]	6.0	~1 mL
irritating to eyes, respiratory system and skin; combustible liquid; may be harmful by inhalation, ingestion, or skin absorption		
acetylacetone [123-54-6]	6.0	~0.8 mL
flammable; toxic if swallowed; irritating to eyes, respiratory system and skin; neurological hazard; target organs: thymus, nerves		
methyl 4-methoxyacetoacetate [41051-15-4]	6.0	~1 mL
irritating to eyes, respiratory system and skin; combustible; vapor or mist is irritating to the eyes, mucous membranes, and upper respiratory tract; may be harmful by inhalation, ingestion, or skin absorption		

(Continued)

Table 6.21 *(Continued)*

Reagent [CAS No.]	mmol	Quantity
methyl 3-oxo-pentanoate [30414-53-0]	6.0	~950 μL
acetoacetanilide [102-01-2] harmful; harmful in contact with skin and if swallowed	6.0	1.06 g
2-chlorobenzaldehyde [89-98-5] corrosive; causes burns; combustible; material is extremely destructive to the tissue of the mucous membranes and upper respiratory tract; may be harmful if swallowed or inhaled	2×4.0	~1.1 mL
2-thiophenecarboxaldehyde [98-03-3] harmful if swallowed; irritating to eyes, respiratory system and skin; combustible	4.0	~550 μL
4-fluorobenzaldehyde [459-57-4] irritant; irritating to eyes, respiratory system and skin	4.0	~650 μL
o-tolualdehyde [529-20-4] harmful if swallowed; irritating to respiratory system and skin; risk of serious damage to eyes; combustible	4.0	~650 μL
acetaldehyde [75-07-0] flammable liquid; harmful by ingestion; skin sensitizer; irritant; carcinogen; target organ effect; target organs: blood, kidney, lungs, cardiovascular system, liver, central nervous system	4.0	~450 μL
m-tolualdehyde [620-23-5] combustible; may cause eye or skin irritation; may be irritating to mucuous membranes and upper respiratory tract; may be harmful if swallowed, inhaled, or absorbed through the skin	4.0	~650 μL
3-nitrobenzaldehyde [99-61-6] irritating to eyes, respiratory system and skin; may be harmful by inhalation, ingestion, or skin absorption	2×4.0	1.21 g
2,3-dichlorobenzaldehyde [6334-18-5] corrosive; causes burns	4.0	700 mg
benzaldehyde [100-52-7] harmful; harmful in contact with skin and if swallowed; irritating to skin; may cause sensitization by inhalation and skin contact; combustible; readily adsorbed through skin; target organs: central nervous system; liver	4.0	~600 μL
3-hydroxybenzaldehyde [100-83-4] irritating to eyes, respiratory system and skin; may be harmful if swallowed, inhaled, or absorbed through the skin	4.0	488 mg
acetic acid, glacial [64-19-7]/ethanol 3 : 1 combustible; corrosive; harmful in contact with skin; causes severe burns; lachrymator; target organs: kidneys, teeth		~10 mL
ethanol [64-17-5] flammable; irritant; irritating to eyes, respiratory system and skin; target organs: nerves, liver		~5 mL

Table 6.21 (Continued)

Reagent [CAS No.]	mmol	Quantity
acetonitrile [75-05-8]		~5 mL
flammable liquid; delayed target organ effects (lungs, blood, kidney, liver, central nervous system); mild skin irritant; severe eye irritant		
water/ethanol 2 : 1		90 mL
water/ethanol 1 : 1		60 mL
water/acetonitrile 1 : 2		12 mL
water		30 mL

^{a}For the liquid reagents only an approximate amount is given since the liquid handler will dispense the exact amount.

Equipment	Biotage Emrys Liberator
Method programming	The library is programmed using the EmrysWorkflowManager software. All chemicals needed are entered by drawing the structure. The mmol-amount of all reagents is entered: 4.0 mmol of aldehydes, 4.0 mmol of urea/thiourea, 6.0 mmol of active methylene compound and 0.4 mmol of ytterbium triflate. The software automatically calculates the corresponding amounts. The amount of solvent (for X = O: AcOH/EtOH 3 : 1, for X = S: MeCN) is entered so that a total reaction volume of about 2.5 mL is reached. Then the dispensing procedure of the liquid components is defined.
Vial preparation	Into the corresponding 10 mL process vials equipped with a stirring bar are placed urea (240 mg, 4.0 mmol) or thiourea (304 mg, 4.0 mmol) and ytterbium triflate (248 mg, 0.4 mmol). Any solid building blocks are added as well (see Table 6.21). For the liquid dispensing, fill all the liquid reagents and solvents into the appropriate microwave storage vials according to the software instruction. All the vials are sealed by capping with a Teflon septum fitted in an aluminum crimp top.

Microwave processing		
Time:	600 s (for X = S: 1200 s)
Temperature:	120 °C
Pre-stirring:	15 s
Absorption level:	high
Fixed hold time:	on

For any variations in the microwave protocol, see Table 6.22.

Caution: Any remaining overpressure after the cooling step has to be removed by piercing the septum with a syringe needle.

Work-up and isolation	The products are either purified by filtration (**A**) in the case of already precipitated product or by precipitation and filtration (**B**) if no product has precipitated after the cooling step.
	(**A**) The vials are placed in the fridge for about 2 h. The precipitate is filtered and washed with 10–15 mL of a mixture of cold water/EtOH 2 : 1 in the case of (**a**)–(**f**) or with 15 mL of cold water in the case of (**k**).

(B) The reaction mixture is poured onto about 20 g of crushed ice and is stirred for 2 h vigorously. The resulting precipitates are filtered and washed with 20 mL of a mixture of cold water/EtOH 1 : 1 in the case of (g)–(i), about 12 mL of a mixture of cold water/MeCN 1 : 2 in the case of (j) or 15 mL of cold water in the case of (l).
All products are dried overnight at 50 °C.

Yield and physical data

See Table 6.22 for yield data.

(a) light yellow solid; mp 210–212 °C; ^{1}H NMR (360 MHz, DMSO-d_6) $\delta = 9.26$ (br s, 1H), 7.70 (br s, 1H), 7.40 (d, $J = 7.5$ Hz, 1H), 7.24–7.31 (m, 3H), 5.62 (d, $J = 2.7$ Hz, 1H), 3.89 (q, $J = 7.1$ Hz, 2H), 2.29 (s, 3H), 0.99 (t, $J = 7.1$ Hz, 3H).

(b) light tan solid; mp 214–217 °C, ^{1}H NMR (360 MHz, DMSO-d_6) $\delta = 9.31$ (br s, 1H), 7.90 (br s, 1H), 7.34 and 7.36 (dd, $J = 0.9 + 5.0$ Hz, 1H), 6.88–6.95 (m, 2H), 5.41 (d, $J = 3.5$ Hz, 1H), 4.06 (q, $J = 7.1$ Hz, 2H), 2.21 (s, 3H), 1.16 (t, $J = 7.1$ Hz, 3H).

(c) white solid; mp 188–190 °C; ^{1}H NMR (360 MHz, DMSO-d_6) $\delta = 9.25$ (br s, 1H), 7.77 (br s, 1H), 7.23–7.24 (m, 2H), 7.14 (t, $J = 8.8$ Hz, 2H), 5.14 (d, $J = 3.2$ Hz, 1H), 3.52 (s, 3H), 2.25 (s, 3H).

(d) off-white solid; mp 235–237 °C; ^{1}H NMR (360 MHz, DMSO-d_6) $\delta = 9.18$ (br s, 1H), 7.63 (br s, 1H), 7.11–7.17 (m, 4H), 5.39 (d, $J = 2.7$ Hz, 1H), 3.45 (s, 3H), 2.41 (s, 3H), 2.29 (s, 3H).

(e) white solid; mp 194 °C; ^{1}H NMR (360 MHz, DMSO-d_6) $\delta = 8.98$ (br s, 1H), 7.20 (br s, 1H), 4.02–4.14 (m, 3H), 2.15 (s, 3H), 1.19 (t, $J = 7.1$ Hz, 3H), 1.09 (d, $J = 6.3$ Hz, 3H).

(f) yellow solid; mp 250–252 °C; ^{1}H NMR (360 MHz, DMSO-d_6) $\delta = 9.14$ (br s, 1H), 7.77 (br s, 1H), 7.20 (t, $J = 7.8$ Hz, 1H), 7.01–7.07 (m, 3H), 5.21 (d, $J = 3.3$ Hz, 1H), 2.28 (s, 6H), 2.09 (s, 3H).

(g) light yellow solid; mp 199–201 °C; ^{1}H NMR (360 MHz, DMSO-d_6) $\delta = 9.34$ (br s, 1H), 8.07–8.15 (m, 2H), 7.88 (br s, 1H), 7.64–7.71 (m, 2H), 5.28 (d, $J = 3.0$ Hz, 1H), 4.78–4.85 (m, 1H), 2.26 (s, 3H), 1.17 (d, $J = 6.2$ Hz, 3H), 0.98 (d, $J = 6.2$ Hz, 3H).

(h) light yellow solid; mp 182–184 °C; ^{1}H NMR (360 MHz, DMSO-d_6) $\delta = 8.89$ (br s, 1H), 8.10–8.16 (m, 2H), 8.00 (br s, 1H), 7.64–7.72 (m, 2H), 5.34 (d, $J = 3.3$ Hz, 1H), 4.45–4.58 (m, 2H), 3.57 (s, 3H), 3.31 (s, 3H).

(i) white solid; mp 212–213 °C, ^{1}H NMR (360 MHz, DMSO-d_6) $\delta = 8.36$ (br s, 1H), 7.77 (br s, 1H), 7.54 and 7.56 (dd, $J = 1.4$ and 7.9 Hz, 1H), 7.36 (t, $J = 7.9$ Hz, 1H), 7.25 and 7.27 (dd, $J = 1.4$ and 7.8 Hz, 1H), 5.66 (d, $J = 3.1$ Hz, 1H), 3.45 (s, 3H), 2.69 (d, $J = 7.3$ Hz, 2H), 1.15 (t, $J = 7.3$ Hz, 3H).

(j) off-white solid; mp 195–197 °C; ^{1}H NMR (360 MHz, DMSO-d_6) $\delta = 10.05$ (br s, 1H), 9.86 (br s, 1H), 9.34 (br s, 1H), 7.01–7.51 (m, 9H), 5.77 (s, 1H), 2.01 (s, 3H).

(k) light yellow solid; mp 205 °C; ^{1}H NMR (360 MHz, DMSO-d_6) $\delta = 10.33$ (br s, 1H), 9.65 (br s, 1H), 7.20–7.37 (m, 5H), 5.17 (d, $J = 3.5$ Hz, 1H), 4.01 (q, $J = 7.0$ Hz, 2H), 2.29 (s, 3H), 1.10 (t, $J = 7.1$ Hz, 3H).

(l) light tan solid; mp 179 °C; ^{1}H NMR (360 MHz, DMSO-d_6) $\delta = 9.60$ (br s, 1H), 9.44 (br s, 1H), 7.12 (t, $J = 7.6$ Hz, 1H), 6.63–6.65 (m, 3H), 1.12 (t, $J = 7.0$ Hz, 3H), 2.28 (s, 3H), 4.02 (q, $J = 7.0$ Hz, 2H), 5.08 (d, $J = 3.3$ Hz, 1H).

Literature

[31]

Table 6.22 Isolated yields.

Entry	R^1	R^2	R^3	X	Yield (%)
a	OEt	Me	2-Cl-Ph	O	74
b	OEt	Me	2-thienyl	O	57
c	OMe	Me	4-F-Ph	O	69
d	OMe	Me	2-tolyl	O	79
e^a	OEt	Me	Me	O	67
f^b	Me	Me	3-tolyl	O	53
g	OiPr	Me	3-NO_2-Ph	O	70
h	OMe	CH_2OMe	3-NO_2-Ph	O	69
I	OMe	Et	2,3-$(Cl)_2$-Ph	O	72
j^c	NHPh	Me	2-Cl-Ph	S	51
k	OEt	Me	Ph	S	70
l	OEt	Me	3-OH-Ph	S	78

^{a}EtOH as solvent; 120 °C, 20 min.
^{b}EtOH as solvent.
^{c}Total reaction volume: 3600 µL.

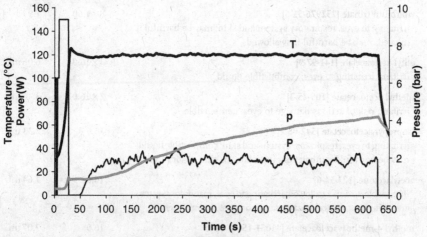

Figure 6.23 Representative power/pressure/temperature profile for the Biginelli library synthesis on a 4 mmol scale.

Background The Biginelli synthesis, the cyclocondensation of β-keto esters, aldehydes and (thio)urea, is another MCR useful for generating libraries relevant to medicinal chemists which is exceptionally flexible in creating chemical diversity about the dihydropyrimidine (DHPM) scaffold. This protocol for the synthesis of a 12-membered DHPM library under fully automated conditions was taken from a literature example [31]. The preparation of this library including work-up can be carried out within about 9 h.

6.3.2
Parallel Library Generation

6.3.2.1 Biginelli Synthesis

8 examples
(35–72%)

Table 6.23 Reagents and hazard information for the Biginelli synthesis.

Reagent [CAS No.]	mmol	Quantity
urea [57-13-6] may be harmful if inhaled; may cause respiratory tract irritation; may be harmful if absorbed through skin; may cause skin irritation	8 × 16.0	7.69 g
ytterbium triflate [252976-51-5] irritating to eyes, respiratory system and skin; may be harmful if inhaled; may be harmful if swallowed	8 × 1.60	7.94 g
ethyl acetoacetate [141-97-9] irritant; irritating to eyes; combustible liquid	3 × 16.0	6.06 mL
methyl acetoacetate [105-45-3] irritant; risk of serious damage to eyes; combustible	2 × 16.0	3.46 mL
isopropyl acetoacetate [542-08-5] irritating to eyes, respiratory system and skin; combustible liquid; may be harmful by inhalation, ingestion, or skin absorption ·	16.0	2.33 mL
acetylacetone [123-54-6] flammable; toxic if swallowed; irritating to eyes, respiratory system and skin; neurological hazard; target organs: thymus, nerves	16.0	1.64 mL
methyl 4-methoxyacetoacetate [41051-15-4] irritating to eyes, respiratory system and skin; combustible; vapor or mist is irritating to the eyes, mucous membranes, and upper respiratory tract; may be harmful by inhalation, ingestion, of skin absorption	16.0	2.07 mL
2-chlorobenzaldehyde [89-98-5] corrosive; causes burns; combustible; material is extremely destructive to the tissue of the mucous membranes and upper respiratory tract; may be harmful if swallowed or inhaled	16.0	1.81 mL
2-thiophenecarboxaldehyde [98-03-3] harmful if swallowed; irritating to eyes, respiratory system and skin; combustible	16.0	1.49 mL
4-fluorobenzaldehyde [459-57-4] irritant; irritating to eyes, respiratory system and skin	16.0	1.71 mL

Table 6.23 (Continued)

Reagent [CAS No.]	mmol	Quantity
o-tolualdehyde [529-20-4] harmful if swallowed; irritating to respiratory system and skin; risk of serious damage to eyes; combustible	16.0	1.85 mL
3-nitrobenzaldehyde [99-61-6] irritating to eyes, respiratory system and skin; may be harmful by inhalation, ingestion, or skin absorption	2 × 16.0	4.84 mg
acetaldehyde [75-07-0] flammable liquid; harmful by ingestion; skin sensitizer; irritant; carcinogen; target organ effect; target organs: blood, kidney, lungs, cardiovascular system, liver, central nervous system	16.0	0.90 mL
m-tolualdehyde [620-23-5] combustible; may cause eye or skin irritation; may be irritating to mucuous membranes and upper respiratory tract; may be harmful if swallowed, inhaled, or absorbed through the skin	16.0	1.88 mL
acetic acid, glacial/ethanol 3:1 [64-19-7]/[64-17-5] combustible; corrosive; harmful in contact with skin; causes severe burns; lachrymator; target organs: kidneys, teeth flammable; irritant; irritating to eyes, respiratory system and skin; target organs: nerves, liver		34.4 mL
water/ethanol 2:1		90 mL
water/ethanol 1:1		40 mL

Equipment	Milestone MicroSYNTH, PRO-16 rotor
Vial preparation	Into each of eight 70 mL reaction vessels equipped with stirring bars are placed urea (960 mg, 16.0 mmol), ytterbium triflate (992 mg, 1.60 mmol), the active methylene component (16.0 mmol), the aldehyde component (16.0 mmol), and enough of a 3:1 acetic acid/ethanol mixture to bring the total volume up to 10 mL. For the reagent and solvent amounts see Table 6.24. The vials are sealed with the proper closures and placed symmetrically in the 16-position rotor.

Microwave processing	Mode:	High pressure reactor	
	Step 1:	Temperature:	120 °C
		Time:	1 min
		Max. power:	400 W
	Step 2:	Temperature:	120 °C
		Time:	10 min
		Max. power:	300 W
		Venting:	10 min

Work-up and isolation	For the work-up procedure, see Section 6.3.1.2.
Yield and physical data	See Table 6.25 for yield data. See Section 6.3.1.2 for physical data.
Literature	[32]

Table 6.24 Reagent charge amounts.

Entry	Active methylene	Amount	Aldehyde	Amount	Solvent
a	ethyl acetoacetate	2.02 mL	2-chlorobenzaldehyde	1.81 mL	4.22 mL
b	ethyl acetoacetate	2.02 mL	2-thiophenecarboxaldehyde	1.49 mL	4.54 mL
c	methyl acetoacetate	1.73 mL	4-fluorobenzaldehyde	1.71 mL	4.62 mL
d	methyl acetoacetate	1.73 mL	o-tolualdehyde	1.85 mL	4.48 mL
e	isopropyl acetoacetate	2.33 mL	3-nitrobenzaldehyde	2.42 g	3.30 mL
f	ethyl acetoacetate	2.02 mL	acetaldehyde	0.90 mL	5.13 mL
g	acetylacetone	1.64 mL	m-tolualdehyde	1.88 mL	4.53 mL
h	methyl 4-methoxyacetoacetate	2.07 mL	3-nitrobenzaldehyde	2.42 g	3.56 mL

Table 6.25 Isolated yields.

Entry	R^1	R^2	R^3	Yield (%)
a	OEt	Me	2-Cl-Ph	68
b	OEt	Me	2-thienyl	59
c	OMe	Me	4-F-Ph	72
d	OMe	Me	2-tolyl	60
e	OEt	Me	Me	48
f	Me	Me	3-tolyl	69
g	Oi-Pr	Me	3-NO$_2$-Ph	60
h	OMe	CH$_2$OMe	3-NO$_2$-Ph	35

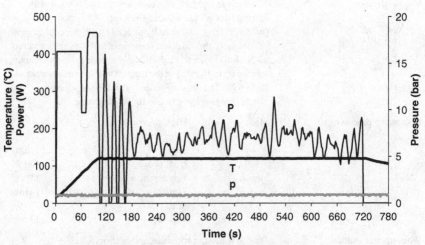

Figure 6.24 Power/pressure/temperature profile for the Biginelli library synthesis on a 16 mmol scale. Fiber-optic temperature measurement in one reference vessel is shown.

Background   The Biginelli library generation using sequential techniques has been reported [32], and the current example adapts these conditions for parallel processing. In this way, multi-gram quantities of eight derivatives can be synthesized within a half-hour period.

6.4
Reaction Scale-Up

For more information on scale-up in MAOS see Section 4.8.

6.4.1
Single Batch Mode

6.4.1.1 Synthesis of Aspirin on a 10 and 100 mmol Scale

10 mmol: 81%
100 mmol: 84%

Table 6.26 Reagents and hazard information for the synthesis of aspirin.

Reagent [CAS No.]	mmol	Quantity
salicylic acid [69-72-7]	(A) 10.0	1.38 g
harmful if swallowed; risk of serious damage to eyes; target organs: central nervous system, blood	(B) 100	13.8 g
acetic anhydride [108-24-7]	12.6	1.19 mL
toxic if inhaled; material is extremely destructive to the tissue of the mucous membranes and upper respiratory tract; may cause respiratory tract irritation; may be harmful if absorbed through skin; causes skin burns; may cause eye irritation; causes eye burns; harmful if swallowed; causes burns	126	11.9 mL
water		15 mL
		160 mL

(A) Single-mode synthesis on a 10 mmol scale

Equipment	Biotage Initiator Eight 2.0 EXP
Vial preparation	Into a 10 mL process vial equipped with a stirring bar are placed salicylic acid (1.38 g, 10.0 mmol) and acetic anhydride (1.19 mL, 12.6 mmol). The vial is sealed by capping with a Teflon septum fitted in an aluminum crimp top.

Microwave processing	Time:	3 min
	Temperature:	140 °C
	Pre-stirring:	15 s
	Absorption level:	normal
	Fixed hold time:	on

Work-up and isolation	The reaction vessel is placed in an ice-bath for 30 min, during which the reaction mass crystallizes. The product is thoroughly mixed with water (5.0 mL) and stirred magnetically for 30 min at room temperature. The precipitate is then isolated by filtration, washed with cold water (10 mL) and dried overnight at 50 °C.
Yield and physical data	Acetylsalicylic acid is obtained as a white solid (1.46 g, 81%). mp 131.5–132.2 °C; ^1H NMR (DMSO-d_6, 360 MHz) $\delta = 13.1$ (br s, 1H), 7.93 (d, $J = 8.0$ Hz, 1H), 7.63 (t, $J = 7.6$ Hz, 1H), 7.37 (t, $J = 7.6$ Hz, 1H), 7.19 (d, $J = 8.0$ Hz, 1H), 2.2 (s, 3H).

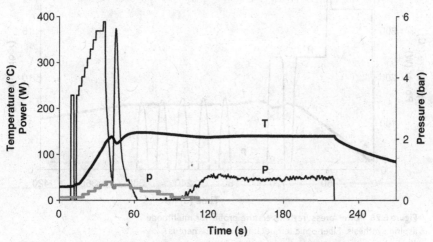

Figure 6.25 Power/pressure/temperature profile for single-mode aspirin synthesis.

(B) Multimode synthesis on a 100 mmol scale

Equipment	Milestone MicroSYNTH, MonoPREP module
Vial preparation	Into the 50 mL glass vessel equipped with a stirring bar are placed salicylic acid (13.8 g, 100 mmol) and acetic anhydride (11.9 mL, 126 mmol). The vessel is assembled in the MonoPREP module. The reaction mixture is stirred for 30 s before irradiation.

Microwave processing	Mode:		High pressure reactor	
	Step 1:		Temperature:	140 °C
			Time:	1 min
			Max. power:	600 W
	Step 2:		Temperature:	140 °C
			Time:	3 min
			Max. power:	300 W
			Venting:	10 min

Work-up and isolation	The reaction vessel is placed in an ice-bath for 30 min, during which the reaction mass crystallizes. The product is thoroughly mixed with water (60 mL) and stirred magnetically for 1 h at room temperature. The precipitate is then isolated by filtration, washed with cold water (100 mL) and dried overnight at 50 °C.
Yield and physical data	Acetylsalicylic acid is obtained as a white solid (15.1 g, 84%). mp 132.2–133.8 °C; for NMR data, see above.
Literature	[34]

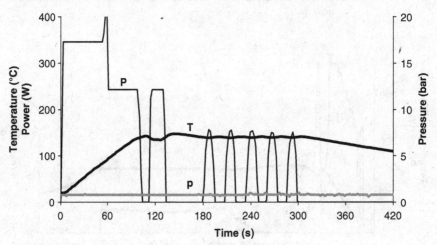

Figure 6.26 Power/pressure/temperature profile for multimode aspirin synthesis. Fiber-optic temperature measurement is shown.

Background Instructional protocols for the aspirin synthesis have emerged using microwave-mediated techniques [33], but they involve the use of modified domestic microwave ovens whereas Ondruschka and coworkers have employed dedicated instruments [34]. The presented procedure was adapted from the latter protocol.

6.4.1.2 Suzuki Reaction under Closed- and Open Conditions (1.5, 15, 150 mmol)

1.5 mmol: 96%
15 mmol: 98%
150 mmol: 96%

Table 6.27 Reagents and hazard information for the Suzuki reaction.

Reagent [CAS No.]	mmol	Quantity
4-bromoanisole [104-92-7]	(A) 1.5	0.19 mL
may be harmful by inhalation, ingestion, or skin absorption; may	(B) 15.0	19 mL
cause irritation	(C) 150	19.0 mL
phenylboronic acid [98-80-6]	1.8	220 mg
may be harmful if inhaled; may cause respiratory tract irritation;	18.0	2.2 g
may be harmful if absorbed through the skin; may cause skin	180	22 g
irritation; may cause eye irritation; harmful if swallowed		
palladium stock solution; 1000 μg mL^{-1} in 5% HCl [Aldrich	1×10^{-5}	1.0 μL
207349]	0.003	319 μL
corrosive; causes burns; harmful if swallowed, inhaled, or	0.03	3.19 mL
absorbed through skin		
ethanol [64-17-5]		1.5 mL
flammable; irritant; irritating to eyes, respiratory system and skin;		28 mL
target organs: nerves, liver		300 mL
sodium carbonate [497-19-8]	1.50	159 mg
may be harmful if inhaled; may cause respiratory tract irritation;	15.0	1.59 g
may be harmful if absorbed through skin; may cause skin	150	15.9 g
irritation; may cause eye irritation		
water		0.90 mL
		18.0 mL
		180 mL
sodium hydroxide, 1N [1310-73-2]		15 mL
corrosive; causes burns		70 mL
		200 mL
ethyl acetate [141-78-6]		30 mL
flammable liquid; irritant; target organ effect; target organs:		240 mL
blood, kidney, liver, central nervous system		600 mL
magnesium sulfate [7487-88-9]		0.5 g
harmful if swallowed; toxic in contact with skin and if swallowed		4.0 g
		10 g

(A) Closed-vessel synthesis on a 1.5 mmol scale

Equipment	Biotage Initiator Eight 2.0 EXP
Vial preparation	Into a 10 mL process vial equipped with a stirring bar are placed 4-bromoanisole (0.19 mL, 1.5 mmol), phenyl-boronic acid (220 mg, 1.8 mmol), sodium carbonate (159 mg, 1.5 mmol), ethanol (1.5 mL), water (0.9 mL), and palladium stock solution (1.0 μL, 0.01 μmol). The vial is sealed by capping with a Teflon septum fitted in an aluminum crimp top.

Microwave processing

Time:	7 min
Temperature:	150 °C
Pre-stirring:	10 s
Absorption level:	high
Fixed hold time:	on

Work-up and isolation	The reaction mixture is transferred to a separatory funnel and partitioned between 1 N sodium hydroxide (15 mL) and ethyl acetate (15 mL). The organic phase is collected and the aqueous phase is extracted again with ethyl acetate (15 mL). The combined organic portions are dried over magnesium sulfate (0.5 g) and concentrated *in vacuo* to give a white solid, which is dried overnight at 50 °C.
Yield and physical data	4-Methoxybiphenyl is obtained as a white solid (264 mg, 96%). mp 88 °C; ^{1}H NMR (DMSO-d_6, 360 MHz) δ = 7.57–7.65 (m, 4H), 7.43 (t, J = 7.3 Hz, 2H), 7.30 (t, J = 7.3 Hz, 1H), 7.02 (d, J = 8.6 Hz, 2H), 3.79 (s, 3H).
Literature	[19, 36]

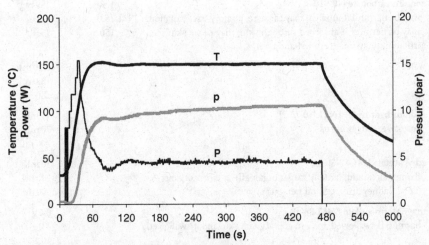

Figure 6.27 Power/pressure/temperature profile for the closed-vessel Suzuki reaction on a 1.5 mmol scale.

(B) Open-vessel synthesis on a 15 mmol scale

Equipment	CEM Discover LabMate
Vial preparation	Into a 100 mL round-bottom flask equipped with a stirring bar are placed 4-bromoanisole (1.90 mL, 15.0 mmol), phenylboronic acid (2.20 g, 18.0 mmol), ethanol (28 mL), and palladium stock solution (319 µL, 3.0 µmol). In a separate flask are combined sodium carbonate (1.59 g, 15.0 mmol) and water (18 mL). Immediately before heating, the aqueous solution is added to the contents of the round-bottom flask, to which is attached an extension adapter and reflux condenser. The assembly is placed in the microwave cavity and secured with the appropriate attenuator.

Microwave processing		
	Mode:	open-vessel
	Temperature:	78 °C
	Power:	120 W
	Time:	30 min

Work-up and isolation	The reaction mixture is transferred to a separatory funnel and partitioned between 1 N sodium hydroxide (70 mL) and ethyl acetate (120 mL). The organic phase is collected and the aqueous phase is extracted again with ethyl acetate (120 mL). The combined organic portions are dried over magnesium sulfate (4.0 g) and concentrated *in vacuo* to give a white solid, which is dried overnight at 50 °C.
Yield and physical data	4-Methoxybiphenyl is obtained as a white solid (2.60 g, 98%). For NMR data, see above.
Literature	[19, 36]

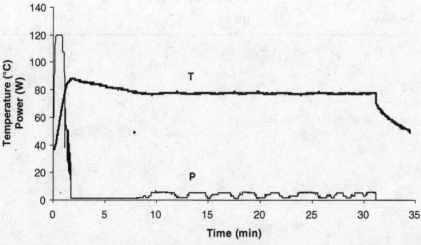

Figure 6.28 Power/temperature profile for the open-vessel Suzuki reaction on a 15 mmol scale.

(C) Open-vessel synthesis on a 150 mmol scale

Equipment	Milestone MicroSYNTH
Vial preparation	Into a 1 L three-necked round-bottom flask equipped with a stirring bar are placed 4-bromoanisole (19 mL, 150 mmol), phenylboronic acid (22.0 g, 180 mmol), ethanol (300 mL), and palladium stock solution (3.19 mL, 30 µmol). In a separate flask are combined sodium carbonate (15.9 g, 150 mmol) and water (180 mL). Immediately before heating, the aqueous solution is added to the contents of the round-bottom flask, to which is attached an extension adapter and reflux condenser. The assembly is placed in the microwave cavity and equipped with a fiber-optic thermometer.

Microwave processing

Mode:	Glass reactor	
Step 1:	Temperature:	78 °C
	Time:	30 s
	Max. power:	570 W
Step 2:	Temperature:	78 °C
	Time:	40 min
	Max. power:	330 W
	Venting:	10 min

Work-up and isolation	The reaction mixture is transferred to a separatory funnel and partitioned between 1 N sodium hydroxide (200 mL) and ethyl acetate (300 mL). The organic phase is collected and the aqueous phase is extracted again with ethyl acetate (300 mL). The combined organic portions are dried over magnesium sulfate (10 g) and concentrated *in vacuo* to give a white solid, which is dried overnight at 50 °C.
Yield and physical data	4-Methoxybiphenyl is obtained as a white solid (27.1 g, 96%). For NMR data, see above.
Literature	[19, 36]

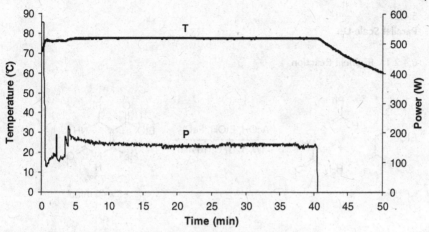

Figure 6.29 Power/temperature profile for the open-vessel Suzuki reaction on a 150 mmol scale. Fiber-optic temperature measurement is shown.

Background Leadbeater and coworkers have developed a microwave-mediated Suzuki coupling protocol with ultra-low loading of the palladium catalyst [35], and have also described the optimization and scale-up of this procedure under open-vessel microwave conditions [36]. The current example is essentially taken from Leadbeater's optimized conditions, using an equimolar amount of sodium carbonate and 4-bromoanisole. While the stoichiometric ratios of components remain constant the catalyst loading has to be increased to 0.02 mol% for the scale-up reactions. A change from a sealed-vessel superheated environment to an open-vessel reflux assembly has to be conducted for the scale-up purpose. Longer reaction times (30–40 min versus 7 min) are required under open-vessel reflux conditions.

6.4.2
Parallel Scale-Up

6.4.2.1 Biginelli Reaction

$$\text{EtO}\overset{O}{\underset{Me}{\bigvee}}\overset{O}{\bigvee} + \overset{Ph}{\underset{H}{\bigvee}}\overset{O}{H} + \overset{NH_2}{\underset{H_2N}{\bigvee}}\overset{O}{\xrightarrow[\text{MW, 120 °C, 10 min}]{\text{AcOH, EtOH, FeCl}_3}}$$

72%
8 x 40 mmol

Table 6.28 Reagents and hazard information for the Biginelli scale-up in parallel.

Reagent [CAS No.]	mmol	Quantity
urea [57-13-6] may be harmful if inhaled; may cause respiratory tract irritation; may be harmful if absorbed through skin; may cause skin irritation	8×40	19.2 g
iron(III) chloride hexahydrate [10025-77-1] harmful by ingestion; corrosive; may be harmful if inhaled; material is extremely destructive to the tissue of the mucous membranes and upper respiratory tract; may cause respiratory tract irritation; may be harmful if absorbed through skin; causes skin burns; may cause eye irritation; causes eye burns	8×4.0	8.64 g
acetic acid, glacial [64-19-7]/ethanol 3:1 combustible; corrosive; harmful in contact with skin; causes severe burns; lachrymator; target organs: kidneys, teeth		120 mL
ethyl acetoacetate [141-97-9] irritant; irritating to eyes; combustible liquid	8×40	40.5 mL
benzaldehyde [100-52-7] harmful; harmful in contact with skin and if swallowed; irritating to skin; may cause sensitization by inhalation and skin contact; combustible; readily adsorbed through skin; target organs: central nervous system; liver	8×40	32.5 mL
ethanol [64-17-5] flammable; irritant; irritating to eyes, respiratory system and skin; target organs: nerves, liver		56 mL

Equipment	Anton Paar Synthos 3000, 16-vessel rotor (16MF100)
Vial preparation	Into each of eight 100 mL reaction vessels equipped with stirring bars are added urea (2.4 g, 40 mmol), iron(III) chloride hexahydrate (1.08 g, 4.0 mmol) and a 3:1 mixture of acetic acid/ethanol (16 mL). The mixtures are allowed to stir for 15 min and then treated with ethyl acetoacetate (5.06 mL, 40 mmol), and benzaldehyde (4.06 mL, 40 mmol). The vessels are sealed with the appropriate closures and placed symmetrically into a 16-vessel rotor.

Microwave processing	Step 1:	Temperature:	120 °C
		Ramp:	5 min
		Hold time:	10 min
		Fan:	0
		Max power:	1000 W
	Step 2:	Temperature:	20 °C
		Ramp:	0 min
		Hold time:	15 min
		Fan:	3

Work-up and isolation	The vessels are placed in an ice-bath for 1 h, after which each precipitate is isolated by filtration, washed with cold ethanol (7 mL), and dried overnight at 50 °C.
Yield and physical data	Ethyl 6-methyl-4-phenyldihydropyrimidone-5-carboxylate is obtained as a tan solid (59.9 g combined, 72%). mp 204 °C; ^{1}H NMR (DMSO-d_6, 360 MHz) $\delta = 9.20$ (s, 1H), 7.74 (s, 1H), 7.20–7.36 (m, 5H), 5.14 (d, $J = 3.3$ Hz, 1H), 3.98 (q, $J = 7.0$ Hz, 2H), 2.25 (s, 3H), 1.09 (t, $J = 7.0$ Hz, 3H).
Literature	[13]

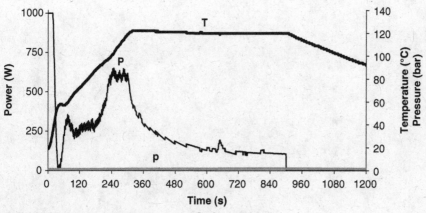

Figure 6.30 Power/pressure/temperature profile for the Biginelli scale-up. Internal gas balloon temperature measurement in one reference vessel is shown.

Background   In contrast to Section 6.3.2.1 where an eight-member DHPM library was synthesized under parallel conditions in a multimode instrument, in this example only one DHPM derivative is obtained in a parallel fashion by performing the reaction in eight individual vessels on a 40 mmol scale each. The current experiment is taken from a literature case study for this technique [13].

6.4.3
Flow Mode

6.4.3.1 Fischer Indole Synthesis

Table 6.29 Reagents and hazard information for the Fischer indole synthesis.

Reagent [CAS No.]	mmol	Quantity
cyclohexanone [108-94-1] combustible; corrosive; harmful in contact with skin and if swallowed; causes burns; possible risk of impaired fertility; readily absorbed through skin; target organs: liver, kidneys	10.1	990 mg
phenylhydrazine [100-63-0] toxic; dangerous for the environment; may cause cancer; also toxic by inhalation, in contact with skin and if swallowed; irritating to eyes and skin; may cause sensitization by skin contact; also toxic: danger of serious damage to health by prolonged exposure through inhalation, in contact with skin and if swallowed; possible risk of irreversible effects; very toxic to aquatic organisms; Calif. Prop. 65 carcinogen; combustible; target organs: blood, liver	11.1	1.2 g
acetic acid, glacial [64-19-7] combustible; corrosive; harmful in contact with skin; causes severe burns; lachrymator; target organs: kidneys, teeth		100 mL
isopropanol [67-63-0] highly flammable; irritant; irritating to eyes; vapors may cause drowsiness and dizziness		50 mL

Equipment	CEM Voyager$_{CF}$
Vial preparation	A flow cell is prepared by filling a 10 mL process vial with glass beads, inserting an inlet tube reaching to the bottom of the vessel, and securing the assembly to the specially designed flow cell attenuator closure. This apparatus is connected to the microwave cavity and the flow cell inlet tube is attached to the pump outlet. In a separate 30 mL Erlenmeyer flask are combined cyclohexanone (990 mg, 10.1 mmol), phenylhydrazine (1.2 g, 11.1 mmol), and a mixture of acetic acid/isopropanol 2 : 1 (20 mL).

The flow system is primed with acetic acid/isopropanol 2 : 1 until solvent begins to elute from the flow cell outlet, at which time a back-pressure valve is attached and the flow rate is set to 1 mL min^{-1}. After a stable system pressure (15–17 bar) is achieved, the pump inlet tubing is transferred from the acetic acid/isopropanol to the flask containing the starting materials, and microwave processing is initiated using the conditions below.

After a 3 mL forerun, collection of the product mixture is begun. When all the starting material mixture has been consumed, the flask and flow cell are rinsed forward with acetic acid/isopropanol 2 : 1 (10 mL), which requires an additional 10 min.

Microwave processing	Priming:	1 min
	Temperature:	160 °C
	Time:	40 min
	Max power:	200 W
	Cooling:	off
	Flow rate:	1 mL min^{-1}

Work-up and isolation

The processed reaction mixture is treated with ice-water (400 mL) and the resulting slurry is stirred for 30 min. The precipitate is isolated by filtration, washed with water (300 mL), and dried overnight at 50 °C.

Yield and physical data

2,3,4,9-Tetrahydro-1H-carbazole is obtained as a white solid (1.69 g, 96%).
mp 118–120 °C; ^{1}H NMR (CDCl$_3$, 360 MHz) δ = 7.57 (s, 1H), 7.45 (m, 1H), 7.04–7.25 (m, 3H), 2.67–2.73 (m, 4H), 1.82–1.94 (m, 4H).

Literature

[37]

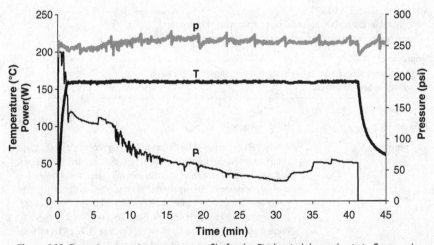

Figure 6.31 Power/pressure/temperature profile for the Fischer indole synthesis in flow mode.

Background   The current example is adapted from a continuous-flow protocol where a custom-built sand-filled flow cell was employed for the Fischer indole synthesis [37].

6.5
Special Processing Techniques

6.5.1
Fractional Product Distilliation (Open-Vessel)

For more information on performing reactions under open-vessel conditions see Section 4.3.

6.5.1.1 Pyranoquinoline Synthesis

81%

Table 6.30 Reagents and hazard information for the pyranoquinoline synthesis.

Reagent [CAS No.]	mol	Quantity
N-methylaniline [100-61-8] toxic by inhalation, in contact with skin and if swallowed; irritating to eyes, respiratory system and skin; combustible; readily absorbed through skin	0.2	22 mL
diethyl malonate [105-53-3] combustible; may be irritating to skin, eyes, mucous membranes and upper respiratory tract; may be harmful if inhaled, swallowed, or absorbed through the skin	0.4	61 mL
diphenyl ether [101-84-8] caution: avoid contact and inhalation; unpleasant odor; target organs: liver; kidneys		100 mL
dioxane [123-91-1] highly flammable; harmful; may form explosive peroxides; irritating to eyes and respiratory system; limited evidence of a carcinogenic effect; repeated exposure may cause skin dryness or cracking		200 mL
diethyl ether [60-29-7] toxic; extremely flammable; may form explosive peroxides; harmful if swallowed; irritating to eyes, respiratory system, and skin; repeated exposure may cause skin dryness or cracking; vapors may cause drowsiness and dizziness; target organs: central nervous system, kidneys		100 mL

Equipment	Milestone MicroSYNTH
Vial preparation	To a 500 mL 2-necked round-bottom flask equipped with a stirring bar are added *N*-methylaniline (22 mL, 0.2 mol), diethyl malonate (61 mL, 0.4 mol), and diphenyl ether (100 mL). The flask is placed in the microwave cavity and fitted with a 60 cm Vigreux column and distillation head.

Microwave processing

Mode:	Glass reactor	
Step 1:	Temperature:	193 °C
	Time:	5 min
	Max power:	850 W
Step 2:	Start temp:	193 °C
	Final temp:	230 °C
	Time:	8.5 min
	Max power:	800 W
Step 3:	Start temp:	230 °C
	Final temp:	233 °C
	Time:	7.5 min
	Max power:	700 W
Step 4:	Start temp:	233 °C
	Final temp:	259 °C
	Time:	55 min
	Max power:	600 W
Step 5:	Start temp:	259 °C
	Final temp:	259 °C
	Time:	6 min
	Max power:	600 W

Work-up and isolation	The reaction mixture is cooled to about 120 °C and treated with dioxane (100 mL) which is added with intense stirring through the Vigreux column. After standing at room temperature for 12 h the resulting precipitate is collected by filtration, washed with dioxane (100 mL) and diethyl ether (100 mL) and dried overnight at 50 °C.
Yield and physical data	4-Hydroxy-6-methyl-2*H*-pyrano[3,2-*c*]quinoline-2,5(6*H*)-dione is obtained as a light orange solid (39.4 g, 81%). mp 255–256 °C; ^{1}H NMR (CDCl$_3$, 360 MHz) $\delta = 13.16$ (s, 1H), 8.33, 8.36 (dd, $J = 10.8$ Hz, 1H), 7.79–7.84 (m, 1H), 7.45–7.54 (m, 2H), 5.97 (s, 1H), 3.80 (s, 3H).
Literature	[38]

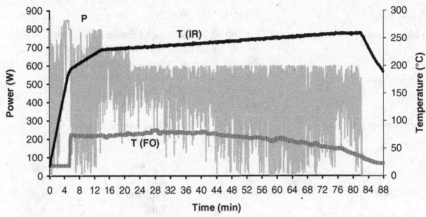

Figure 6.32 Power/temperature profile for the pyranoquinoline synthesis. T (IR) = temperature via IR; T (FO) = distillation head temperature via fiber-optic.

Background Virtually no pyranoquinoline derivative is formed under typical sealed-vessel conditions for the cyclocondensation of N-methylaniline with diethyl malonate. To drive the reaction forward, it is necessary to remove the by-product ethanol by distillation, the accumulation of which also serves as a convenient measure of the extent of reaction. Thus, this reaction must be carried out using an instrument with open-vessel capabilities. The current example is taken essentially from the literature report for this procedure [38], which results in higher yields and shorter reaction times compared to conventional methodology.

6.5.2
Reactions with Gaseous Components

For more information on performing reactions with gaseous reagents under sealed-vessel conditions, see Section 4.4.

6.5.2.1 Ethoxycarbonylation with Carbon Monoxide

78%

Table 6.31 Reagents and hazard information for the ethoxycarbonylation.

Reagent [CAS No.]	mmol	Quantity
4-iodoanisole [696-62-8] irritating to eyes, respiratory system and skin; may be harmful if inhaled, swallowed, or absorbed through skin	4 × 1.0	936 mg
1,8-diazabicyclo[5.4.0]undec-7-ene [6674-22-2] corrosive; harmful if swallowed; causes burns; harmful to aquatic organisms; may cause long-term adverse effects in the aquatic environment; unpleasant odor	4 × 1.1	656 µL[a]
palladium(II) acetate [3375-31-3] irritant; causes severe eye irritation; risk of serious damage to eyes; may be harmful if inhaled, swallowed, or absorbed through skin	4 × 0.001	0.92 mg[a]
ethanol [64-17-5] flammable; irritant; irritating to eyes, respiratory system and skin; target organs: nerves, liver		40 mL[a]
chloroform [67-66-3] harmful by ingestion; irritant; carcinogen; target organ effect; target organs: central nervous system, blood, liver, cardiovascular system, kidney		500 mL

[a] For a better handling due to the very low catalyst loading a stock solution containing 1.64 mL DBU, 2.3 mg Pd(OAc)$_2$ and 100 mL EtOH can be prepared. To each vessel, 10 mL of this stock solution are transferred.

Equipment	Anton Paar Synthos 3000, 8-vessel rotor (8SXQ80)
Vial preparation	Into each of four 80 mL quartz reaction vessels equipped with a stirring bar are added 4-iodoanisole (234 mg, 1.0 mmol), 1,8-diazabicyclo[5.4.0]undec-7-ene (DBU, 164 µL, 1.1 mmol), paladium(II) acetate (0.23 mg, 0.001 mmol), and ethanol (10 mL). Alternatively, 10 mL of the stock solution (see Table 6.31) can be added to each vessel containing 4-iodoanisole. The vessels are sealed with the proper closures, placed symmetrically in the 8-position rotor, flushed and charged with carbon monoxide (10 bar).

Microwave processing	Step 1:	Temperature:	125 °C
		Ramp:	5 min
		Hold time:	20 min
		Fan:	1
		Max power:	1000 W
	Step 2:	Temperature:	20 °C
		Ramp:	0 min
		Hold time:	20 min
		Fan:	3

Work-up and isolation	Excess carbon monoxide is carefully vented in a fume hood. The contents are then transferred to a round-bottom flask and concentrated *in vacuo*. The remaining residue is purified by dry flash column chromatography using chloroform as eluent (500 mL).
Yield and Physical Data	Ethyl 4-methoxybenzoate is obtained as a clear oil (141 mg for one representative vessel, 78%). ^{1}H NMR (CDCl$_3$, 360 MHz) $\delta = 8.01$ (d, $J = 8.9$ Hz, 2H), 6.93 (d, $J = 8.9$ Hz, 2H), 4.36 (q, $J = 7.1$ Hz, 2H), 3.87 (s, 3H), 1.39 (t, $J = 7.1$ Hz, 3H).
Literature	[39]

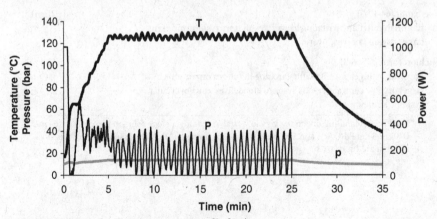

Figure 6.33 Power/pressure/temperature profile for the ethoxycarbonylation. Gas balloon temperature measurement for one reference vessel is shown.

Background The Anton Paar Synthos 3000 offers pressure vessels which enable pre-pressurization with a gaseous component before microwave irradiation. The present carbonylation reaction using gaseous carbon monoxide is adapted from the literature procedure of Kormos and Leadbeater [39]. The same authors also reported on alkoxycarbonylations using near-stoichiometric quantities of carbon monoxide [40].

6.5.2.2 Aminocarbonylation with Molybdenum Hexacarbonyl

Table 6.32 Reagents and hazard information for the aminocarbonylation.

Reagent [CAS No.]	mmol	Quantity
4-iodoanisole [696-62-8] irritating to eyes, respiratory system and skin; may be harmful if inhaled, swallowed, or absorbed through skin	1.0	234 mg
molybdenum hexacarbonyl [13939-06-5] highly toxic; very toxic by inhalation, in contact with skin and if swallowed; may be fatal if inhaled, swallowed, or absorbed through skin	0.5	132 mg
sodium carbonate [497-19-8] may be harmful if inhaled; may cause respiratory tract irritation; may be harmful if absorbed through skin; may cause skin irritation; may cause eye irritation	3.0	318 mg
palladium(II) acetate [3375-31-3] irritant; causes severe eye irritation; risk of serious damage to eyes; may be harmful if inhaled, swallowed, or absorbed through skin	0.05	11 mg
***n*-butylamine [109-73-9]** flammable; highly toxic; harmful in contact with skin and if swallowed; toxic by inhalation; causes severe burns	5.0	495 µL
water		2.0 mL
dichloromethane [75-09-2] toxic; harmful if swallowed; irritating to eyes, respiratory system and skin; may cause cancer; OSHA carcinogen; California Prop. 65 carcinogen; readily absorbed through skin; target organ: heart because methylene chloride is converted to carbon monoxide in the body; target organ: central nervous system because of possible dizziness, headache, loss of consciousness and death at high concentrations		24 mL
ethyl acetate [141-78-6] flammable liquid; irritant; target organ effect; target organs: blood, kidney, liver, central nervous system		35 mL
petroleum ether [101316-46-5] flammable liquid; may be harmful if inhaled; may cause respiratory tract irritation; vapors may cause drowsiness and dizziness; may cause skin irritation; may be harmful if absorbed through skin; target organs: central nervous system; lungs; liver; ears		35 mL

Equipment	Biotage Initiator Eight 2.0 EXP
Vial preparation	To a 10 mL process vial equipped with a stirring bar are added 4-iodoanisole (234 mg, 1.0 mmol), molybdenum hexacarbonyl (132 mg, 0.5 mmol), sodium carbonate (318 mg, 3.0 mmol), palladium(II) acetate (11 mg, 0.05 mmol), *n*-butylamine (495 µL, 5.0 mmol), and water (2.0 mL). The vial is sealed by capping with a Teflon septum fitted in an aluminum crimp top.

Microwave processing		
	Time:	10 min
	Temperature:	110 °C
	Pre-stirring:	10 s
	Absorption level:	high
	Fixed hold time:	on

Work-up and isolation	The contents of the vial are transferred to a separatory funnel and extracted with dichloromethane (4 × 6.0 mL). The combined organic extracts are filtered through Celite and concentrated *in vacuo*. The residue is subjected to column chromatography using 10 g of silica, eluting first with dichloromethane (about 10 mL) and then with petroleum ether/ethyl acetate 1 : 1.
Yield and physical data	*N*-butyl-4-methoxybenzamide is obtained as an off-white solid (181 mg, 87%).
	^{1}H NMR (CDCl$_3$, 360 MHz) δ = 7.74 (d, *J* = 8.7 Hz, 2H), 7.74 (d, *J* = 8.7 Hz, 2H), 6.92 (d, *J* = 8.8 Hz, 2H), 6.07 (br s, 1H), 3.85 (s, 3H), 3.45 (q, *J* = 7.0 Hz, 2H), 1.60 (quint, *J* = 7.3 Hz, 2H), 1.43 (sext, *J* = 7.4 Hz, 2H), 0.96 (t, *J* = 7.3 Hz, 3H).
Literature	[41]

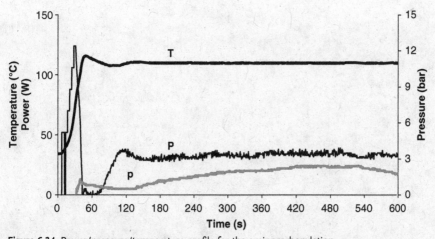

Figure 6.34 Power/pressure/temperature profile for the aminocarbonylation.

Background   Another approach for the introduction of gaseous components is to use a solid reagent such as molybdenum hexacarbonyl. Upon microwave irradiation the inorganic complex slowly releases carbon monoxide into the solution, where it is consumed in the reaction. The present example uses the readily available palladium (II) acetate as a catalyst and is adapted from a protocol published by Larhed and coworkers [41]. Aminocarbonylations using aryl bromides can be performed with a more complex palladacycle catalyst [42].

6.5.3
High-Temperature Processing

For more information on performing reactions at extremely high temperatures and/
or pressures see Section 4.5.1.

6.5.3.1 Hydrolysis of Ethyl Benzoate in Near-Critical Water (NCW)

Table 6.33 Reagents and hazard information for the hydrolysis of ethyl benzoate in NCW.

Reagent [CAS No.]	mmol	Quantity
ethyl benzoate [93-89-0] combustible; may cause skin irritation; may be harmful if absorbed through the skin; may cause eye irritation; may be harmful if inhaled; material may be irritating to mucous membranes and upper respiratory tract; may be harmful if swallowed	6.97	1.0 mL
sodium chloride, 0.03 M solution [7646-14-5] mild skin irritant; mild eye irritant; may be harmful if inhaled; may cause respiratory tract irritation; may be harmful if absorbed through skin; may cause skin irritation; may cause eye irritation; may be harmful if swallowed		15 mL
sodium carbonate, 20% solution [497-19-8] may be harmful if inhaled; may cause respiratory tract irritation; may be harmful if absorbed through skin; may cause skin irritation; may cause eye irritation		10 mL
chloroform [67-66-3] harmful by ingestion; irritant; carcinogen; target organ effect; target organs: central nervous system, blood, liver, cardiovascular system, kidney		80 mL
diethyl ether [60-29-7] toxic; extremely flammable; may form explosive peroxides; harmful if swallowed; irritating to eyes, respiratory system, and skin; repeated exposure may cause skin dryness or cracking; vapors may cause drowsiness and dizziness; target organs: central nervous system, kidneys		80 mL
magnesium sulfate [7487-88-9] harmful if swallowed; toxic in contact with skin and if swallowed		1.5 g

Equipment	Anton Paar Synthos 3000, 8-vessel rotor (8SXQ80)
Vial preparation	Into an 80 mL quartz reaction vessel equipped with a stirring bar are placed ethyl benzoate (1.0 mL, 6.97 mmol), and a 0.03 M sodium chloride solution (15 mL). Additionally for a symmetrical loading of the rotor, three vessels equipped with a stirring bar are filled with a 0.03 M sodium chloride solution (15 mL). The vessels are sealed with the appropriate closures and placed in diametrically opposed slots in the 8-vessel rotor.

Microwave processing			
	Step 1:	Temperature:	295 °C
		Ramp:	8 min
		Hold time:	60 min
		Fan:	1
		Max power:	1000 W
	Step 2:	Temperature:	295 °C
		Ramp:	0 min
		Hold time:	60 min
		Fan:	1
	Step 3:	Temperature:	20 °C
		Hold time:	20 min
		Fan:	3

Work-up and isolation	After cooling, the reaction mixture is basified with a 20% sodium carbonate solution (10 mL) whereupon the precipitate dissolves. The basic solution is then extracted with chloroform (4 × 20 mL). After acidification of the aqueous phase (pH about 5–6), benzoic acid precipitates and the acidic aqueous phase is subsequently extracted with diethyl ether (4 × 20 mL). The ether phases are combined, dried over magnesium sulfate and are concentrated *in vacuo*.
Yield and physical data	Benzoic acid is obtained as a white solid (804 mg, 95%). mp 123 °C; ^{1}H NMR (DMSO-d_6, 360 MHz) $\delta = 12.9$ (br s, 1H), 7.95 (d, $J = 8$ Hz, 2H), 7.63 (t, $J = 8$ Hz, 1H), 7.49 (t, $J = 8$ Hz, 2H).
Literature	[43]

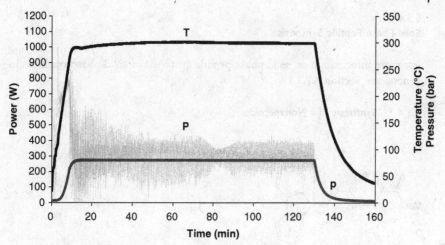

Figure 6.35 Power/pressure/temperature profile for the ethyl benzoate hydrolysis (NCW).

Background In the present showcase example taken from the literature [43], ethyl benzoate is completely hydrolyzed in pure water in just 2 h. The Anton Paar Synthos 3000, equipped with special thick-walled quartz pressure vessels and operation limits of 300 °C and 80 bar, was used for this experiment.

6.5.4
Solid-Phase Peptide Synthesis

For more information on solid-phase peptide synthesis or solid-phase synthesis in general see Section 4.7.2.1.

6.5.4.1 Synthesis of a Nonapeptide

H-Trp-Asp-Thr-Val-Arg-Ile-Ser-Phe-Lys-OH

Table 6.34 Reagents and hazard information for the peptide synthesis.

Reagent [CAS No.]	mmol	Quantity
Rink amide MBHA resin, 0.72 mmol g^{-1} (Nova Biochem 01-64-0037)	0.108	150 mg
irritating to eyes, respiratory system and skin; may be harmful if swallowed, inhaled, or absorbed through the skin		
dichloromethane [75-09-2]		72 mL
toxic; harmful if swallowed; irritating to eyes, respiratory system and skin; may cause cancer; OSHA carcinogen; California Prop. 65 carcinogen; readily absorbed through skin; target organ: heart because methylene chloride is converted to carbon monoxide in the body; target organ: central nervous system because of possible dizziness, headache, loss of consciousness and death at high concentrations		

Table 6.34 (Continued)

Reagent [CAS No.]	mmol	Quantity
N,N-dimethylformamide [68-12-2] combustible liquid; irritant; teratogen; target organs: kidney, liver, blood, cardiovascular system, central nervous system; may be harmful if inhaled; causes respiratory tract irritation		183 mL
20% piperidine [110-89-4] in DMF solution flammable; toxic; harmful if swallowed; toxic by inhalation and in contact with skin; causes burns		40 mL
Fmoc-Lys(Boc)-OH [71989-26-9] may cause skin or eye irritation; may be harmful if swallowed, inhaled, or absorbed through the skin; may be irritating to mucous membranes and upper respiratory tract	0.324	152 mg
Fmoc-Phe-OH [35661-40-6] may cause skin or eye irritation; may be harmful if swallowed, inhaled, or absorbed through the skin; may be irritating to mucous membranes and upper respiratory tract	0.324	126 mg
Fmoc-Ser(tBu)-OH [71989-33-8] may cause skin or eye irritation; may be harmful if swallowed, inhaled, or absorbed through the skin; may be irritating to mucous membranes and upper respiratory tract	0.324	124 mg
Fmoc-Ile-OH [71989-23-6] may cause skin or eye irritation; may be harmful if swallowed, inhaled, or absorbed through the skin; may be irritating to mucous membranes and upper respiratory tract	0.324	115 mg
Fmoc-Arg(Pbf)-OH [154445-77-9] may cause skin or eye irritation; may be harmful if swallowed, inhaled, or absorbed through the skin; may be irritating to mucous membranes and upper respiratory tract	0.324	210 mg
Fmoc-Val-OH [68858-20-8] may cause skin or eye irritation; may be harmful if swallowed, inhaled, or absorbed through the skin; may be irritating to mucous membranes and upper respiratory tract	0.324	110 mg
Fmoc-Thr(tBu)-OH [71989-35-0] may cause skin or eye irritation; may be harmful if swallowed, inhaled, or absorbed through the skin; may be irritating to mucous membranes and upper respiratory tract	0.324	129 mg
Fmoc-Asp(OtBu)-OH [71989-14-5] may cause skin or eye irritation; may be harmful if swallowed, inhaled, or absorbed through the skin; may be irritating to mucous membranes and upper respiratory tract	0.324	133 mg
Fmoc-Trp(Boc)-OH [143824-78-6] may cause skin or eye irritation; may be harmful if swallowed, inhaled, or absorbed through the skin; may be irritating to mucous membranes and upper respiratory tract	0.324	171 mg

(Continued)

Table 6.34 (Continued)

Reagent [CAS No.]	mmol	Quantity
1-hydroxybenzotriazole hydrate [123333-53-9] highly flammable; heating may cause an explosion; may cause skin or eye irritation; may be harmful if swallowed, inhaled, or absorbed through the skin; may be irritating to mucous membranes and upper respiratory tract	9×0.324	450 mg
N-methyl-2-pyrrolidinone [872-50-4] combustible liquid; irritant; target organ effect; target organs: spleen, lymphatic system, thymus, bone marrow		63 mL
N,N'-diisopropylcarbodiimide [693-13-0] flammable liquid; highly toxic by inhalation; target organ effect; target organs: nerves, eyes; skin and respiratory sensitizer	9×0.324	450 μL
methanol [67-56-1] flammable liquid; toxic by inhalation, ingestion, and skin absorption; irritant; target organs: eyes, kidney, liver, heart, central nervous system		9.0 mL
water		2.2 mL
thioanisole [100-68-5] harmful if swallowed; irritating to eyes, respiratory system and skin; combustible liquid; stench		400 μL
1,2-ethanedithiol [540-63-6] combustible; toxic; harmful by inhalation, in contact with skin and if swallowed; irritating to eyes, respiratory system and skin; stench		400 μL
trifluoroacetic acid [76-05-1] delayed target organ effects; target organ: liver; corrosive; may be harmful if inhaled; material is extremely destructive to the tissue of the mucous membranes and upper respiratory tract; may cause respiratory tract irritation; may be harmful if absorbed through skin; causes skin burns; causes eye burns; may cause eye irritation; may be harmful if swallowed; causes burns		10 mL
diethyl ether [60-29-7] toxic; extremely flammable; may form explosive peroxides; harmful if swallowed; irritating to eyes, respiratory system, and skin; repeated exposure may cause skin dryness or cracking; vapors may cause drowsiness and dizziness; target organs: central nervous system, kidneys		123 mL

Equipment	CEM Discover SPS
Vial preparation, part I	Into a CEM Discover SPS reaction vessel is placed Rink amide MBHA resin (150 mg, 0.108 mmol, 0.72 mmol g^{-1}) and a 1 : 1 mixture of dichloromethane/N,N-dimethylformamide (DCM/DMF, 5.0 mL). The vessel is allowed to stand at room temperature for 20 min.
Fmoc deprotection	The solvent is removed by filtration and the resin is washed with DMF (2×3.0 mL). To the resin is added a 20% solution of piperidine in DMF (2.0 mL).

Microwave processing		
Mode:		SPS
Max power:		20 W
Time:		30 s
Max temp:		60 °C
Delta temp:		3 °C

The solvent is removed by filtration and the resin is washed with DMF (2×3.0 mL). To the resin is added a 20% solution of piperidine in DMF (2.0 mL).

Microwave processing		
Mode:		SPS
Max power:		20 W
Time:		150 s
Max temp:		60 °C
Delta temp:		3 °C

Coupling step	The solvent is removed by filtration and the resin is washed with DMF (2×3.0 mL), DCM (2×3.0 mL) and N-methyl-2-pyrrolidinone (NMP, 2×3.0 mL). In a separate vial are combined the Fmoc-protected amino acid (0.324 mmol, see Table 6.34), 1-hydroxybenzotriazole hydrate (50 mg, 0.324 mmol), NMP (1.0 mL), and N,N'-diisopropylcarbodiimide (50 µL, 0.324 mmol). The solution is allowed to stand for 2 min at room temperature and added to the resin.

Microwave processing		
Mode:		SPS
Max power:		10 W
Time:		5 min
Max temp:		60 °C
Delta temp:		3 °C

After the coupling step, Fmoc deprotection has to be performed. The Fmoc deprotection and coupling step are repeated for each amino acid residue.

Vial preparation, part II	The solvent is removed by filtration and the resin is washed with DMF (2×3.0 mL), dichloromethane (5×3.0 mL), methanol (3×3.0 mL), and anhydrous diethyl ether (3.0 mL). The resin is transferred to a glass vial and dried overnight in a vacuum desiccator over potassium hydroxide.

Work-up and isolation	The dried resin is treated with a mixture of deionized water (200 µL), thioanisole (400 µL), 1,2-ethanedithiol (400 µL), and trifluoroacetic acid (9.0 mL). The mixture is stirred at room temperature for 2 h, after which the resin is removed by filtration and washed with trifluoroacetic acid (1.0 mL). The combined filtrates are treated with cold diethyl ether (30 mL) and stored in an ice bath for 15 min. The resulting suspension is centrifuged at 4000 rpm for 5 min and the supernatant is decanted. The precipitate is resuspended in diethyl ether, centrifuged, and decanted three more times. Finally, the precipitate is dissolved in deionized water (20 mL) and lyophilized.
Yield and physical data	The peptide H-Trp-Asp-Thr-Val-Arg-Ile-Ser-Phe-Lys-OH is obtained as an amorphous solid (90 mg, 72%, purity >95%). Mass Spec: calculated 1150.6; found 1150.8.
Literature	[44]

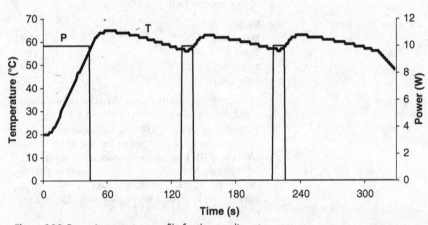

Figure 6.36 Power/temperature profile for the coupling step.

Background The protocol for the synthesis of a nonapeptide was taken from the literature [44], where it was demonstrated that the peptide could be synthesized in a shorter time (about 3.5 h) and with high purity (>95%) under microwave irradiation conditions in comparison with a reference peptide that is obtained by standard methods at room temperature (within 11 h).

References

1 Murphree, S.S. and Kappe, C.O. (2008) *Journal of Chemical Education*, **85**, in press.

2 Raner, K.D. and Strauss, C.R. (1992) *The Journal of Organic Chemistry*, **57**, 6231–6234; Raner, K.D., Strauss, C.R. and Trainor, R.W. (1995) *The Journal of Organic Chemistry*, **60**, 2456–2460; Cablewski, T., Faux, A.F. and Strauss, C.R. (1994) *The Journal of Organic Chemistry*, **59**, 3408–3412.

3 Wilson, N.S., Sarko, C.R. and Roth, G.P. (2004) *Organic Process Research & Development*, **8**, 535–538.

4 Gedye, R., Smith, F., Westaway, K., Ali, H., Baldisera, L., Laberge, L. and Rousell, J. (1986) *Tetrahedron Letters*, **27**, 279–282.

5 Sarju, J., Danks, T.N. and Wagner, G. (2004) *Tetrahedron Letters*, **45**, 7675–7677.

6 Elschenbroich, C. (2006) *Organometallics*, Wiley-VCH, Weinheim, pp. 62–70.

7 Baker, K.V., Brown, J.M., Hughes, N., Skarnulis, A.J. and Sexton, A. (1991) *The Journal of Organic Chemistry*, **56**, 698–703.

8 Mutule, I. and Suna, E. (2005) *Tetrahedron*, **61**, 11168–11176.

9 Gold, H., Larhed, M. and Nilsson, P. (2005) *Synlett*, 1596–1600.

10 Dressen, M.H., Kruijs, B.H., Meuldijk, J., Vekemans, J.A. and Hulshof, L.A. (2007) *Organic Process Research & Development*, **11**, 865–869.

11 Larhed, M., Moberg, C. and Hallberg, A. (2002) *Accounts of Chemical Research*, **35**, 717–727.

12 Arvela, R.K. and Leadbeater, N.E. (2005) *The Journal of Organic Chemistry*, **70**, 1786–1790; Arvela, R.K. Leadbeater, N.E. and Collins, M.J. Jr. (2005) *Tetrahedron*, **61**, 9349–9355.

13 Stadler, A., Yousefi, B.H., Dallinger, D., Walla, P., Van der Eycken, E., Kaval, N. and Kappe, C.O. (2003) *Organic Process Research & Development*, **7**, 707–716.

14 Eynde, V., Jacques, J. and Mayence, A. (2003) *Molecules*, **8**, 381–391.

15 Bagley, M.C. and Lubinu, M.C. (2006) *Synthesis*, 1283–1288.

16 Haines, A.H. (1988) *Methods for the Oxidation of Organic Compounds: Alcohols, Alcohol Derivatives, Alkyl Halides, Nitroalkanes, Alkyl Azides, Carbonyl Compounds, Hydroxyarenes, and Aminoarenes*, Academic Press, London.

17 Thottumkara, A.P., Bowsher, M.S. and Vinod, T.K. (2005) *Organic Letters*, **7**, 2933–2936; Schulze, A. and Giannis, A. (2006) *Synthesis*, 257–260; Jain, S.L., Sharma, V.B. and Sain, B. (2006) *Tetrahedron*, **62**, 6841–6847.

18 Trost, B.M., Vidal, B. and Thommen, M. (1999) *Chemistry – A European Journal*, **5**, 1055–1069.

19 Razzaq, T. and Kappe, C.O. (2008) *Chemistry & Sustainability: Energy & Materials*, **1**, 123–132.

20 Leadbeater, N.E. and Torenius, H.M. (2002) *The Journal of Organic Chemistry*, **67**, 3145–3148.

21 Van der Eycken, E., Appukkuttan, P., de Borggraeve, W., Dehaen, W., Dallinger, D. and Kappe, C.O. (2002) *The Journal of Organic Chemistry*, **67**, 7904–7907.

22 Kremsner, J.M. and Kappe, C.O. (2006) *The Journal of Organic Chemistry*, **71**, 4651–4658.

23 Mirafal, G.A. and Summer, J.M. (2000) *Journal of Chemical Education*, **77**, 356–357; Kwon, P.-S., Kim, J.-K., Kwon, T.-W., Kim, Y.-H. and Chung, S.-K. (1997) *Bulletin of the Korean Chemical Society*, **18**, 1118–1119.

24 Molteni, V., Hamilton, M.M., Mao, L., Crane, C.M., Termin, A.P. and Wilson, D.M. (2002) *Synthesis*, 1669–1674.

25 Paquin, L., Hamelin, J. and Texier-Boullet, F. (2006) *Synthesis*, 1652–1656.

26 Bose, A.K., Ganguly, S.N., Manhas, M.S., Guha, A. and Pombo-Villars, E. (2006) *Tetrahedron Letters*, **47**, 4605–4607.

27 Stadler, A., Pichler, S., Horeis, G. and Kappe, C.O. (2002) *Tetrahedron*, **58**, 3177–3183.

28 Dallinger, D. and Kappe, C.O. (2007) *Nature Protocols*, **2**, 317–321.

29 Glasnov, T.N., Tye, H. and Kappe, C.O. (2008) *Tetrahedron*, **64**, 2035–2041.

30 Öhberg, L. and Westman, J. (2001) *Synlett*, 1296–1298.

31 Dallinger, D. and Kappe, C.O. (2007) *Nature Protocols*, **2**, 1713–1721.

32 Stadler, A. and Kappe, C.O. (2001) *Journal of Combinatorial Chemistry*, **3**, 624–630; Stadler, A. and Kappe, C.O. (2003) *Methods in Enzymology*, **369**, 197–223.

33 Mirafzal, G.A. and Summer, J.M. (2000) *Journal of Chemical Education*, **77**, 356–357; Montes, I., Sanabria, D., Garcia, M., Castro, J. and Fajardo, J. (2006) *Journal of Chemical Education*, **83**, 628–631.

34 Nüchter, M., Ondruschka, B., Weiß D., Beckert, R., Bonrath, W. and Gum, A. (2005) *Chemical Engineering & Technology*, **28**, 871–881.

35 Leadbeater, N.E. and Marco, M. (2003) *The Journal of Organic Chemistry*, **68**, 888–892; Arvela, R.K., Leadbeater, N.E., Sangi, M.S., Williams, V.A., Granados, P. and Singer, R.D. (2005) *The Journal of Organic Chemistry*, **70**, 161–168; Arvela, R.K., Leadbeater, N.E. and Collins, M.J. Jr (2005) *Tetrahedron*, **61**, 9349–9355.

36 Leadbeater, N.E., Williams, V.A., Barnard, T.M. and Collins, M.J. Jr (2006) *Organic Process Research & Development*, **10**, 833–837.

37 Bagley, M.C., Jenkins, R.L., Lubinu, M.C., Mason, C. and Wood, R. (2005) *The Journal of Organic Chemistry*, **70**, 7003–7006; Glasnov, T.N. Vugts, D.J., Koningstein, M.M., Desai, B., Fabian, W.M.F., Orru, R.V.A. and Kappe, C.O. (2006) *The QSAR & Combinatorial Science*, **25**, 509–518.

38 Razzaq, T. and Kappe, C.O. (2007) *Tetrahedron Letters*, **48**, 2513–2517.

39 Kormos, C.M. and Leadbeater, N.E. (2007) *Organic and Biomolecular Chemistry*, **5**, 65–68.

40 Kormos, C.M. and Leadbeater, N.E. (2007) *Synlett*, 2006–2010.

41 Wu, X., Ekegren, J.K. and Larhed, M. (2006) *Organometallics*, **25**, 1434–1439.

42 Wu, X. and Larhed, M. (2005) *Organic Letters*, **7**, 3327–3329.

43 Kremsner, J.M. and Kappe, C.O. (2005) *European Journal of Organic Chemistry*, 3672–3679.

44 Bacsa, B. and Kappe, C.O. (2007) *Nature Protocols*, **2**, 2222–2227.

Index

Practical Microwave Synthesis for Organic Chemists: Strategies, Instruments, and Protocols
C. Oliver Kappe, Doris Dallinger, and S. Shaun Murphree
Copyright © 2009 WILEY-VCH Verlag GmbH & Co. KGaA, Weinheim
ISBN: 978-3-527-32097-4

Printed in the United States
By Bookmasters